OpenStack 云计算平台部署与运维

主　编　杨美霞　李　银
副主编　单　峰　丁宪杰

北京理工大学出版社
BEIJING INSTITUTE OF TECHNOLOGY PRESS

内容简介

OpenStack 作为开源云计算技术，有着广泛的受众、活跃的社区和良好的传播，本书从理论到实践操作，由浅入深，带领读者认识 OpenStack 云计算的全貌，轻松步入 OpenStack 云计算的世界。本书主要介绍基于 OpenStack 的云平台的搭建与运维，从云计算的概念入手，介绍云计算的基本概念、虚拟化技术、OpenStack 云平台部署、云主机运行及 OpenStack 云平台运维。本书共分为 5 个项目，项目一主要介绍云计算和 OpenStack 的基本概念；项目二主要介绍虚拟化的技术特点和体系架构，并以 KVM 为例讲解虚拟机管理器的使用；项目三基于 OpenStack（Rocky）版本，介绍 Keystone、Glance、Nova、Neutron、Horizon、Cinder 等 OpenStack 核心组件的手工部署；项目四介绍以命令行和 Web 端两种方式从云平台上创建云主机；项目五介绍云平台的运维管理。本书以人才岗位需求为目标，突出知识与技能的有机融合，让学生在学习过程中举一反三，创新思维，以适应高等职业教育人才建设需求。

图书在版编目(CIP)数据

OpenStack 云计算平台部署与运维 / 杨美霞,李银主编. --北京:北京理工大学出版社,2022.11
ISBN 978-7-5763-1881-4

Ⅰ.①O… Ⅱ.①杨…②李… Ⅲ.①云计算 Ⅳ.①TP393.027

中国版本图书馆 CIP 数据核字(2022)第 229504 号

责任编辑：王玲玲　　**文案编辑**：王玲玲
责任校对：刘亚男　　**责任印制**：施胜娟

出版发行 / 北京理工大学出版社有限责任公司
社　　址 / 北京市丰台区四合庄路 6 号
邮　　编 / 100070
电　　话 / (010) 68914026（教材售后服务热线）
(010) 68944437（课件资源服务热线）
网　　址 / http://www.bitpress.com.cn

版 印 次 / 2022 年 11 月第 1 版第 1 次印刷
印　　刷 / 河北盛世彩捷印刷有限公司
开　　本 / 787 mm×1092 mm　1/16
印　　张 / 19.25
字　　数 / 407 千字
定　　价 / 68.00 元

前言

随着新一代信息技术的快速发展，云计算作为一种新兴的 IT 服务越来越受到人们的关注。云计算将计算、服务和应用作为一种公共设施提供给公众，使人们能够像使用水、电、燃气那样使用计算机资源，企业和个人消费者通过使用网络终端即可获取存储、计算和带宽等资源。

OpenStack 是一个开源的云计算管理平台项目，是一系列软件开源项目的组合。本书基于 OpenStack Rocky 版本，从 OpenStack 的 Keystone、Nova、Glance、Neutron、Horizon、Cinder 几个主要组件入手，介绍云计算平台的部署与运维。全书共分为 5 个项目：项目一讲述 OpenStack 初识，项目二讲述虚拟化技术，项目三讲述 OpenStack 云平台部署，项目四讲述云主机运行，项目五讲述 OpenStack 云平台运维。

为了适应国家培养复合型技术技能人才的需要，本书力求内容新颖，叙述简练，应用性强。特色如下：

1. 本书以毕业生小李初入职场需要完成一个云平台的项目为出发点，记录该学生学习和工作的过程，以及对工作过程的一些反思，以实际项目案例培养学生成为一个云计算实施工程师。

2. 本书坚持正确的政治方向和价值导向，秉持和践行立德树人的教学理念，在“小李的反思”中将云计算技术与素养元素相结合，将“科技强国”“自主可控”“树立远大理想”“职业素养”“网络安全”等内容融入课堂，素养元素贯穿教育教学全过程，做到“润物细无声”。

3. 本书对标云计算开发与运维“1 + X”证书制度标准，融入全国职业院校技能大赛云计算赛项比赛内容，通过若干任务的具体操作，使学生掌握云平台搭建与运维技能的同时，也能考取“1 + X”证书，实现书证融通、课证融通。

4. 本书采用项目式编写体例，实施任务驱动，每个项目的开始包含项目导入和项目目标，每个任务包括“任务描述”“知识要点”“任务实施”“任务工单”“知识巩固”“小李的反思”等模块，框架清晰，循序渐进，层次分明。

5. 本书为校企共同开发，华为技术有限公司、新华三技术有限公司等企业一线人员为本书的编写提供了大量的案例素材与宝贵建议。

本书适合作为本科层次、专科层次职业教育院校计算机类、电子信息类专业的教学用书，也可作为云计算初学者及相关开发人员的自学参考书。

本书由天津现代职业技术学院的杨美霞、李银担任主编，新华三技术有限公司单峰、中国戏曲学院丁宪杰担任副主编。

为了方便读者学习，本书还开通了“云视界”公众号，读者可以关注该公众号，获取相关技术文档。

由于编者水平所限，书中不足之处难免，敬请读者批评指正。

编　者

目录

项目一

OpenStack 初识

OpenStack 是一个开源的云计算管理平台项目，是一个云操作系统框架，OpenStack 本身不支持各种资源的管理功能，但通过在 OpenStack 内部署各服务组件以实现对计算、存储、网络等资源的管理，实现云操作系统的功能。OpenStack 技术是云计算领域非常重要的一项开源技术，是所有云计算从业者必须掌握的一项核心技术。

项目导入

小李是一名刚毕业的云计算专业的学生，应聘到一家 IT 公司做云计算助理工程师，公司交给他一个云平台搭建项目，该项目需要基于 OpenStack 搭建私有云，进行云上资源的下发和管理，能够根据工作需求，分配不同资源规格的虚拟机。

小李接到项目后，决定将上学期间学到的知识梳理串联起来形成体系化的知识框架，学中做、做中学，将理论与实践相结合，首先他想深入地了解 OpenStack 这一关键开源技术框架。

本项目将随着小李的学习脚步，首先认识云计算，包括云计算的发展历程、云计算的概念、3 种服务模式和 4 种部署模式；其次解构 OpenStack，包括 OpenStack 的概念、发展、作用、技术框架及各核心组件。

项目目标

【素质目标】

- 培养学生科技报国的使命感
- 培养学生任务交付的综合能力
- 激发学生学习热情

【知识目标】

- 了解云计算的概念及发展历史
- 掌握云计算的 3 种服务模式
- 掌握云计算的 4 种部署模式
- 了解 OpenStack 的概念及发展历史
- 掌握 OpenStack 的技术框架
- 掌握 OpenStack 各关键组件的作用

【能力目标】

- 具备新知识的学习能力
- 具备关键知识的信息获取能力

任务1 认识云计算

【任务描述】

小李已经确定了他的第一步计划是了解云计算，他围绕着几个问题开始了他的云计算之旅：什么是云计算？云计算有哪些特点？云计算的部署模式有哪些？云计算的服务模式有哪些？

【知识要点】

美国国家标准与技术研究院（NIST）定义，云计算是一种按使用量付费的模式，这种模式提供可用的、便捷的、按需的网络访问，并可以通过网络访问可配置的计算资源共享池（网络、服务器、存储和应用软件等资源），这些资源能够被快速提供，只需投入很少的管理工作或与服务商进行很少的交互。

云计算具有五大基本特征：资源池化、快速弹性、自助服务、可计量、广泛的网络访问。

云计算有 3 种服务模式：IaaS（Infrastructure as a Service）、PaaS（Platform as a Service）、SaaS（Software as a Service）

云计算有 4 种部署模式：私有云、公有云、混合云和社区云。

【任务实施】

1. 云计算发展历程

云计算被视为计算机网络领域的一次革命，因为它的出现，社会的工作方式和商业模式也在发生巨大的改变。

云计算的根源可以追溯到 1956 年，Christopher Strachey 发表了一篇有关虚拟化的论文，正式提出了虚拟化的概念。虚拟化是今天云计算基础架构的核心，是云计算发展的基础。而后随着网络技术的发展，逐渐孕育了云计算的萌芽。

在 20 世纪 90 年代，计算机网络出现了大爆炸，出现了以思科为代表的一系列公司，随即网络出现泡沫时代。

在 2004 年，Web2. 0 会议举行，Web2. 0 成为当时的热点，这也标志着互联网泡沫破灭，计算机网络发展进入了一个新的阶段。在这一阶段，让更多的用户方便快捷地使用网络服务成为会联网发展亟待解决的问题，与此同时，一些大型公司也开始致力于开发大型计算能力的技术，为用户提供了更加强大的计算处理服务。

在 2006 年 8 月 9 日，Google 首席执行官埃里克·施密特（Eric Schmidt）在搜索引擎大会（SESSanJose2006）首次提出“云计算”（Cloud Computing）的概念。这是云计算发展史上第一次正式地提出这一概念，有着巨大的历史意义。

2007 年以来，“云计算”成为计算机领域最令人关注的话题之一，同样也是大型企业、互联网建设着力研究的重要方向。因为云计算的提出，互联网技术和 IT 服务出现了新的模式，引发了一场变革。

在 2008 年，微软发布其公共云计算平台（Windows Azure Platform），由此拉开了微软的云计算大幕。同样，云计算在国内也掀起一场风波，许多大型网络公司纷纷加入云计算的阵列。

2009 年 1 月，阿里软件在江苏南京建立首个“电子商务云计算中心”。同年 11 月，中国移动云计算平台“大云”计划启动。到现阶段，云计算已经发展到较为成熟的阶段。

2017 年，华为提出了南贵北乌的云数据中心布局，在贵安新区和乌兰察布大数据产业园均规划建设超大型数据中心基地。

2019 年 8 月 17 日，北京互联网法院发布《互联网技术司法应用白皮书》。发布会上，北京互联网法院互联网技术司法应用中心揭牌成立。

2020 年，我国云计算市场规模达到 1 781 亿元，增速为 33.6%。其中，公有云市场规模达到 990.6 亿元，同比增长 43.7%，私有云市场规模达 791.2 亿元，同比增长 22.6%。

2. 云计算概念

云计算（Cloud Computing）是分布式计算的一种，指的是通过网络“云”将巨大的数据计算处理程序分解成无数个小程序，然后，通过多部服务器组成的系统进行处理和分析这些小程序，得到结果并返回给用户。云计算早期，简单地说，就是简单的分布式计算，解决任务分发，并进行计算结果的合并。因而，云计算又称为网格计算。通过这项技术，可以在很短的时间内（几秒钟）完成对数以万计的数据的处理，从而达到强大的网络服务。

现阶段所说的云服务已经不单单是一种分布式计算，而是分布式计算、效用计算、负载均衡、并行计算、网络存储、热备份冗杂和虚拟化等计算机技术混合演进并跃升的结果。

根据美国国家标准与技术研究院（NIST）定义，云计算是一种按使用量付费的模式，这种模式提供可用的、便捷的、按需的网络访问，并可以通过网络访问可配置的计算资源共享池（网络、服务器、存储和应用软件等资源），这些资源能够被快速提供，只需投入很少的管理工作或与服务商进行很少的交互。

传统的 IT 系统需要自建机房、网络部署、购买服务器、部署操作系统、中间件及应用软件等一系列工作，这样的部署需要较长的建设周期，同时，需要更高的前期建设成本和后期的运维成本；在云计算中，IT 业务不再部署在本地计算机或服务器上，而是运行在远端的分布式系统上，只需要根据使用量从云服务提供商那里购买相应的云服务即可。云服务作为一种商品进行流通，就像水、电、燃气一样，取用方便、按量付费、费用低廉，只是云服务需要通过网络去访问。

3. 云计算的特征

云计算具有五大基本特征：资源池化、快速弹性、自助服务、可计量、广泛的网络访问。

（1）资源池化。云端计算资源需要被池化，以便通过多租户形式共享给多个消费者，也只有池化，才能根据消费者的需求动态分配或再分配各种物理的和虚拟的资源。消费者通常不知道自己正在使用的计算资源的确切位置，但是在自助申请时，允许指定大概的区域范围（比如在哪个国家、哪个省或者哪个数据中心）。

（2）快速弹性。消费者能方便、快捷地按需获取和释放计算资源，也就是说，需要时能快速获取资源从而扩展计算能力，不需要时能迅速释放资源以便降低计算能力，从而减少资源的使用费用。对于消费者来说，云端的计算资源是无限的，可以随时申请并获取任意数量的计算资源。但是我们一定要消除一个误解，那就是一个实际的云计算系统不一定是投资巨大的工程，不一定要购买成千上万台计算机，也不一定具备超大规模的运算能力。云端建设方案一般采用可伸缩性策略，刚开始时采用几台计算机，以后根据用户数量规模来弹性增减计算机数量。

（3）自助服务。消费者不需要或很少需要云服务提供商的帮助，就可以单方面按需获取并使用云端的计算资源。

（4）可计量。消费者使用云端计算资源是要付费的，付费的计量方法有很多，比如根据某类资源（如存储、CPU、内存、网络带宽等）的使用量和时间长短计费，也可以按照每使用一次多少钱来计费。但不管如何计费，对消费者来说，价码要清楚，计量方法要明确，而云服务提供商需要监视和控制资源的使用情况，并及时输出各种资源的使用报表，做到供/需双方费用结算清楚明白、准确无误。

（5）广泛的网络访问。消费者可以随时随地使用任何云终端设备接入网络并使用云端的计算资源。常见的云终端设备包括手机、平板电脑、笔记本电脑、PDA 掌上电脑和台式计算机等。

4. 云计算服务模式

企业要部署一套 IT 系统，需要经过机房建设、网络布线、购买存储设备、购买服务器、安装操作系统、安装中间件和数据库、部署应用软件等一系列过程，如图 1－1 所示，IT 环境从逻辑上划分为基础设施层、平台层、应用软件层，云服务提供商针对 IT 环境的三个逻辑分层提供三种服务，分别是 IaaS、PaaS、SaaS。

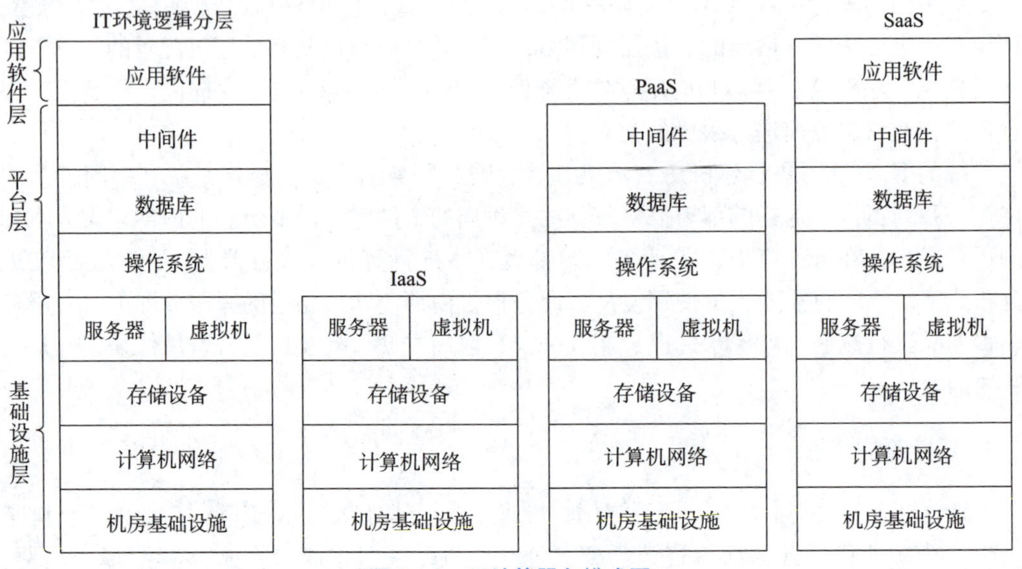

图 1－1　云计算服务模式图

IaaS 是 Infrastructure as a Service 的首字母缩写，意思是基础设施即服务，即把 IT 系统的基础设施层作为服务出租出去。由云服务提供商把 IT 系统的基础设施建设好，并对计算设备进行池化，然后直接对外出租硬件服务器、虚拟主机、存储或网络设施（负载均衡器、防火墙、公网 IP 地址及诸如 DNS 等基础服务）等。云服务提供商负责管理机房基础设施、计算机网络、磁盘柜、服务器和虚拟机，租户自己安装和管理操作系统、数据库、中间件和运行库、应用软件和数据信息。

PaaS 是 Platform as a Service 的首字母缩写，意思是平台即服务，即把 IT 系统的平台层作为服务出租出去。相比 IaaS 云服务提供商，PaaS 云服务提供商要做的事情增加了，它们需要准备机房，布好网络，购买设备，安装操作系统、数据库和中间件，即把基础设施层、平台和软件层都搭建好，然后在平台软件层上划分“小块”（习惯称之为容器）并对外出租。PaaS 云服务提供商也可以从其他 IaaS 云服务提供商那里租赁计算资源，然后自己部署平台和软件层。另外，为了让消费者能直接在云端开发调试程序，PaaS 云服务提供商还得安装各种开发调试工具。相反，租户要做的事情比 IaaS 要少很多，租户只要开发和调试软件或者安装、配置和使用应用软件即可。

SaaS 是 Software as a Service 的首字母缩写，意为软件即服务。简言之，就是软件部署在云端，让用户通过网络来使用它，即云服务提供商把 IT 系统的应用软件层作为服务出租，而消费者可以使用任何云终端设备接入计算机网络，然后通过网页浏览器或者编程接口使用云端的软件。这进一步降低了租户的技术门槛，应用软件也无须自己安装了，而是直接使用软件。

5. 云计算部署模式

根据云服务的消费者来源不同，将云计算的部署模型分为 4 种：私有云、公有云、混合云和社区云，区分方法如下：

（1）私有云的所有消费者只来自一个特定的单位组织。

（2）公有云的所有消费者来自社会公众。

（3）混合云的资源来自两个或两个以上的云。

（4）社区云的所有消费者来自两个或两个以上特定的单位组织。

私有云是为一个客户（公司）单独使用而构建的，因而提供对数据、安全性和服务质量的最有效控制。该公司拥有基础设施，并可以控制在此基础设施上部署应用程序的方式。私有云可部署在企业数据中心的防火墙内，也可以将它们部署在一个安全的主机托管场所，私有云的核心属性是专有资源。

公有云通常指第三方提供商为用户提供的能够使用的云，云端资源面向社会大众开放，符合条件的任何个人或单位组织都可以租赁并使用云端资源。公有云一般可通过 Internet 使用，可能是免费或成本低廉的，公有云的核心属性是共享资源服务。

混合云融合了公有云和私有云，是近年来云计算的主要模式和发展方向。我们已经知道私有云主要是面向企业用户，出于安全考虑，企业更愿意将数据存放在私有云中，但是同时又希望可以获得公有云的计算资源，在这种情况下，混合云被越来越多地采用，它将公有云和私有云进行混合和匹配，以获得最佳的效果。这种个性化的解决方案，达到了既省钱又安全的目的。

社区云是由几个组织共享的云端基础设施，它们支持特定的社群，有共同的关切事项，例如使命任务、安全需求、策略与法规遵循考量等。管理者可能是组织本身，也可能是第三方；管理位置可能在组织内部，也可能在组织外部。

【任务工单】

工单号：1－1

<table>
<tr><td colspan="2">项目名称：OpenStack 初识</td><td colspan="2">任务名称：认识云计算</td></tr>
<tr><td colspan="2">班级：</td><td>学号：</td><td>姓名：</td></tr>
<tr><td>任务安排</td><td colspan="3">□描述云计算的概念
□绘制云计算服务模式示意图
□说明云计算部署模式，并描述如何区分
□自行通过搜索引擎、书籍、公众号等多种形式了解云计算与虚拟化的联系
□将教师所讲内容和自学获取到的知识整理笔记并在班级内汇报</td></tr>
<tr><td>成果交付形式</td><td colspan="3">将任务内容整理形成笔记文档，上传至教学平台</td></tr>
<tr><td>任务实施总结</td><td colspan="3">任务自评（0～10 分）：
项目收获：

改进点：</td></tr>
<tr><td>成果验收</td><td colspan="3">□完全满足任务要求
□基本满足任务要求
要求全部完成，存在一些知识理解的小问题，需要改进的地方：

□不能满足需求
需求未全部完成，存在一些影响知识理解的问题，需要改进的地方：</td></tr>
</table>

【知识巩固】

1. 什么是云计算?

2. 以下是云计算服务模式的是（　　）。

A. IaaS　　B. DaaS　　C. SaaS　　D. PaaS

3. 以下是云计算部署模式的是（　　）。

A. 私有云　　B. 公有云　　C. 混合云　　D. 社区云

【小李的反思】

大学之道，在明明德，在亲民，在止于至善。

出自《大学》，意思是大学的宗旨在于弘扬光明正大的品德，在于使人弃旧图新，在于使人达到至善至美的最高境界。

国无德不兴，人无德不立。人类是推动社会历史不断发展的主体因素，作为社会个人，我们要从自身做起，静心修身，提高素养，做一个有益于他人、有益于社会的人，这是历史前进所呼唤的，更是人类文明所要求的。作为一名当代大学生，无论学习何专业，以后从事何职业，重要的是要有进取的意识，有敬业的精神，努力学习科学文化知识，以自身的技术技能助力国家发展，实现科技报国，共同努力解决某些关键核心领域的“卡脖子”问题。

20 世纪 40 年代，钱学森在美国已是知名的火箭导弹专家。1949 年，在得知中华人民共和国成立后，他决定辞去美国的一切职务全家回国报效祖国，然而回国之路却充满艰难险阻。1950 年，他在港口登船前被联邦调查局以莫须有的罪名抓进了监狱，随后被软禁限制自由长达五年。当时美国海军次长丹尼·金布尔甚至声称：“钱学森抵得上 5 个师的兵力。我宁可把这个家伙枪毙了，也不能放他回中国去！”1955 年 8 月 4 日，钱学森终于收到了美国移民局允许他回国的通知。1955 年 10 月 1 日清晨，钱学森一家终于回到了魂牵梦绕的祖国。钱学森在登上回国轮船前，向媒体发表了讲话，他说：“我将竭尽努力，和中国人民一道建设自己的国家，让我的同胞过上有尊严的幸福生活。”他配备有世界第一流搞科研的技术设备，享有非常优裕的生活条件。如果从追求个人的科研成果来说，那真是“得天独厚”。但他毅然冲破美国的种种阻挠，回到祖国，在“一穷二白”的土地上创造中国人的火箭、导弹事业。有人问他为什么归心似箭，他说：“因为我是一个中国人，我的事业在中国，我的归宿在中国。”有人问他中国既无人才又无设备，搞火箭导弹能行吗？他回答是：“外国人能干的，中国人为什么不能干！”钱学森的誓言实现了，中国卫星上天了，洲际导弹可以同外国“比武”了。

任务 2 解构 OpenStack

【任务描述】

小李已经了解了云计算的基本概念，他开始着手进入正餐：了解 OpenStack。小李想要使用 OpenStack 搭建私有云，他着手开始研究 OpenStack 是什么，OpenStack 的技术框架是什么样，OpenStack 有哪些组件、作用是什么。

【知识要点】

OpenStack 是一个云平台管理的项目，是一个云操作系统框架。OpenStack 采用按字母顺序编排的单词来描述版本新旧，2010 年发布了该项目的第一个版本 Austin，截至目前，最新版本是第 25 版本 Yoga。OpenStack 是一个云操作系统框架，本身并不具备计算、存储、网络等服务功能，而这些功能是通过 OpenStack 框架内的各个服务组件提供的。OpenStack 核心组件包括 Keystone、Nova、Glance、Neutron、Cinder 等，其中，Keystone 提供身份认证服务、Nova 提供计算服务、Glance 提供镜像服务、Neutron 提供网络服务、Cinder 提供块存储服务。

【任务实施】

1. 什么是 OpenStack

OpenStack 是一个云平台管理的项目，它不是一个软件，它是由几个主要的组件组合起来，为公有云、私有云和混合云的建设与管理提供软件的开源项目。现在已经有来自 100 多个国家的数万名个人和 200 多家企业参与到 OpenStack 的开发，如华为、NASA、谷歌、惠普、Intel、IBM、微软等。这些个人与企业将 OpenStack 作为基础设施即服务资源的通用前端。OpenStack 项目的首要任务是简化云的部署过程并为其带来良好的可扩展性。OpenStack 系统或其演变版本目前被广泛应用在各行各业，包括自建私有云、公共云、租赁私有云及混合云，用户包括思科、英特尔、IBM、华为、希捷等，OpenStack 支持 KVM、Xen、LVC、Docker 等虚拟机软件或容器。

2. OpenStack 发展历史

OpenStack 是由美国国家航空航天局（NASA）和 Rackspace 合作研发，Apache 许可的开源项目。项目自 2010 年成立，发布了该项目的第一个版本 Austin，该版本是当时第一个开源的云计算平台项目。此后，OpenStack 基本每半年发行一个新版本，截至目前，最新版本是第 25 版本 Yoga，不同于其他软件的版本号采用数字编码，OpenStack 采用一个单词来描述不同的版本，其中，单词首字母指明版本的新旧。比如目前的版本 Yoga 就比之前的 Xena 要新，同时，“Y”在 26 个字母中排行第 25，所以称第 25 版本。各个版本的发行时间表参考网站 https://releases.openstack.org/。

3. OpenStack 能做什么

OpenStack 的主要目标是管理数据中心的资源，简化资源分配。它管理三部分资源，分别是：

- 计算资源：OpenStack 可以规划并管理大量虚拟机，从而允许企业或服务提供商按需提供计算资源；开发者可以通过 API 访问计算资源，从而创建云应用，管理员与用户则可以通过 Web 访问这些资源。
- 存储资源：OpenStack 可以为云服务或云应用提供所需的对象及块存储资源。因为对性能及价格有需求，很多组织已经不能满足于传统的企业级存储技术，因此，OpenStack 可以根据用户需要提供可配置的对象存储或块存储功能。
- 网络资源：如今的数据中心存在大量的配置工作，如服务器、网络设备、存储设备、安全设备均需要配置，而它们还将被划分成更多的虚拟设备或虚拟网络，这会导致 IP 地址的数量、路由配置、安全规则将爆炸式增长；传统的网络管理技术无法真正的可高扩展、高自动化地管理下一代网络，因而 OpenStack 提供了插件式、可扩展、API 驱动型的网络及 IP 管理功能。

4. OpenStack 服务组件

OpenStack 是一个云操作系统框架，本身并不具备计算、存储、网络等服务功能，而这些功能是通过 OpenStack 框架内的各个服务组件提供的。通过各个组件的协作构建完整的云操作系统，组件间通过公共 API 相互交互，每个服务都至少有一个 API 进程，该进程侦听 API 请求，对其进行预处理并将其传递给服务的其他部分。OpenStack 的技术架构如图 1-2 所示。

以下列举 OpenStack 的 6 个最重要的核心项目。

Keystone：Keystone 是 OpenStack 的认证服务，Keystone 为所有的 OpenStack 组件提供认证和访问策略服务，它依赖自身 REST 系统进行工作，主要对 Swift、Glance、Nova 等进行认证与授权，它对动作消息中的来源者进行合法性鉴定。Keystone 采用两种授权方式：一种基于用户名/密码；另一种基于令牌（Token）。除此之外，Keystone 提供以下 3 种服务：

- 令牌服务：令牌中含有授权用户或群组的授权信息，授权给合法用户或群组。
- 目录服务：目录中含有合法用户或群组的可用服务列表。
- 策略服务：利用 Keystone 具体指定用户或群组的某些访问权限。

Nova：Nova 是一套控制器，用于管理虚拟机实例的整个生命周期，根据用户需求来提供虚拟服务。Nova 负责管理整个云的计算资源、网络资源、授权及测度。虽然 Nova 本身并不提供任何虚拟能力，但是它将使用 libvirt API 与虚拟机的宿主机进行交互。Nova 通过 Web 服务 API 来对外提供处理接口。

Glance：Glance 负责 OpenStack 的镜像服务，Glance 是一套虚拟机镜像发现、注册、检索系统，它提供虚拟机镜像的存储、查询和检索功能，为 Nova 服务，依赖于存储服务和数据库服务。

Swift：Swift 为 OpenStack 提供了一种分布式、持续虚拟对象存储。Swift 具有跨节点的存储能力。Swift 组件有冗余和失效备援管理功能，也能够处理归档和媒体流，特别是对大数

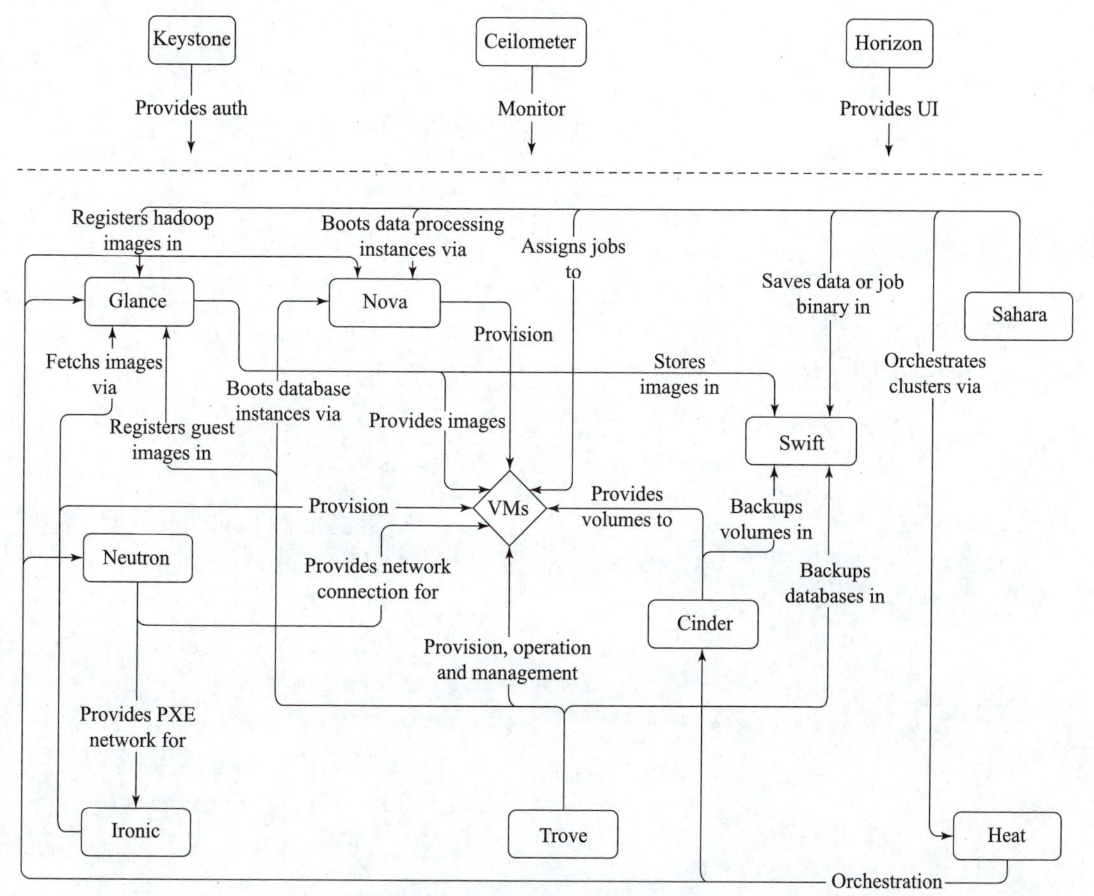

图 1-2　OpenStack 技术架构

据和大容量的测度非常高效。

Cinder：Cinder 是 OpenStack 的块存储服务组件，它管理所有块存储设备，为 VM 服务。

Neutron：Neutron 为 OpenStack 提供虚拟的网络功能，为每个不同的租户建立独立的网络环境。

【任务工单】

工单号：1 - 2

<table>
<tr><td colspan="2">项目名称：OpenStack 初识</td><td colspan="2">任务名称：解构 OpenStack</td></tr>
<tr><td colspan="2">班级：</td><td>学号：</td><td>姓名：</td></tr>
<tr><td>任务安排</td><td colspan="3">□描述什么是 OpenStack
□进入 https://releases. openstack. org/网址查看 OpenStack 的发行版本时间表
□描述 OpenStack 的技术框架
□说明 OpenStack 的 Keystone、Nova、Neutron 等核心组件的作用
□自行通过官方网站、搜索引擎、书籍、公众号等多种形式了解 Heat、Horizon 组件的作用及各组件的协作关系
□将教师所讲内容和自学获取到的知识整理笔记并在班级内汇报</td></tr>
<tr><td>成果交付形式</td><td colspan="3">将任务内容整理形成笔记文档，上传至教学平台</td></tr>
<tr><td>任务实施总结</td><td colspan="3">任务自评（0～10 分）：
项目收获：______
______
______
改进点：______
______
______</td></tr>
<tr><td>成果验收</td><td colspan="3">□完全满足任务要求
□基本满足任务要求
要求全部完成，存在一些知识理解的小问题，需要改进的地方：
______
______
______
□不能满足需求
需求未全部完成，存在一些影响知识理解的问题，需要改进的地方：
______
______
______</td></tr>
</table>

【知识巩固】

1. 描述什么是 OpenStack。

2. 下列选项是 OpenStack 身份认证服务的是（　　）。

A. Cinder　　B. Neutron

C. Nova　　D. Keystone

3. 下列选项是 OpenStack 计算服务的是（　　）。

A. Cinder　　B. Neutron

C. Nova　　D. Keystone

【小李的反思】

道虽迩，不行不至；事虽小，不为不成。

出自《荀子·修身》，意思是说，即使路程再近，不走也不会到达；即使事情再小，不做也不会成功，强调了踏实笃行的意义。

作为云计算专业的学生，最初听到 OpenStack 感觉是一个很高深的技术，充满了畏惧心理，但踏实下来一点一点学习了解，就会发现只要肯开始学，任何技术难题都会迎刃而解。作为当代大学生，我们在学习过程中不能因为事情简单而轻视不去实践，也不能因为事情困难而畏惧不前，对于学习，最早的日子就是今天，最早的时刻就是此刻。

汉朝信都人孙敬，年少好学，博闻强记，而且嗜书如命。孙敬读书时，常常一直看到后半夜，有时不免打起瞌睡来。一觉醒来，又懊悔不已。有一天，他找来一根绳子，将绳子的一头拴在房梁上，下边这头就跟自己的头发拴在一起。这样，每当他累了、困了、想打瞌睡时，只要头一低，绳子就会猛地拽一下他的头发，一疼就会惊醒而赶走睡意。年复一年地刻苦学习，使孙敬饱读诗书，博学多才，成为一名通晓古今的大学问家。这就是头悬梁的故事。

战国时期出名的政治家苏秦，在年轻时，由于学问不深，曾到很多地方做事，都不受重视。回家后，家人对他也很冷淡，瞧不起他。这对他的刺激很大，所以，他决心要发奋读书。他常常读书到深夜，想睡觉时，就拿一把锥子，一打瞌睡，就用锥子往大腿上刺一下。这样，猛然间感到疼痛，使自己醒来，再坚持读书。这就是锥刺股的故事。

“头悬梁，锥刺股”的故事至今广为流传，也教育了一代又一代人，但凡学识渊博之人，都勤奋读书；要想有所成就，刻苦勤奋是必不可少的，这也正是踏实笃行的经典案例。正所谓：道虽迩，不行不至；事虽小，不为不成。亦可说：路虽远，行则将至；事虽难，做则可成。

项目评价

项目名称：OpenStack 初识					
班级：		学号：	姓名：		
评价指标		评价等级及分值	学生自评	组内互评	教师评分
素质目标达成情况（30%）	科技报国使命感（10%）	A（10分）：充分认识到科技发展的重要性，立志学好技术，为社会发展做贡献 B（7分）：充分认识到科技发展的重要性，但对学习技术没有明确目标 C（3分）：对技术的发展和使用不太关心			
	任务交付的职业综合能力（10%）	A（10分）：任务交付及时，任务实施过程完整，结果正确 B（7分）：任务交付比较及时，任务实施过程相对完整，结果正确 C（3分）：能够上交任务结果，完成部分任务实施过程，结果部分正确			
	自我学习热情（10%）	A（10分）：自我学习热情高涨 B（7分）：自我学习热情较好 C（3分）：自我学习热情一般			
知识目标达成情况（40%）	任务实施完成情况（20%）	A（20分）：任务实施全部完成 B（16分）：任务实施大部分完成 C（10分）：任务实施部分完成			
	测验作业完成情况（10%）	A（10分）：测验作业全部完成 B（7分）：测验作业大部分完成 C（3分）：测验作业部分完成			
	课上活动（10%）	A（10分）：积极参与课上抢答、提问、主题讨论等 B（7分）：能够参与课上抢答、提问、主题讨论等 C（3分）：部分课上抢答、提问、主题讨论等			
能力目标达成情况（30%）	任务实施完成质量（20%）	A（20分）：任务实施完成质量优秀 B（16分）：任务实施完成质量良好 C（10分）：任务实施完成质量一般			
	超凡脱俗（10%）	A（10分）：能够自主获取讲授外的课程知识，并传递给身边的同学 B（7分）：能够自主获取讲授外的课程知识 C（3分）：能够吸收教师所讲授的知识			

项目总结

近年来，以“云大物智”（云计算、大数据、物联网、人工智能）为代表的新一波数字技术浪潮席卷各行各业，云计算作为互联网的广泛普及和深度应用，它颠覆了个人计算，开创了崭新的技术领域，云计算实现了从芯片操作系统、应用软件到服务产业链的垂直整合。OpenStack 作为云计算领域的一项关键技术，每一个学习云计算技术的学生，都需要了解 OpenStack 技术。本项目主要包含了认识云计算和解构 OpenStack 两个任务。在认识云计算任务中，主要围绕什么是云计算，云计算有哪些特点，云计算的部署模式有哪些，云计算的服务模式有哪些，这四个问题开展云计算的介绍；在解构 OpenStack 任务中，围绕着 OpenStack 是什么，OpenStack 的技术框架什么样，OpenStack 有哪些组件，作用是什么，这四个问题开展对 OpenStack 的介绍。通过问题的形式，带着问题在任务中寻找答案，加强对云计算和 OpenStack 的理解。

项目二

虚拟化技术

项目导入

小李所做的私有云平台需要将计算、存储、网络等资源池化，从资源池内分发不同规格的虚拟机，那么搭建该平台就需要很好地理解虚拟化技术。为了在搭建云平台时能更好地理解底层架构，方便解决问题，小李决定在了解 OpenStack 的基础上，进一步了解虚拟化技术。

本项目将随着小李的学习脚步，首先，探索虚拟化技术，包括虚拟化技术概念、虚拟化技术特点、服务器虚拟化方式、虚拟化架构及 VMware 上完成虚拟机的安装；其次，掌握 KVM 虚拟机的使用，包括 KVM 安装、虚拟机创建、虚拟机管理、虚拟机存储管理、虚拟机网络管理等。

项目目标

【素质目标】

- 培养学生精益求精的工匠精神
- 激发学生知学 - 好学 - 乐学的学习热情

【知识目标】

- 了解虚拟化技术概念、特点
- 掌握服务器虚拟化方式及虚拟化架构
- 掌握 VMware 创建虚拟机的过程
- 掌握 KVM 安装过程
- 掌握 KVM 虚拟机的创建、管理、存储管理和网络管理

【能力目标】

- 能够在 VMware 上创建 CentOS 7 的虚拟机
- 能够完成 KVM 的部署和 KVM 虚拟机的管理

任务1　探索虚拟化

【任务描述】

小李决定先了解一下虚拟化的一些基本理论知识，包括虚拟化技术的概念、特点、服务器虚拟化方式、虚拟化体系架构等；然后在 VMware 虚拟机管理器上部署一台 Linux 虚拟机并安装 CentOS7 操作系统，通过实际操作加强理论知识的理解。

【知识要点】

1. 虚拟化技术的概念

虚拟化技术是云计算的根基，在计算机技术中，虚拟化（技术）或虚拟技术（Virtualization）是一种资源管理技术，是将计算机的各种实体资源（CPU、内存、磁盘空间、网络适配器等）予以抽象、转换后呈现出来，并可供分割、组合为一个或多个配置环境，打破实体结构间的不可切割的障碍，使用户可以比原本的配置更好的方式来应用硬件资源。虚拟化的本质是将硬件资源池化，按需分配，随意切割硬件资源。

2. 虚拟化技术特点

虚拟化技术的特点包含四个方面：分区、隔离、封装和相对于硬件独立，如图 2－1～图 2－4 所示。

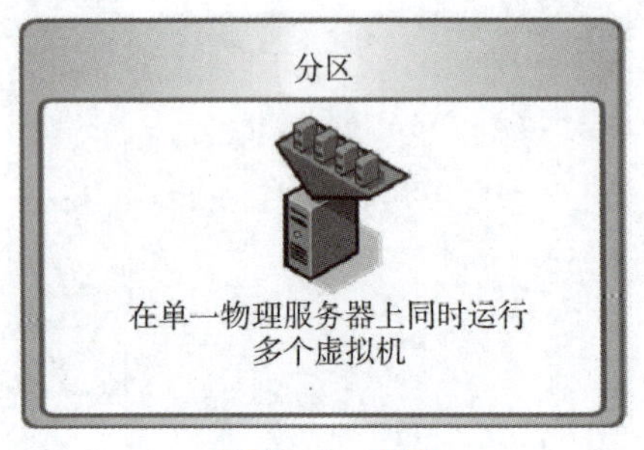

图 2－1　分区

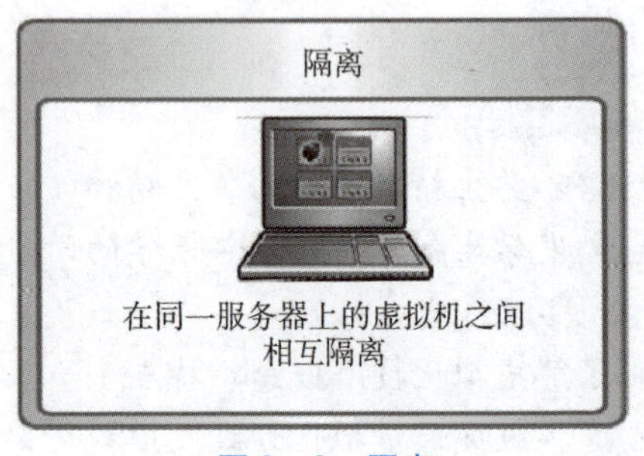

图 2－2　隔离

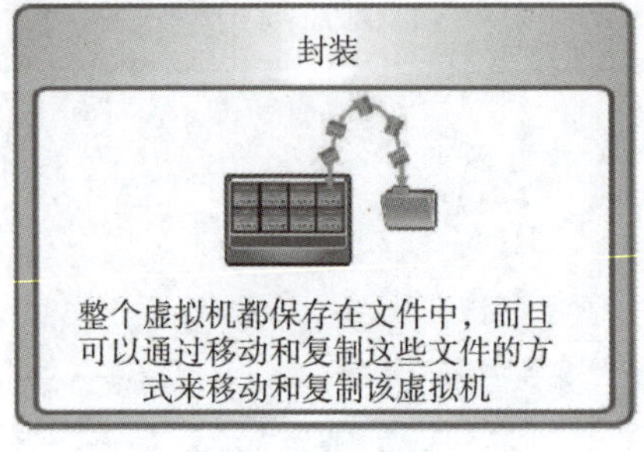

图 2－3　封装

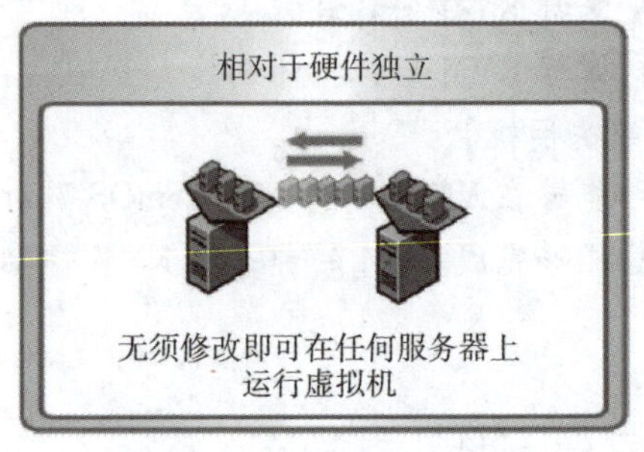

图 2－2　相对于硬件独立

● 分区

在一个单独的物理系统上，可以运行多个操作系统和应用；计算资源可以被放置在资源池中，并能够被有效地控制。

● 隔离

虚拟化能够提供理想化的物理机，每个虚拟机互相隔离；数据不会在虚拟机之间泄露；应用只能在配置好的网络连接上进行通信。

● 封装

虚拟单元的所有环境被存放在一个单独文件中；为应用展现的是标准化的虚拟硬件，确保兼容性；整个磁盘分区被存储为一个文件，易于备份、转移和复制。

● 相对于硬件独立

可以在其他服务器上不加修改地运行虚拟机。虚拟技术支持高可用性、动态资源调整，极大地提高系统的可持续运行能力。

3. 虚拟化技术的优势

● 更高的资源利用率

虚拟化技术支持实现物理资源和资源池的动态共享，提高资源利用率，特别是针对那些平均需求远低于需要为其提供专用资源的不同负载。

● 降低管理成本

虚拟化技术通过以下途径提高工作人员的效率：减少必须进行管理的物理资源的数量；隐藏物理资源的部分复杂性；通过实现自动化、获得更好的信息和实现中央管理来简化公共管理任务；实现负载管理自动化。另外，虚拟化技术还可以支持在多个平台上使用公共的工具。

● 提高使用灵活性

通过虚拟化实现动态的资源部署和重配置，满足不断变化的业务需求。

● 提高安全性

虚拟化技术可以实现较简单的共享机制无法实现的隔离和划分，这些特性可实现对数据和服务进行可控和安全的访问。

● 更高的可用性

虚拟化技术在不影响用户的情况下对物理资源进行删除、升级或改变。

● 更高的可扩展性

根据不同的产品，资源分区和汇聚可支持实现比个体物理资源小得多或大得多的虚拟资源，这意味着可以在不改变物理资源配置的情况下进行规模调整。

● 互操作性和投资保护

虚拟资源可提供底层物理资源无法提供的与各种接口和协议的兼容性。

● 改进资源供应

与个体物理资源单位相比，虚拟化能够以更小的单位进行资源分配。

4. 虚拟化体系架构

目前市场上 x86 架构管理程序（Hypervisor）的架构差异，有三种主要的虚拟化架构：

裸金属虚拟化（Ⅰ型）：裸金属虚拟化架构指直接在硬件上面安装虚拟化软件，再在其上安装操作系统和应用，依赖虚拟层内核和服务器控制台进行管理，如图 2-6 所示。

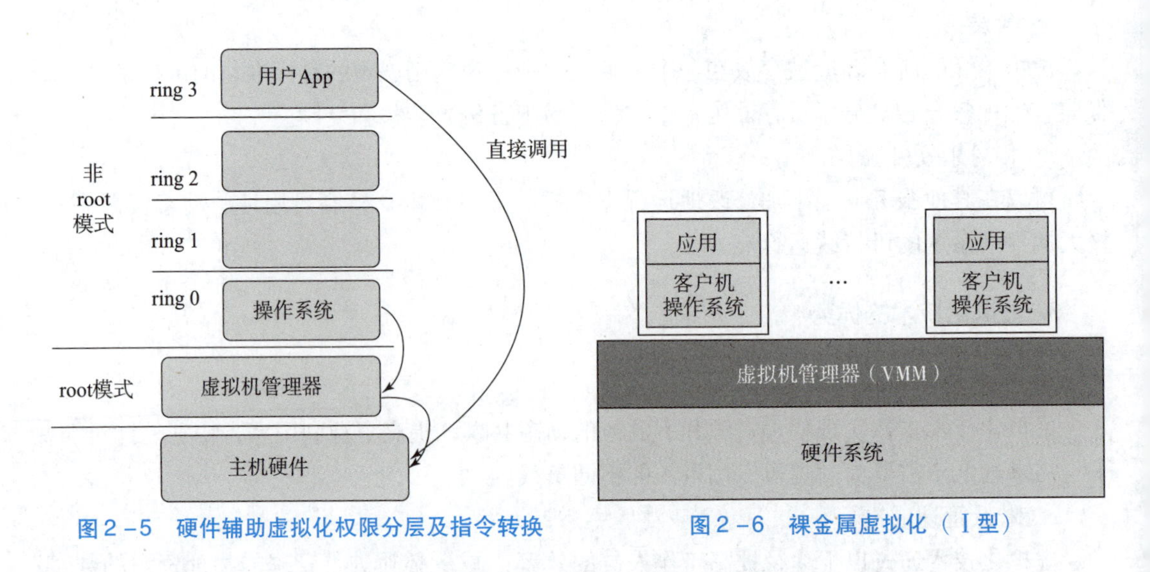

图 2-5　硬件辅助虚拟化权限分层及指令转换　　图 2-6　裸金属虚拟化（Ⅰ型）

宿主虚拟化（Ⅱ型）：寄居虚拟化架构指在宿主操作系统之上安装和运行虚拟化程序，依赖于宿主操作系统对设备的支持和物理资源的管理。类似于 VMware Workstation，如图 2-7 所示。

容器虚拟化：容器虚拟化架构指在宿主机操作系统上安装和运行容器引擎，类似于 Docker。基于容器引擎创建多个容器，通过一个或多个容器组合运行应用，如图 2-8 所示。

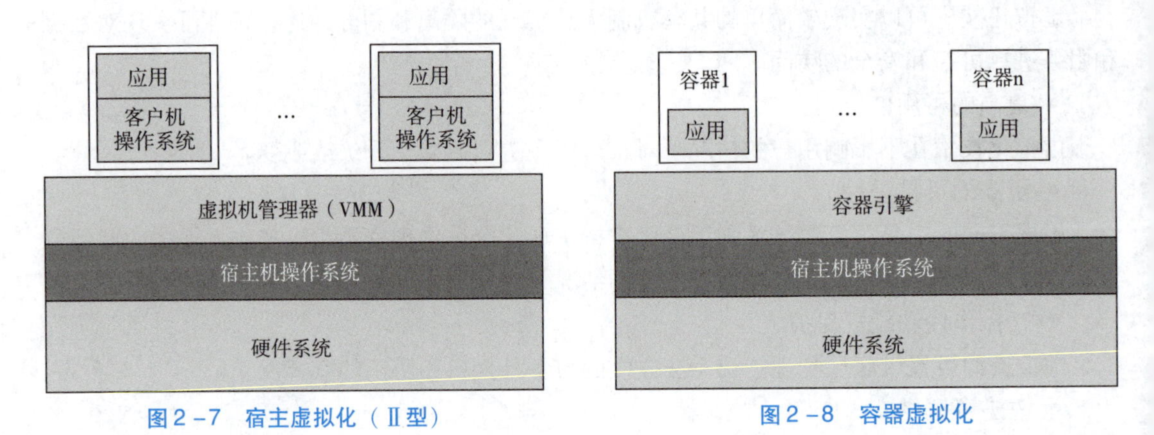

图 2-7　宿主虚拟化（Ⅱ型）　　图 2-8　容器虚拟化

【任务实施】

1. 虚拟机安装

1）安装 VMware Workstation Pro 虚拟化软件

（1）双击 VMware Workstation Pro 安装包进入安装向导欢迎界面，单击“下一步”按钮，如图 2－9 所示。

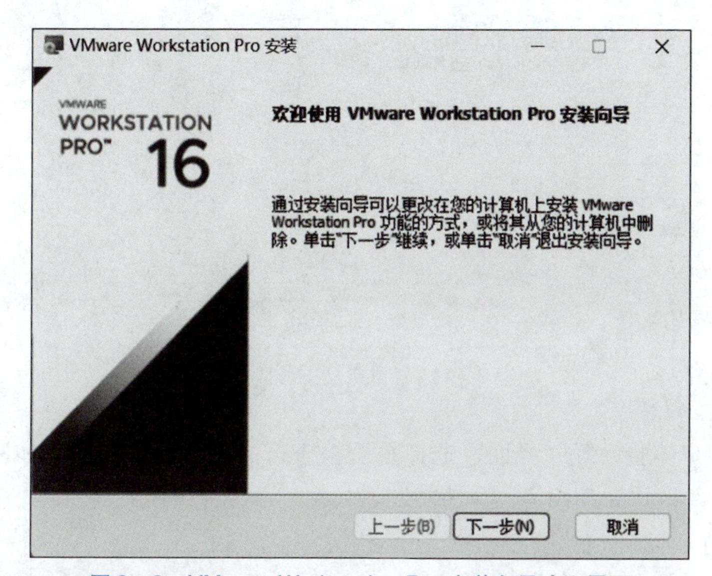

图 2－9　VMware Workstation Pro 安装向导欢迎界面

（2）选择“我接受许可协议中的条款”复选框并单击“下一步”按钮，如图 2－10 所示。

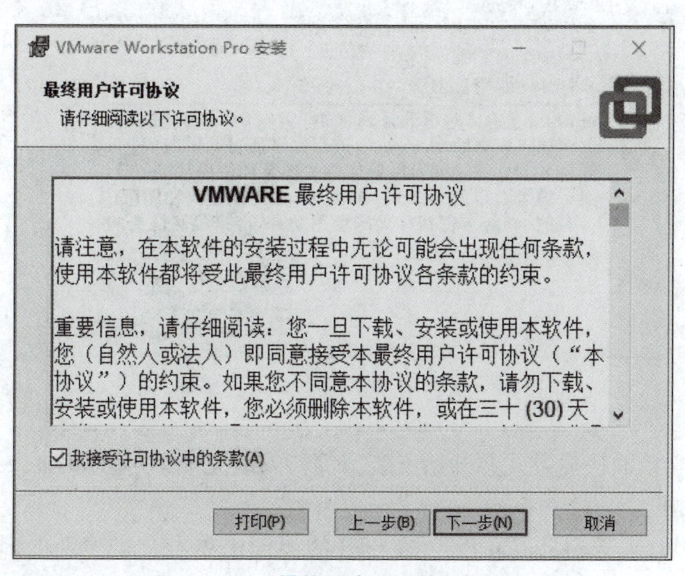

图 2－10　“最终用户许可协议”界面

（3）单击“更改”按钮，自定义软件安装位置，之后单击“下一步”按钮，如图 2-11 所示。

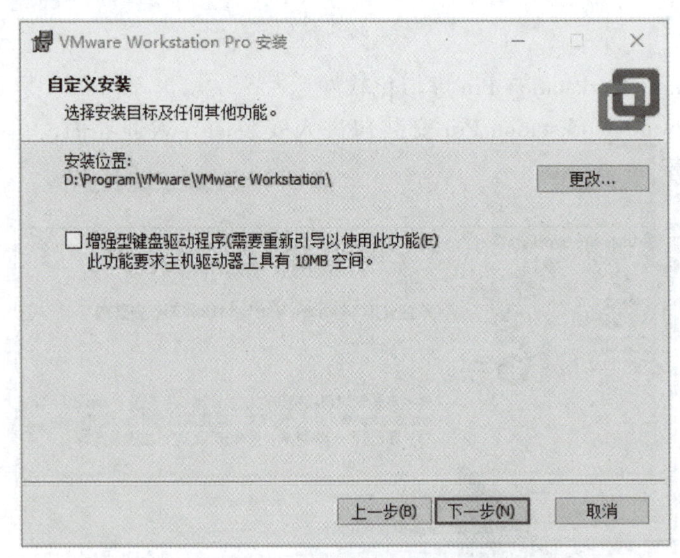

图 2-11 “自定义安装”界面

（4）在“用户体验设置”界面，选择是否检查产品更新及是否加入体验提升计划，之后单击“下一步”按钮，如图 2-12 所示。

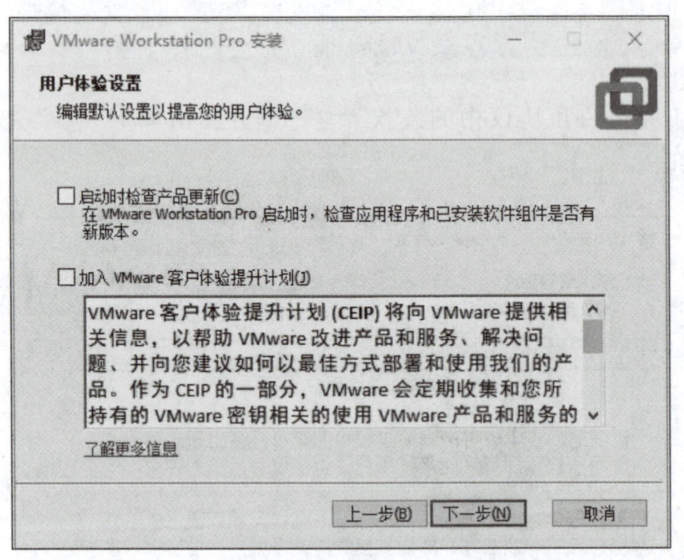

图 2-12 “用户体验设置”界面

（5）选择创建桌面快捷方式并开始安装，等待安装完成即可，如图 2-13 所示。

2）创建虚拟机

（1）打开 VMware Workstation Pro，在打开的界面中单击“创建新的虚拟机”按钮，如图 2-14 所示。

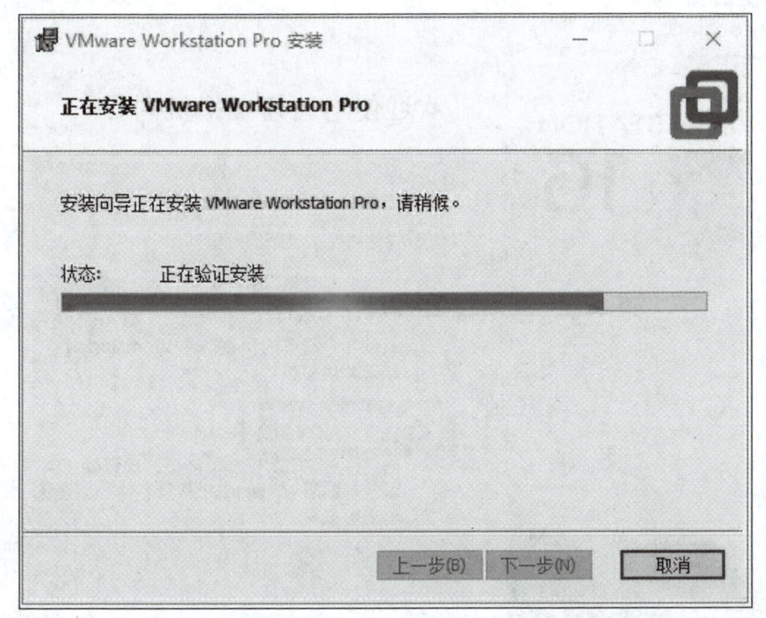

图 2-13　“正在安装 VMware Workstation Pro”界面

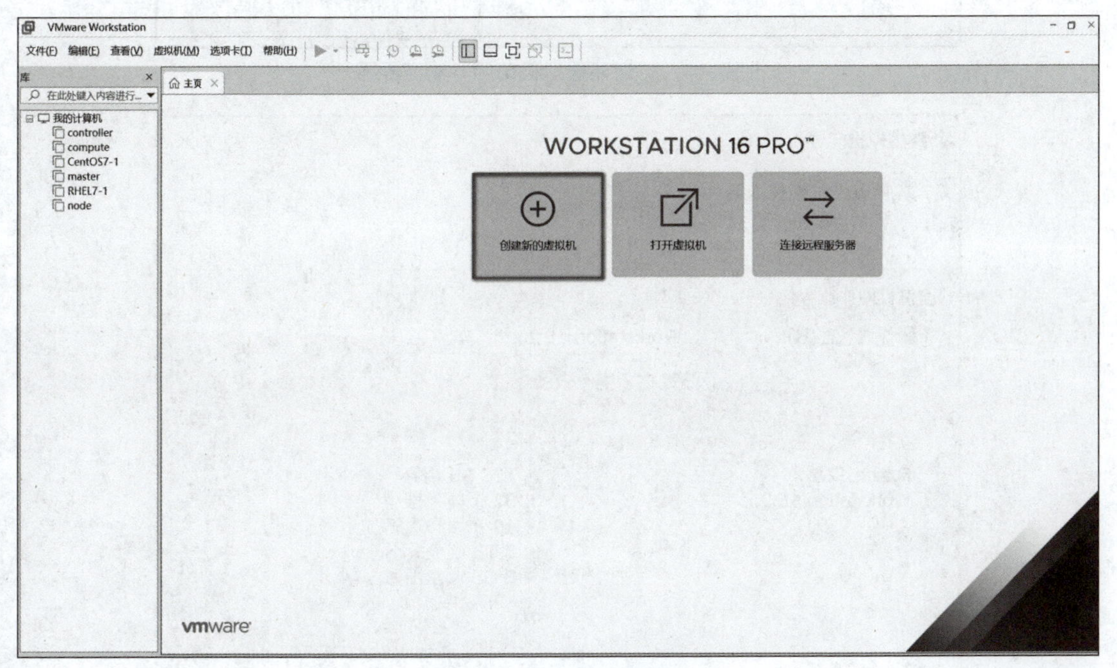

图 2-14　创建新的虚拟机

（2）在“新建虚拟机向导”欢迎界面中选择“自定义（高级）”单选按钮，并单击“下一步”按钮，如图 2-15 所示。

（3）在“选择虚拟机硬件兼容性”界面中，保持默认设置不变，并单击“下一步”按钮，如图 2-16 所示。

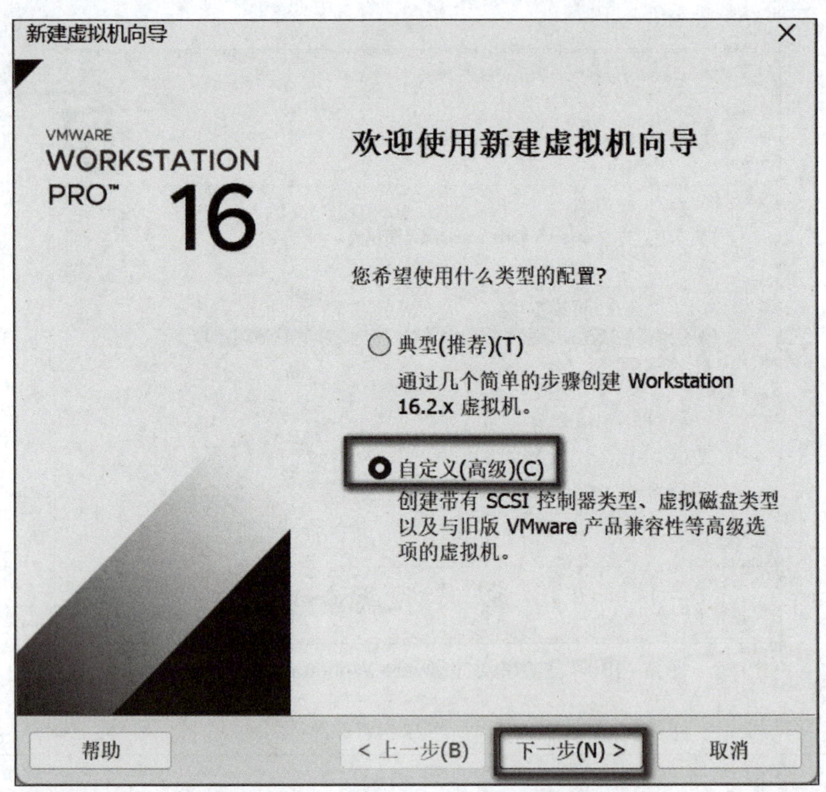

图 2-15 “新建虚拟机向导”欢迎界面

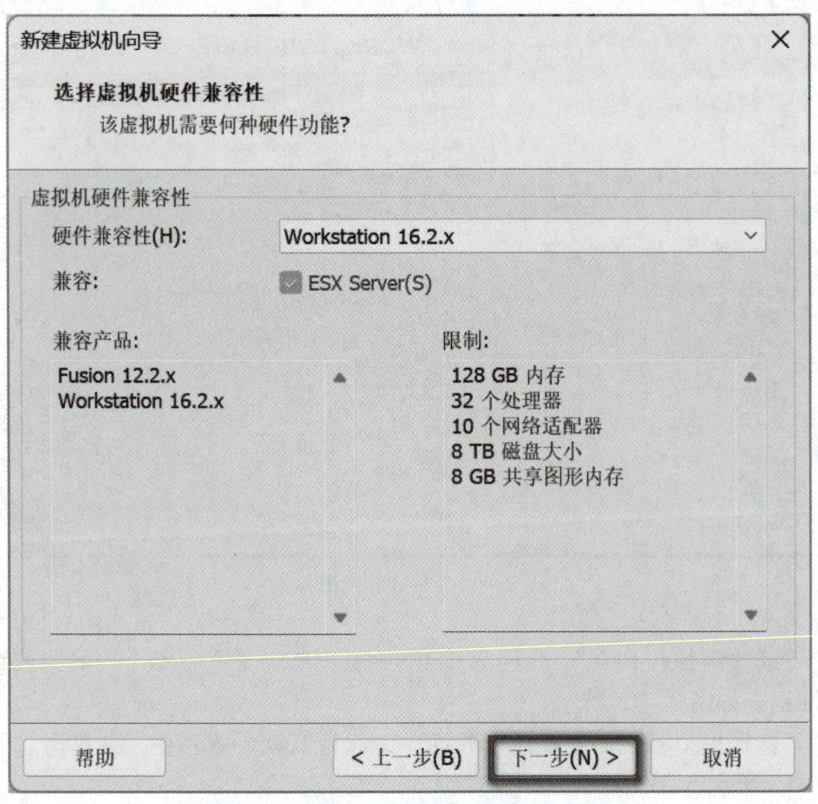

图 2-16 “选择虚拟机硬件兼容性”界面

（4）在“安装客户机操作系统”界面中，选择“稍后安装操作系统”单选按钮并单击“下一步”按钮，如图 2－17 所示。

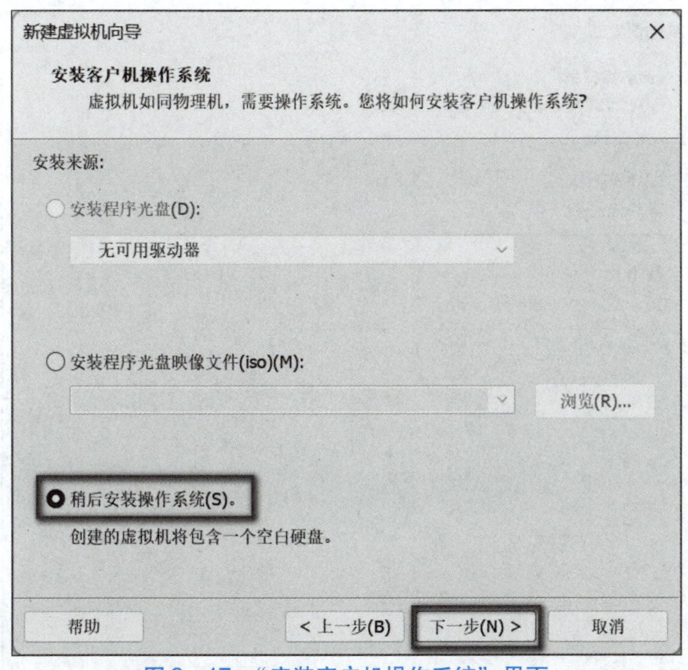

图 2－17　“安装客户机操作系统”界面

（5）在“选择客户机操作系统”界面的“客户机操作系统”列表中，选择“Linux”单选按钮；版本选择“CentOS 7 64 位”，并单击“下一步”按钮，如图 2－18 所示。

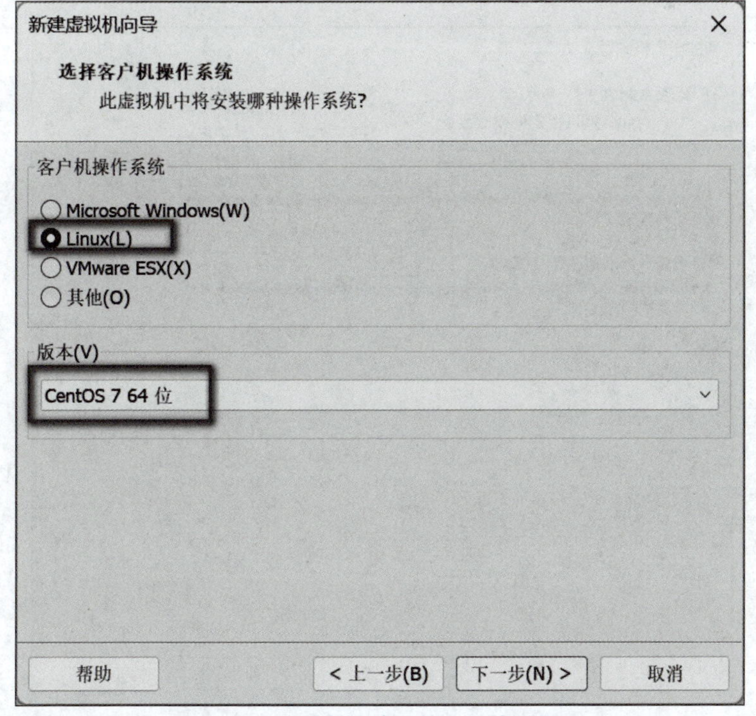

图 2－18　“选择客户机操作系统”界面

（6）在“命名虚拟机”界面中，输入虚拟机的名称，如“vm1 - centos7”，在“位置”编辑框选择虚拟机的存储位置，单击“下一步”按钮，如图 2 - 19 所示。

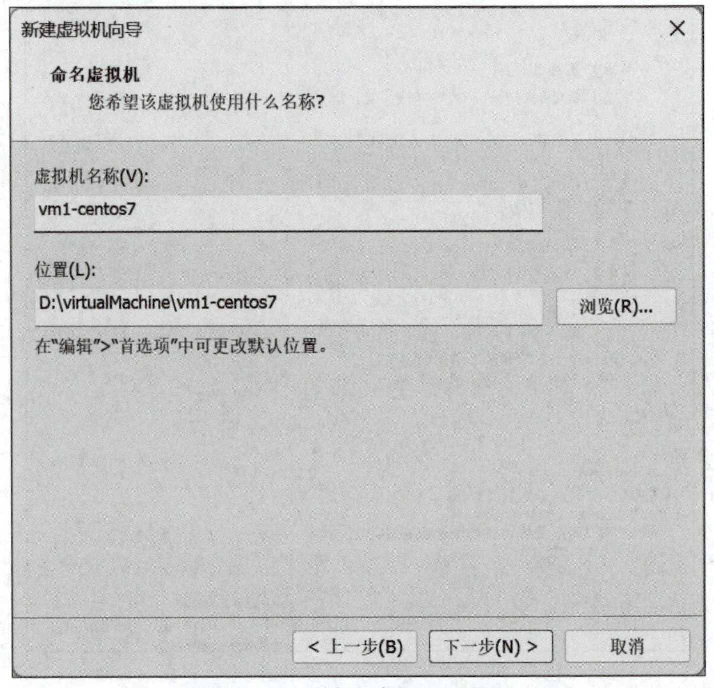

图 2 - 19 “命名虚拟机”界面

（7）在“处理器配置”界面中，设置处理器的数量和每个处理器的内核数量，单击“下一步”按钮，如图 2 - 20 所示。

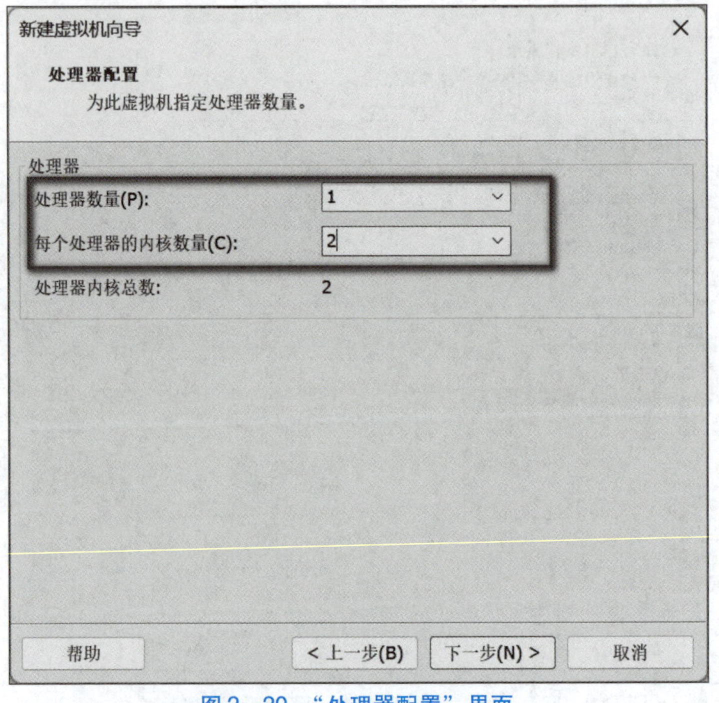

图 2 - 20 “处理器配置”界面

（8）在“此虚拟机的内存”界面中，设置该虚拟机的内存大小，单击“下一步”按钮，如图 2-21 所示。

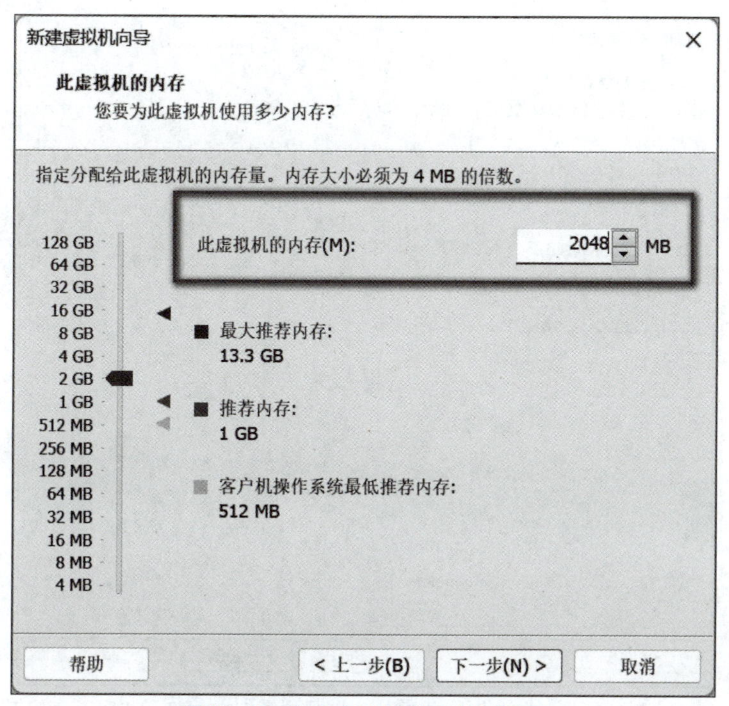

图 2-21 “此虚拟机的内存”界面

（9）在“网络类型”界面中，选择“使用网络地址转换（NAT）”单选按钮并单击“下一步”按钮，如图 2-22 所示。

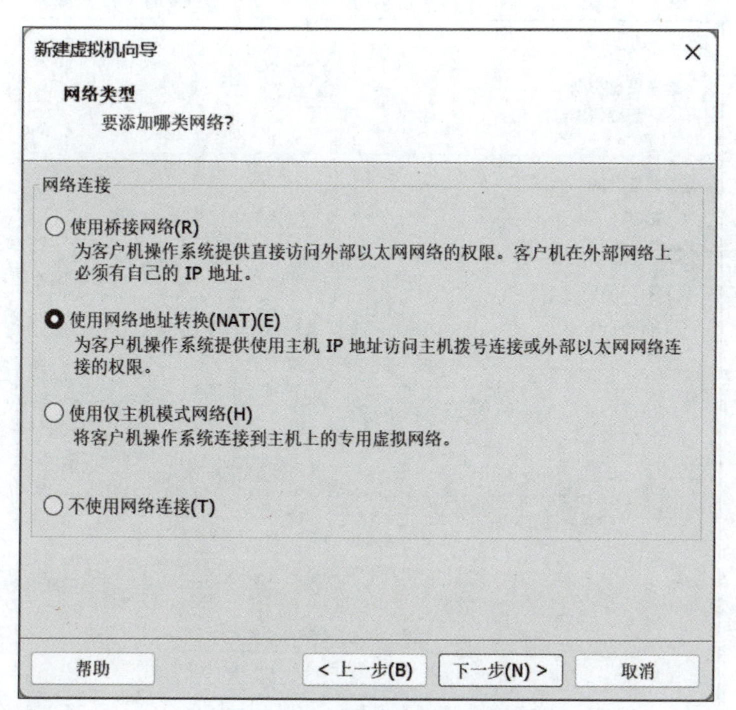

图 2-22 “网络类型”界面

（10）在“选择 I/O 控制器类型”界面中，设置虚拟机的 I/O 控制器类型为“LSI Logic”，单击“下一步”按钮，如图 2－23 所示。

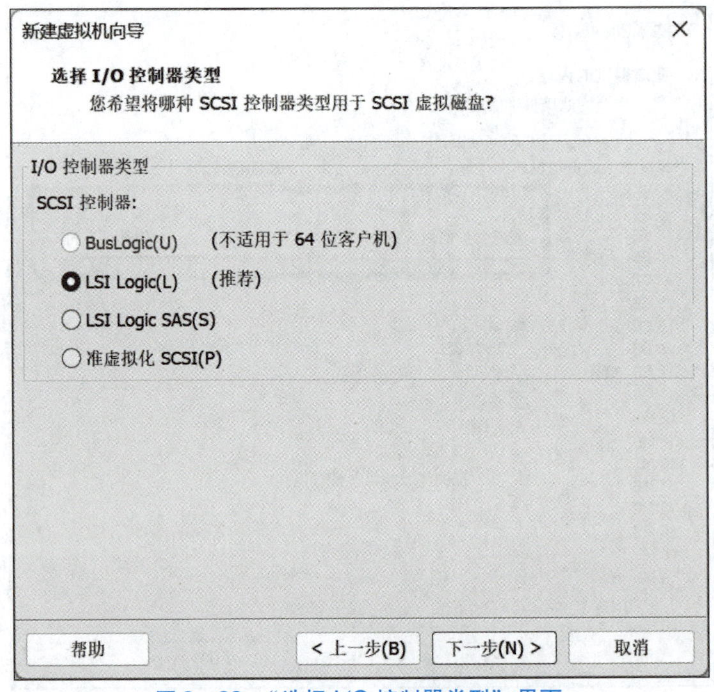

图 2－23 “选择 I/O 控制器类型”界面

（11）在“选择磁盘类型”界面中，设置虚拟磁盘类型为“SCSI”，单击“下一步”按钮，如图 2－24 所示。

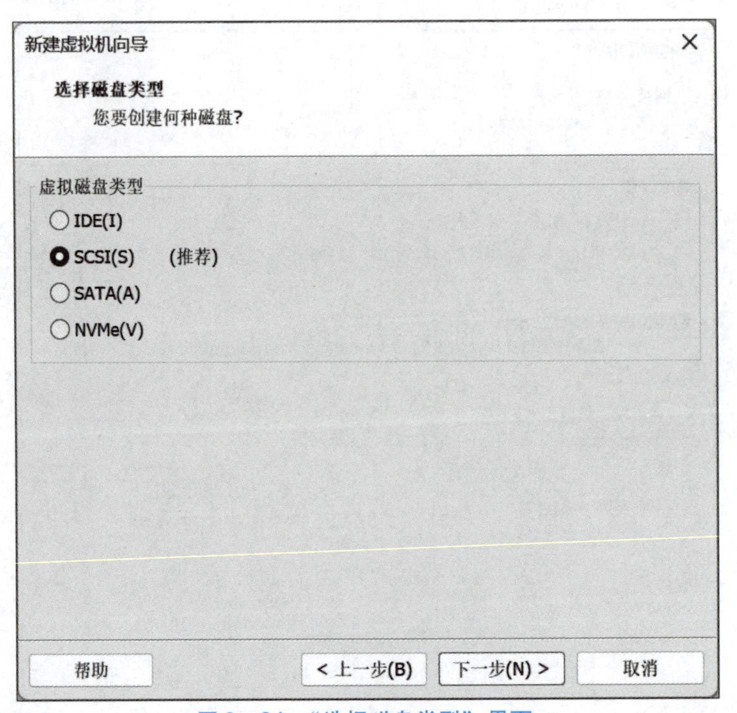

图 2－24 “选择磁盘类型”界面

（12）在“选择磁盘”界面，选择“创建新虚拟磁盘”单选按钮并单击“下一步”按钮，如图 2－25 所示。

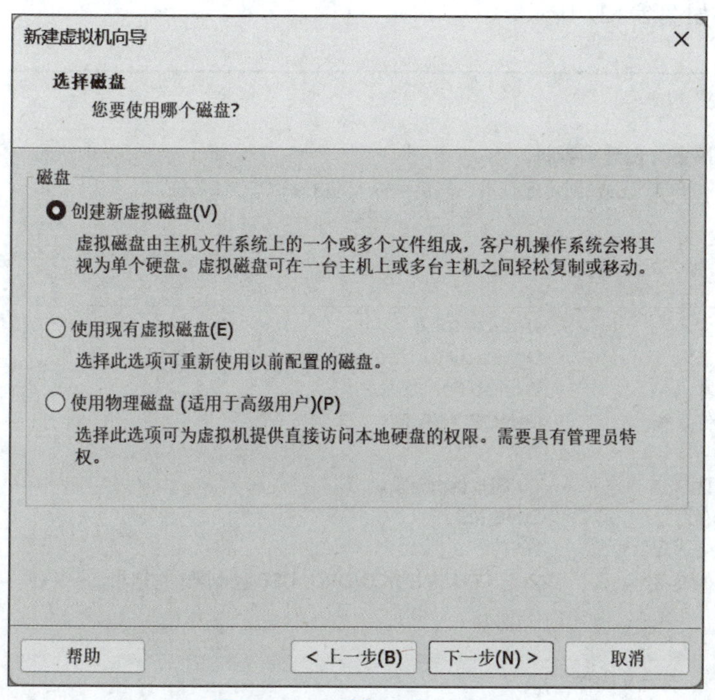

图 2－25 “选择磁盘”界面

（13）在“指定磁盘容量”界面中，设置虚拟机磁盘大小并选择“将虚拟磁盘拆分成多个文件”单选按钮，单击“下一步”按钮，如图 2－26 所示。

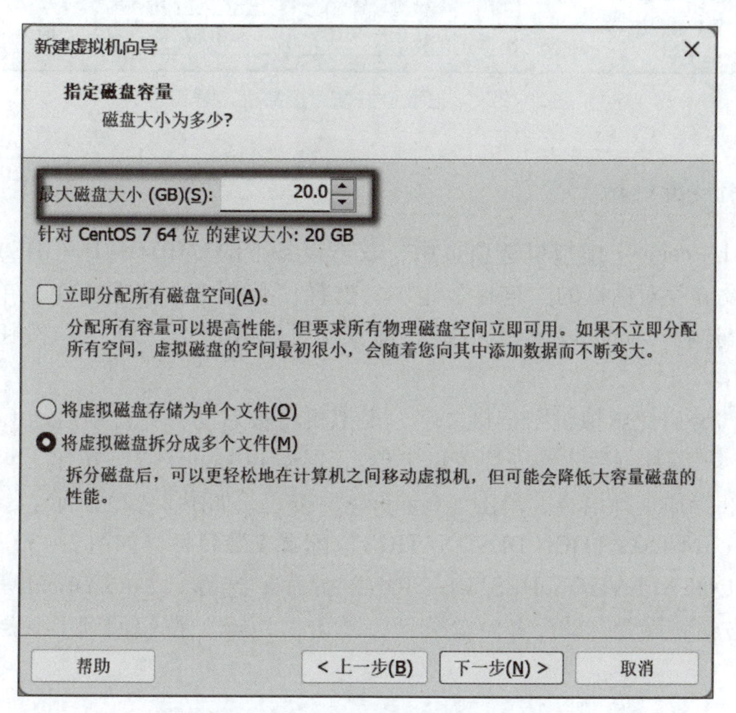

图 2－26 “指定磁盘容量”界面

（14）指定磁盘文件界面保持默认设置不变，并单击“下一步”按钮。

（15）“已准备好创建虚拟机”界面会显示虚拟机的所有配置，单击“完成”按钮完成虚拟机的创建，如图 2－27 所示。

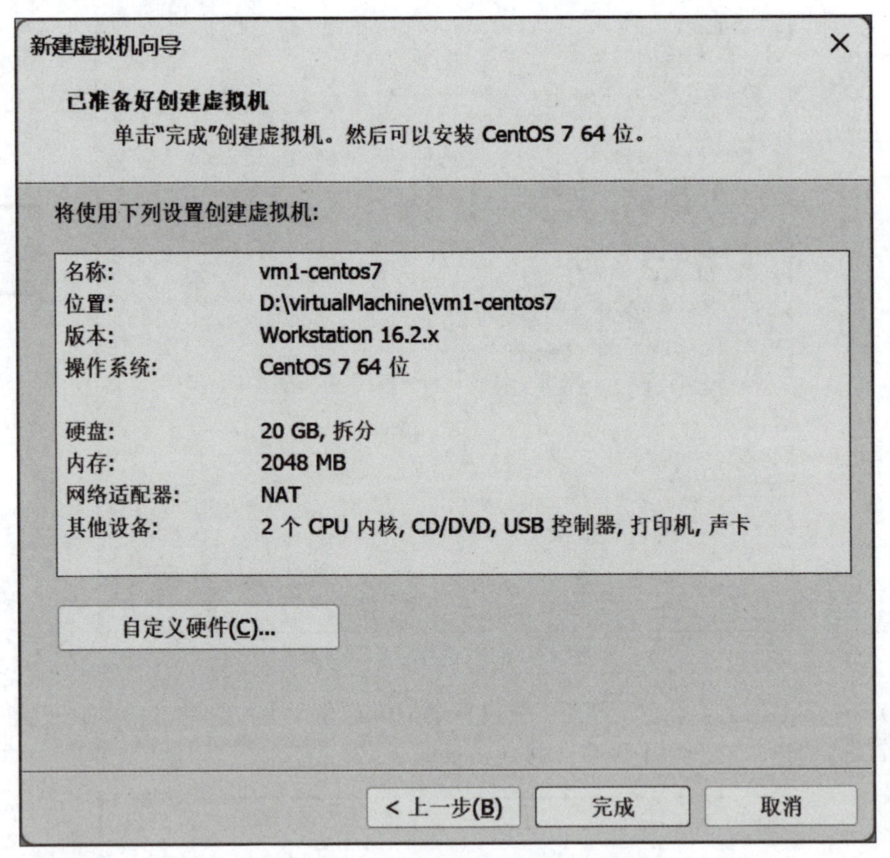

图 2－27 “已准备好创建虚拟机”界面

2. 为虚拟机安装 CentOS 7

（1）在 vm1－centos7 虚拟机选项页中，双击设备下的“CD/DVD（IDE）”选项，在弹出的“虚拟机设置”对话框的“连接”组中，选择“使用 ISO 映像文件”单选按钮，单击“浏览”按钮，选择 CentOS 7 操作系统的镜像文件，并单击“确定”按钮，如图 2－28 所示。

（2）单击“开启此虚拟机”链接文字，虚拟机将被启动，并自动启动 CentOS 7 操作系统的安装程序，稍等片刻后，在安装窗口中单击“Install CentOS 7”按钮。

（3）在安装界面选择语言，单击“Continue”按钮，如图 2－29 所示。

（4）选择“INSTALLATION DESTINATION”配置安装目标，如图 2－30 所示。

（5）在“INSTALLATION DESTINATION”界面，选择“Automatically configure partitioning”单选按钮，设置自动配置磁盘分区，单击“Done”按钮回到上一层配置界面，如图 2－31 所示。

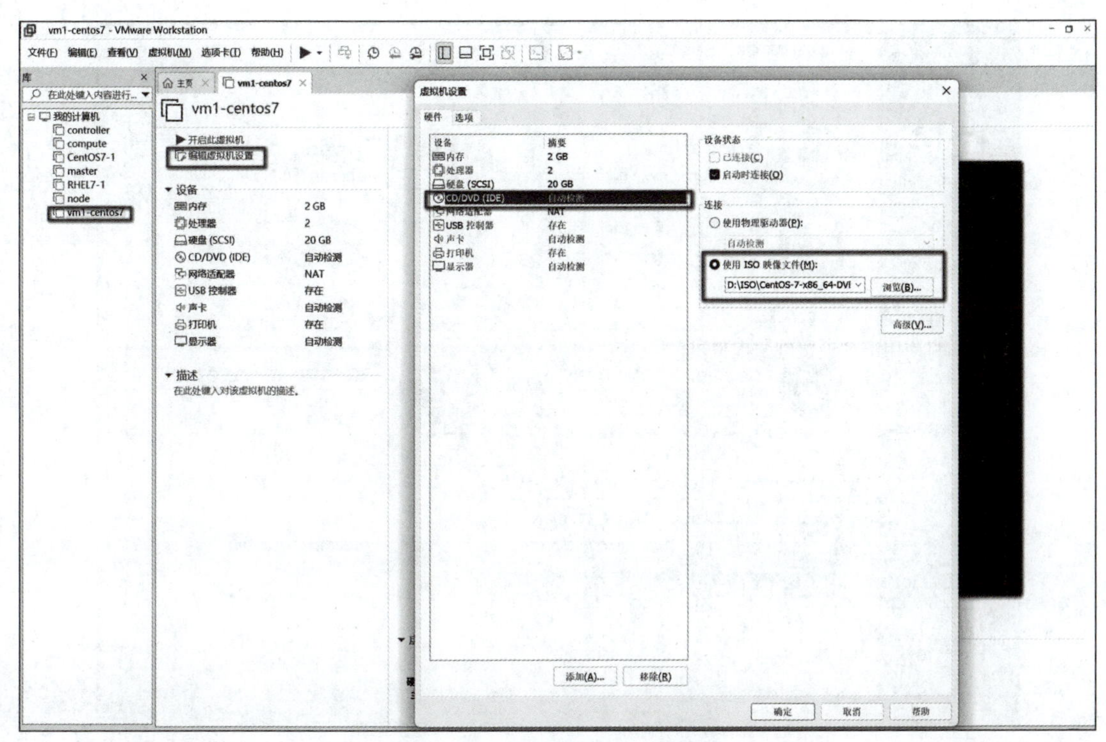

图 2－28　“虚拟机设置”对话框

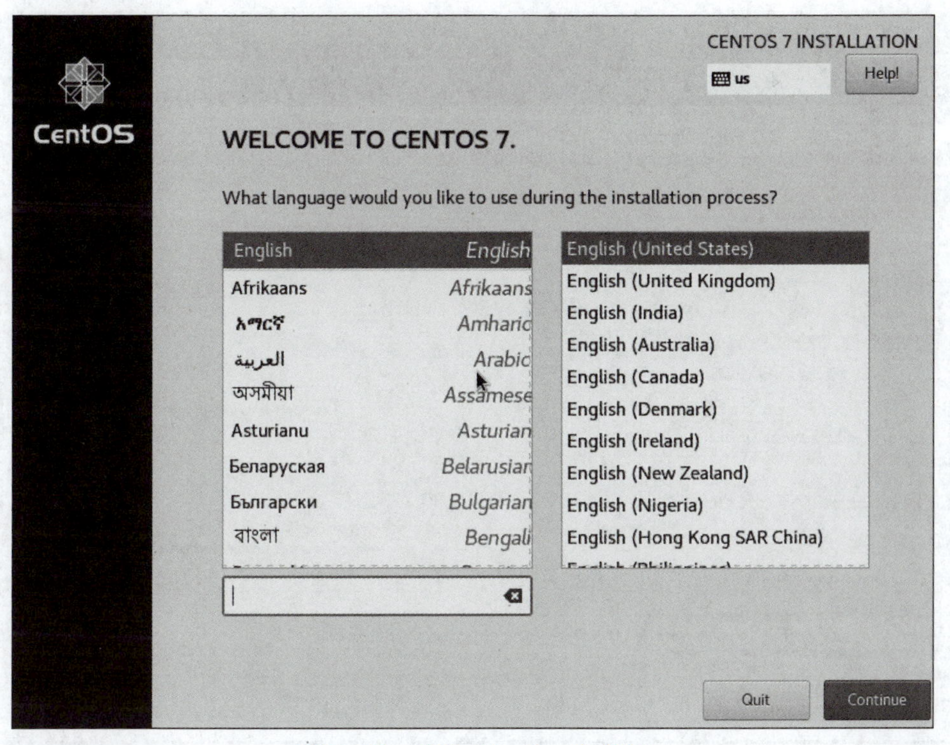

图 2－29　单击“Continue”按钮

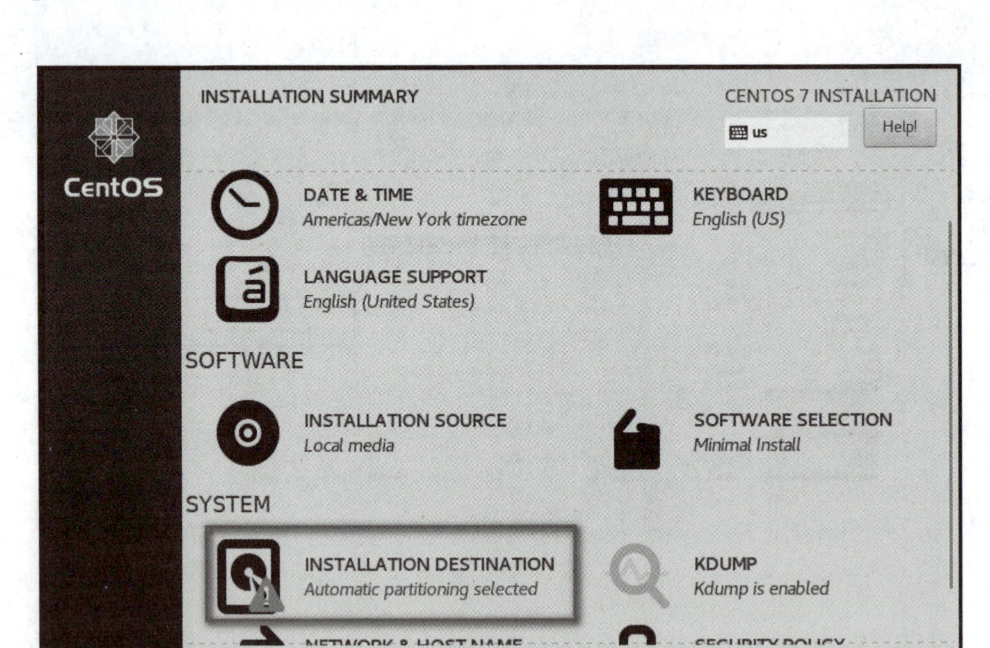

图 2－30 “INSTALLATION SUMMARY” 界面

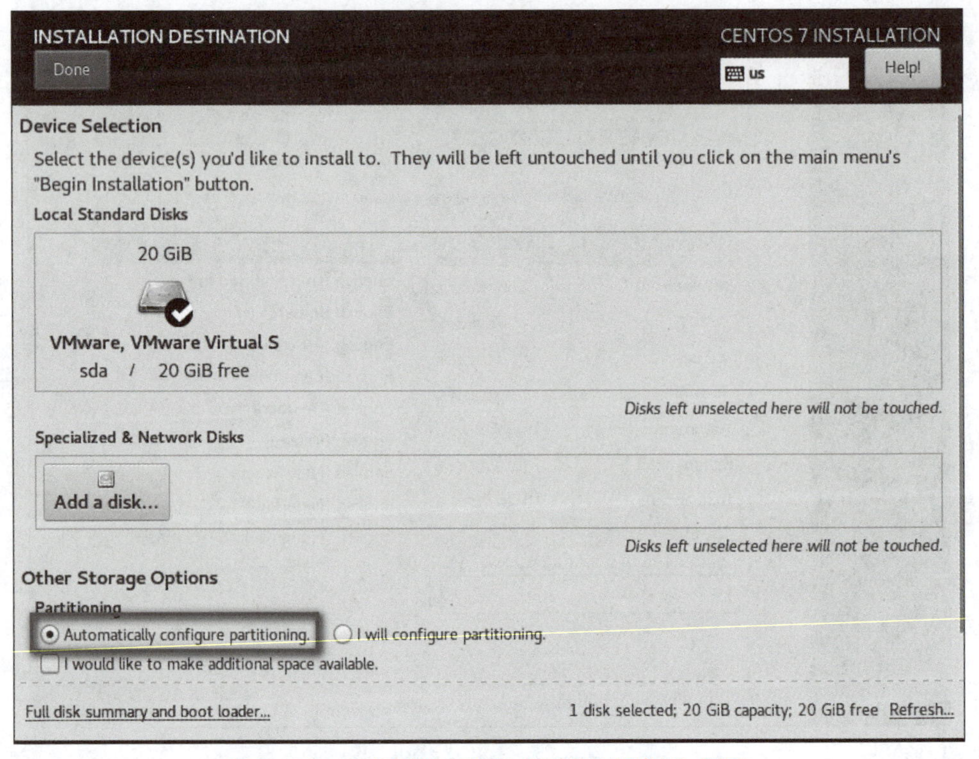

图 2－31 “INSTALLATION DESTINATION” 界面

（6）在“SOFTWARE SELECTION”选项内可以选择虚拟机安装的软件，默认是最小化安装，只安装支撑系统运行的基本软件。如果需要图形界面或更多功能，可以单击选择带GUI的服务器，单击“Begin Installation”按钮开始虚拟机安装，如图2-32所示。

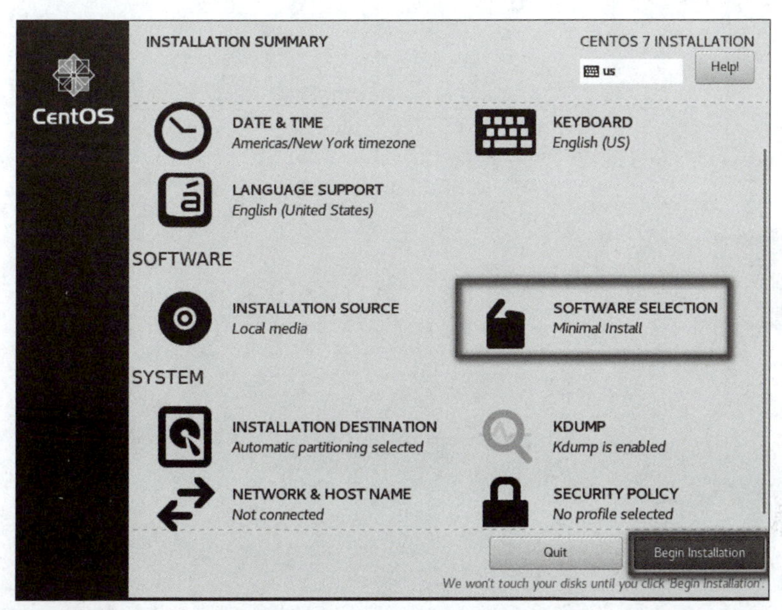

图2-32　开始安装界面

（7）开始安装后，在“CONFIGURATION”界面配置ROOT用户的密码，也可以创建新的用户并设置密码，如图2-33所示。

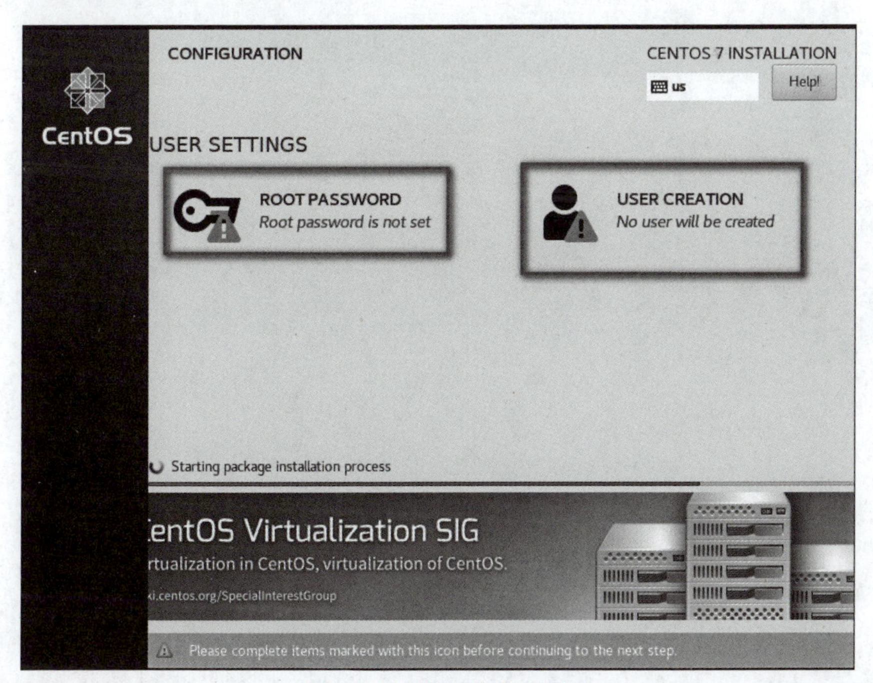

图2-33　配置用户密码界面

（8）此时虚拟机的安装已经完成，单击“Reboot”，重启虚拟机，登录即可使用。

【任务工单】

工单号：2－1

<table>
<tr><td colspan="2">项目名称：虚拟化技术</td><td colspan="2">任务名称：探索虚拟化</td></tr>
<tr><td colspan="2">班级：</td><td>学号：</td><td>姓名：</td></tr>
<tr><td>任务安排</td><td colspan="3">□总结虚拟化的技术特点
□安装 VMware Workstation Pro 虚拟化软件
□在 VMware 上创建双核 4G 规格的 Linux 虚拟机
□对虚拟机安装 CentOS 7 操作系统</td></tr>
<tr><td>成果交付形式</td><td colspan="3">在装好的 CentOS 7 虚拟机上执行 ip addr，查看 IP 地址信息，并截图上传至教学平台</td></tr>
<tr><td>任务实施总结</td><td colspan="3">任务自评（0～10 分）：
项目收获：________
____
____
改进点：________
____
________</td></tr>
<tr><td>成果验收</td><td colspan="3">□完全满足任务要求
□基本满足任务要求
要求全部完成，虚拟机存在一些小的问题，需要改进的地方：
____
____
____
□不能满足需求
需求未全部完成，虚拟机存在一些影响使用的问题，需要改进的地方：
____
____
________</td></tr>
</table>

【知识巩固】

1. 虚拟化技术特点包括（　　）。

A. 分区　　B. 隔离　　C. 封装　　D. 硬件独立

2. 安装虚拟机时，网络连接的选项包括（　　）。

A. 桥接网络　　B. NAT 网络　　C. 不使用网络　　D. 仅主机网络

【小李的反思】

知之愈明，则行之愈笃。行之愈笃，则知之益明。

出自朱熹《朱子语类》，意思是说，理解得越清楚，实践就越扎实；实践越扎实，认识就会更加清晰。

在学习的过程中，要加强理论知识的理解，更要实践出真知，同学们初次接触虚拟化技术，只通过理论的学习理解虚拟化概念是很难做到的，在本任务中，通过部署虚拟机的实践操作，通过实践加强知识的理解。

漫画大师丰子恺先生有一天忽然灵感大发，研墨挥毫画就了一幅题为《卖羊》的漫画：一个农人牵着两只湖羊，到羊肉馆来卖给老板。画好后，丰先生觉得很满意，就带上漫画来到羊肉馆，想让老板和顾客们也欣赏一番。谁知道，一位农民顾客看了却连连摇头发笑。丰先生觉得纳闷儿，就上前虚心地请教他因何而摇头发笑。那农民说，多画了一条绳子。丰先生听了，回过头来又仔细看看自己的画，觉得想不通：两条绳子牵两只羊，哪里多了绳子？这时，那个农民站起来认真地告诉他，牵羊只需牵头羊，不管多少只，只要一条绳子就够了。此时，丰先生才恍然大悟。

丰子恺先生认为两只羊用两根绳子牵没有问题，那是因为他没有真正牵过羊，经常实践的农民却知道牵羊只须牵头羊即可，这正所谓实践出真知。

任务2 使用 KVM

【任务描述】

小李已经通过 VMware 部署虚拟机的过程了解了虚拟化的一些理论知识，这时，小李又了解到，搭建 OpenStack 私有云会用到更多基于 Linux 内核的 KVM 虚拟机管理器。于是，小李开始了解 KVM 虚拟机管理器的使用，包括 KVM 的安装、KVM 虚拟机的安装和管理、虚拟机存储管理和虚拟机网络管理等。

【知识要点】

KVM（Kernel - based Virtual Machine，基于内核的虚拟机）是一个开源的系统虚拟化模块，负责 CPU 和内存的虚拟化。KVM 的虚拟化需要硬件支持（如 Intel VT 技术或者 AMD V 技术），是基于硬件的完全虚拟化。

QEMU 是一个独立的虚拟化解决方案，从这个角度来说它并不依赖 KVM。而 KVM 是另一套虚拟化解决方案，不过因为这个方案实际上只实现了内核中对处理器虚拟化特性的支持，换言之，它缺乏设备虚拟化以及相应的用户空间管理虚拟机的工具，所以它借用了 QEMU 的代码并加以精简，连同 KVM 一起构成了另一个独立的虚拟化解决方案：KVM 是 QEMU 的加速器。

KVM 负责 CPU 虚拟化 + 内存虚拟化，实现了 CPU 和内存的虚拟化，但 KVM 并不能模拟其他设备，还必须有一个运行在用户空间的工具才行。KVM 的开发者选择了比较成熟的开源虚拟化软件 QEMU 作为这个工具。QEMU 模拟 I/O 设备（网卡、磁盘等）对其进行了修改，KVM 加上 QEMU 后，就是完整意义上的服务器虚拟化，最后形成了 QEMU - KVM。

libvirt 是目前使用最为广泛的对 KVM 虚拟机进行管理的工具和应用程序接口。libvirt 屏蔽了不同虚拟化的实现，提供统一管理接口。用户只关心高层的功能，而 VMM 的实现细节，对于最终用户应该是透明的。libvirt 就作为 VMM 和高层功能之间的桥梁，接收用户请求，然后调用 VMM 提供的接口，来完成最终的工作，如图 2 - 34 所示。

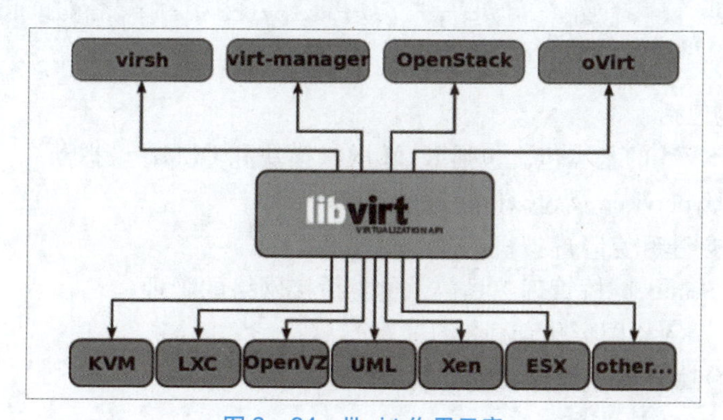

图 2 - 34 libvirt 作用示意

【任务实施】

1. KVM 软件包安装

（1）安装 KVM 虚拟化软件包需要的硬件条件：

CPU：64 位，开启虚拟化功能，使用 egrep 'vmx | svm'/proc/cpuinfo 命令检查 CPU 是否开启了虚拟化功能。输出内容如下所示时，为开启了 CPU 虚拟化；如果没有输出内容，则没有开启其 CPU 虚拟化。

```
[root@ centos ~]# egrep 'vmx |svm'/proc/cpuinfo
flags     :fpu vme de pse tsc msr pae mce cx8 apic sep mtrr pge mca cmov pat pse36
clflush mmx fxsr sse sse2 ss ht syscall nx pdpe1gb rdtscp lm constant_tsc arch_perfmon
nopl xtopology tsc_reliable nonstop_tsc eagerfpu pni pclmulqdq vmx ssse3 fma cx16 pcid
sse4_1 sse4_2 x2apic movbe popcnt tsc_deadline_timer aes xsave avx f16c rdrand hypervi-
sor lahf_lm abm 3dnowprefetch ssbd ibrs ibpb stibp ibrs_enhanced tpr_shadow vnmi ept
vpid fsgsbase tsc_adjust bmi1 avx2 smep bmi2 invpcid rdseed adx smap clflushopt clwb
sha_ni xsaveopt xsavec arat pku ospke spec_ctrl intel_stibp flush_l1d arch_capabili-
ties
flags     :fpu vme de pse tsc msr pae mce cx8 apic sep mtrr pge mca cmov pat pse36
clflush mmx fxsr sse sse2 ss ht syscall nx pdpe1gb rdtscp lm constant_tsc arch_perfmon
nopl xtopology tsc_reliable nonstop_tsc eagerfpu pni pclmulqdq vmx ssse3 fma cx16 pcid
sse4_1 sse4_2 x2apic movbe popcnt tsc_deadline_timer aes xsave avx f16c rdrand hypervi-
sor lahf_lm abm 3dnowprefetch ssbd ibrs ibpb stibp ibrs_enhanced tpr_shadow vnmi ept
vpid fsgsbase tsc_adjust bmi1 avx2 smep bmi2 invpcid rdseed adx smap clflushopt clwb
sha_ni xsaveopt xsavec arat pku ospke spec_ctrl intel_stibp flush_l1d arch_capabili-
ties
```

操作系统：64 位的 CentOS 或 Red Hat，6.0 以上版本。

内存：不少于 2 GB。

存储：不少于 6 GB。

（2）安装软件：

```
[root@ centos ~]# yum install -y qemu-kvm libvirt virt-install bridge-utils qe-
mu-img
```

安装的软件包包括：

qemu-kvm，KVM 的基本包，包括 KVM 内核模块和 QEMU 模拟器。

libvirt，提供 Hypervisor 及虚拟机管理的 API。

virt-install，创建和克隆虚拟机的命令行工具。

bridge-utils，Linux 网桥管理工具包，负责桥接网络的管理。

virt-manager，KVM 图形化管理工具。

qemu-img，QEMU 磁盘镜像管理工具。

（3）启动 libvirt 守护进程：

```
[root@ centos ~]# systemctl start libvirtd
```

```
[root@ centos ~]# systemctl enable libvirtd
```

检查与 KVM 有关的内核模块，如果有输出，则说明 KVM 内核模块已加载。

```
[root@ centos ~]# lsmod |grep kvm
kvm_intel              183621 0
kvm                     586948 1 kvm_intel
irqbypass              13503 1 kvm
```

2. 虚拟机创建

常用来创建虚拟机的方式有三种：virt - manager 可视化工具创建、virt - install 命令行创建和 XML 配置文件创建。

1）使用 virt - manager 可视化工具创建虚拟机

（1）在命令行中执行 virt - manager 命令启动 virt - manager 图形化界面，如图 2 - 35 所示。

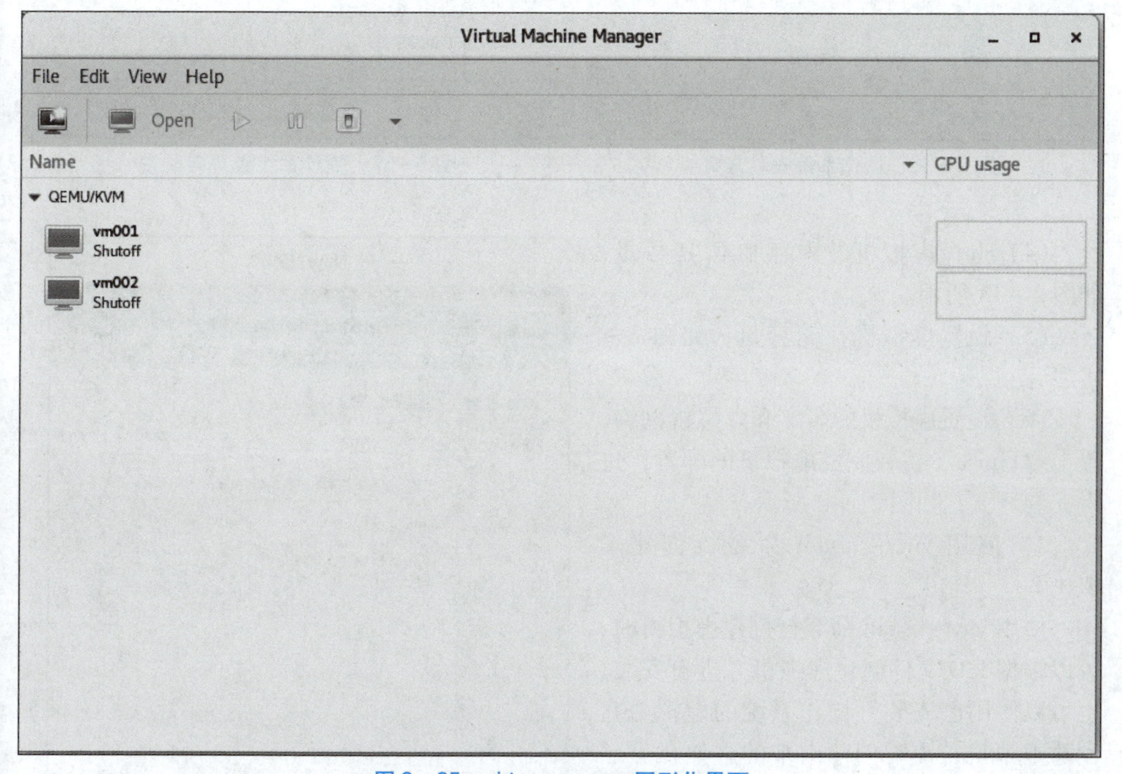

图 2 - 35　virt - manager 图形化界面

（2）菜单栏的“File”选项下选择“New Virtual Machine”选项，此时会弹出“New VM”对话框，选择操作系统安装方式，共包含 4 种安装方式，如图 2 - 36 所示。

- 通过 ISO 文件安装
- 通过网络安装
- 通过 PXE 安装

- 通过现有磁盘镜像文件安装

（3）以通过 ISO 文件安装为例。单击“Forward”按钮之后，选择 ISO 镜像文件以及操作系统类型和版本（也可以自动识别），如图 2－37 所示。

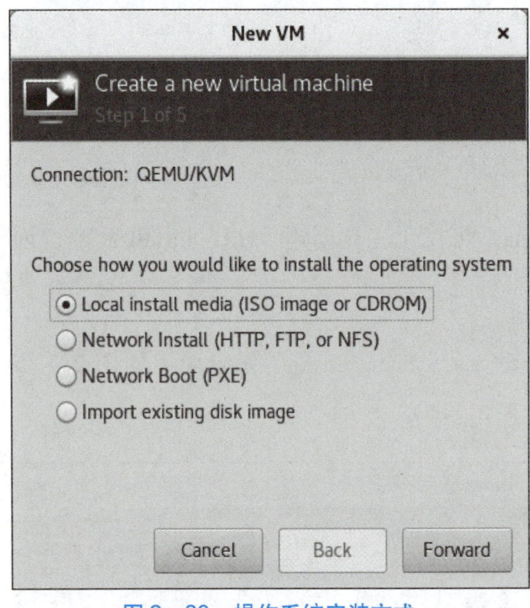

图 2－36　操作系统安装方式

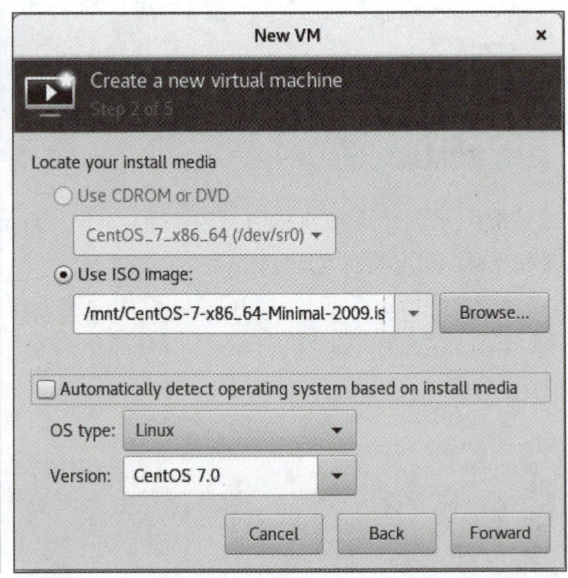

图 2－37　操作系统选择与配置

（4）选择虚拟机的内存和 CPU 数量，如图 2－38 所示。

（5）设置虚拟机存储容量，如图 2－39 所示。

（6）设置虚拟机的名字和虚拟机网络，单击“Finish”按钮完成虚拟机的配置，进入安装环节，如图 2－40 所示。

2）使用 virt－install 命令行创建虚拟机

使用 virt－install 命令行创建虚拟机时，可以根据 ISO 文件创建虚拟机并进行安装，也可以使用已安装好操作系统的虚拟磁盘创建虚拟机，此时创建出来的虚拟机就已经是安装好操作系统的，不需要再次单独进行安装。

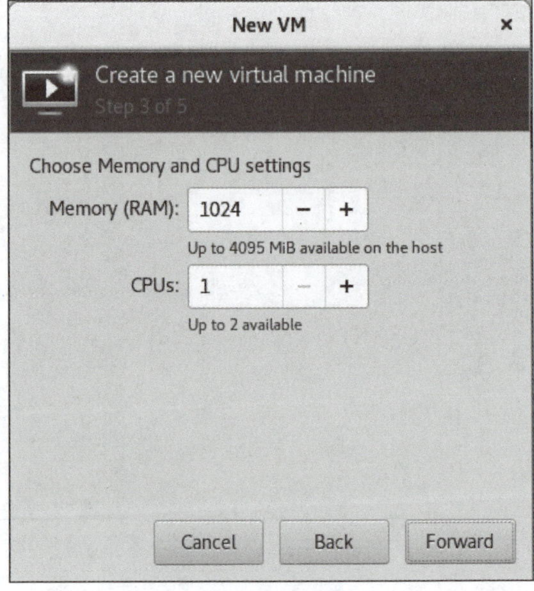

图 2－38　虚拟机规格配置

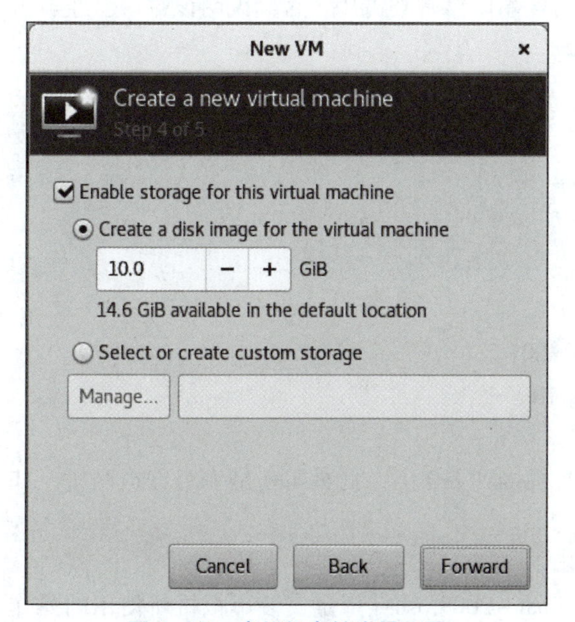

图 2－39　虚拟机存储容量配置

New VM
Create a new virtual machine
Ready to begin the installation
Name: centos7.0
OS: CentOS 7.0
Install: Local CDROM/ISO
Memory: 1024 MiB
CPUs: 1
Storage: 10.0 GiB /var/lib/libvirt/images/centos7.0.qcow2
Customize configuration before install
Network selection
Virtual network 'default' : NAT
Cancel
Back
Finish

图 2－40　虚拟机配置确认

（1）使用 ISO 镜像文件创建虚拟机。

①关闭防火墙：

```
[root@ controller ~]# systemctl stop firewalld.service
[root@ controller ~]# systemctl disable firewalld.service
```

②关闭 SELinux，并将 SELinux 的值修改为 disabled：

```
[root@ controller ~]# setenforce 0
[root@ controller ~]# vi/etc/selinux/config
# This file controls the state of SELinux on the system.
# SELINUX=can take one of these three values:
#      enforcing - SELinux security policy is enforced.
#      permissive - SELinux prints warnings instead of enforcing.
#      disabled - No SELinux policy is loaded.
SELINUX=disabled
# SELINUXTYPE=can take one of three values:
#      targeted - Targeted processes are protected,
#      minimum - Modification of targeted policy. Only selected processes are pro-
tected.
#      mls - Multi Level Security protection.
SELINUXTYPE=targeted
```

③安装虚拟机时，需要使用 VNC 客户端程序连接 <ip>：port，才能完成安装，VNC 端口默认从 5900 开始，此处暂用 5910 端口，安装前需要先检查 5910 端口是否被占用，检查命令如下：

```
[root@ centos ~]# netstat -tuln |grep 5910
```

命令执行后，如果没有输出结果，则说明该端口没有被占用，否则说明该端口已经被占用，需要更换其他端口。

④执行命令安装虚拟机：

```
[root@ centos ~]# virt-install --name vm001 \
--ram 1024 --vcpus 1 --network bridge=br0 \
--disk size=20 --cdrom/mnt/CentOS-7-x86_64-Minimal-2009.iso \
--boot hd,cdrom --graphics vnc,listen=0.0.0.0,port=5910
```

各参数说明如下：

--name vm001：指定虚拟机的名字为 vm001。

--ram 1024：指定虚拟机内存为 1 024 MB。

--vcpus 1：指定 CPU 的数量。

--network bridge=br0：指定虚拟机网络桥接到 br0 上，此处 br0 是一个虚拟网桥，如果不存在，可通过 brctl addbr br0 创建。

--disk size=20：指定磁盘容量为 20 GB。

--cdrom/mnt/CentOS-7-x86_64-Minimal-2009.iso：指定安装虚拟机所使用的操作系统。

--boot hd，cdrom：指定启动顺序，此例中先从磁盘启动，再从 CD-ROM 启动。

--graphics vnc，listen=0.0.0.0，port=5910：指定使用 VNC 参数。

命令执行完毕后，会进入虚拟机安装界面，在/var/lib/libvirt/images 目录下会生成一个 vm001.qcow2 的虚拟磁盘文件，同时，会在/etc/libvirt/qemu 目录下生成一个 vm001.xml 的配置文件。

（2）使用已安装好操作系统的虚拟磁盘创建虚拟机。

①将已安装好操作系统的 vm001.qcow2 虚拟磁盘复制并改名为 vm002.qcow2。

```
[root@ centos ~]# cp/var/lib/libvirt/images/vm001.qcow2 \
/var/lib/libvirt/images/vm002.qcow2
```

②修改 vm002.qcow2 的用户和组为 qemu，并修改权限为 600。

```
[root@ centos ~]# chown qemu:qemu/var/lib/libvirt/images/vm002.qcow2
[root@ centos ~]# chmod 600/var/lib/libvirt/images/vm002.qcow2
```

③执行命令安装虚拟机。

```
[root@ centos ~]# virt-install --name vm002 \
--ram 1024 --vcpus 1 --network bridge=br1 \
--disk/var/lib/libvirt/images/vm002.qcow2 \
--import
```

各参数说明如下：

--disk/var/lib/libvirt/images/vm002.qcow2：指定虚拟磁盘文件。

--import：表示从磁盘导入系统。

（3）使用配置文件创建虚拟机。

每台虚拟机都有一个 XML 配置文件，该配置文件存放在/etc/libvirt/qemu 目录中，libvirtd 启动时，会根据 XML 配置文件生成虚拟机。

①将已安装好操作系统的 vm001. qcow2 虚拟磁盘复制并改名为 vm003. qcow2。

```
[root@ centos ~]# cp/var/lib/libvirt/images/vm001.qcow2 \
/var/lib/libvirt/images/vm003.qcow2
```

②修改 vm003. qcow2 的用户和组为 qemu，并修改权限为 600。

```
[root@ centos ~]# chown qemu:qemu/var/lib/libvirt/images/vm003.qcow2
[root@ centos ~]# chmod 600/var/lib/libvirt/images/vm003.qcow2
```

③创建一个 vm003. xml 的配置文件，为实现简化，此处从现有的 vm001. xml 文件复制并改名为 vm003. xml。

```
[root@ centos ~]# cp/etc/libvirt/qemu/vm001.xml/etc/libvirt/qemu/vm003.xml
```

对该 XML 配置文件进行如下修改：

- 修改虚拟机名称，代码如下：

```
<name>vm001</name>
```

- 修改 UUID，代码如下：

```
<uuid>54e8426d-c72a-4979-b9fb-8d0e6448d1de</uuid>
```

- 修改磁盘文件指向新的磁盘文件，代码如下：

```
<disk type='file' device='disk'>
       #…
       <source file='/var/lib/libvirt/images/vm001.qcow2'/>
#…
</disk>
```

- 修改网卡的 MAC 地址，代码如下：

```
<interface type='bridge'>
        <mac address='52:54:00:44:f1:0c'/>
       #…
</interface>
```

④执行命令创建虚拟机。

```
[root@ centos ~]# virsh define/etc/libvirt/qemu/vm003.xml
```

3. 虚拟机管理

1）虚拟机管理操作命令

（1）查看虚拟机列表。

```
virsh list[--all]
```

不添加任何参数，默认查看的是正在运行的虚拟机，--all 参数表示列出所有的虚拟机，包括处于关闭状态的虚拟机。

（2）开启虚拟机。

```
virsh start <vm-name>
```

（3）重启虚拟机。

```
virsh reboot <vm-name>
```

（4）关闭虚拟机。

```
virsh shutdown <vm-name>
```

（5）强制断电。

```
virsh destroy <vm-name>
```

（6）设置虚拟机随宿主机自启动。

```
virsh autostart <vm-name>
```

（7）取消开机自启动。

```
virsh autostart --disable <vm-name>
```

2）虚拟机管理操作实例

（1）查看虚拟机列表。

```
[root@ centos ~]# virsh list --all
Id      Name                                State
-----------------------------------------------
-vm001 shut                                 off
```

（2）开启 vm001 虚拟机。

```
[root@ centos ~]# virsh start vm001
Domain vm001 started
```

（3）重启 vm001 虚拟机。

```
[root@ centos ~]# virsh reboot vm001
Domain vm001 is being rebooted
```

（4）关闭 vm001 虚拟机。

```
[root@ centos ~]# virsh shutdown vm001
Domain vm001 is being shutdown
```

（5）对虚拟机强制断电。

```
[root@ centos ~]# virsh destroy vm001
Domain vm001 destroyed
```

（6）设置虚拟机随宿主机自启动。

```
[root@ centos ~]# virsh autostart vm001
Domain vm001 marked as autostarted
```

（7）取消开机自启动。

```
[root@ centos ~]# virsh autostart --disable vm001
Domain vm001 unmarked as autostarted
```

4. 虚拟机快照管理

1）虚拟机快照管理操作命令

（1）查看快照信息。

```
virsh snapshot-list <vm-name>
```

（2）创建快照。

```
virsh snapshot-create-as --domain <vm-name> --name <snapshot-name>
```

参数说明如下：

--domain 指定创建快照的虚拟机。

--name 指定快照的名字。

（3）虚拟机恢复到快照。

```
virsh snapshot-revert <vm-name> <snapshot-name>
```

（4）快照的删除。

```
virsh snapshot-delete <vm-name> <snapshot-name>
```

2）虚拟机快照管理操作实例

（1）查看 vm001 虚拟机快照信息。

```
[root@ centos ~]# virsh snapshot-list vm001
Name                    Creation Time                   State
-----------------------------------------
snapshot1               2022-09-26 15:16:54 +0800 running
snapshot2               2022-09-26 15:17:54 +0800 running
```

（2）为 vm001 虚拟机创建 snapshhot3 快照。

```
[root@ centos ~]# virsh snapshot-create-as --domain vm001 --name snapshot3
Domain snapshot snapshot3 created
```

（3）恢复虚拟机到快照 snapshot1。

```
[root@ centos ~]# virsh snapshot-revert vm001 snapshot1
```

（4）删除 vm001 虚拟机的 snapshot3 快照。

```
[root@ centos ~]# virsh snapshot-delete vm001 snapshot3
Domain snapshot snapshot3 deleted
```

5. 虚拟机存储管理

虚拟机使用卷来模拟磁盘，而卷建立在存储池的基础上，在虚拟机存储管理部分包含对存储池的管理和对卷的管理。

1）存储池管理

（1）存储池管理操作命令。

①查看存储池列表。

```
virsh pool-list[ --all]
```

不添加任何参数，默认查看的是激活状态的存储池，--all 参数表示列出所有的存储池，包括未激活的存储池。

②定义存储池，以目录型存储池为例。

```
virsh pool-define-as <pool-name> --type dir --target <targetPath>
```

--type 指定存储池类型。

--target 指定存储池的存储空间路径。

③创建存储池。

```
virsh pool-build <pool-name>
```

④激活存储池。

```
virsh pool-start <pool-name>
```

⑤设置存储池自动启动。

```
virsh pool-autostart <pool-name>
```

⑥查询存储池的基本信息。

```
virsh pool-info <pool-name>
```

⑦查询存储池的 XML 配置文件。

```
virsh pool-dumpxml <pool-name>
```

（2）存储池管理操作实例。

①查看存储池列表。

```
[root@ centos ~]# virsh pool-list --all
Name                          State          Autostart
--------------------------------
default                       active         yes
mnt                            active         yes
```

②以目录型存储为例，定义 data 存储池。

```
[root@ centos ~]# mkdir/data
[root@ centos ~]# virsh pool-define-as data --type dir --target/data
Pool data defined
[root@ centos ~]# virsh pool-list --all
Name                          State          Autostart
-----------------------------------
data                         inactive        no
default                      active         yes
mnt                          active          yes
```

③创建 data 存储池。

```
[root@ centos ~]# virsh pool-build data
Pool data built
```

④启动存储池。

```
[root@ centos ~]# virsh pool-start data
Pool data started
```

⑤设置存储池自动启动。

```
[root@ centos ~]# virsh pool-autostart data
Pool data marked as autostarted
```

⑥查看存储池信息。

```
[root@ centos ~]# virsh pool-info data
Name:           data
UUID:           f93fc226-7a8e-4889-883c-d257d9e1d3ca
State:          running
Persistent:     yes
Autostart:      yes
Capacity:       29.99 GiB
Allocation:     15.91 GiB
Available:      14.08 GiB
```

⑦查看存储池 data 的 xml 配置文件。

```
[root@ centos ~]# virsh pool-dumpxml data
<pool type='dir'>
  <name>data</name>
  <uuid>f93fc226-7a8e-4889-883c-d257d9e1d3ca</uuid>
  <capacity unit='bytes'>32196526080</capacity>
  <allocation unit='bytes'>17081978880</allocation>
  <available unit='bytes'>15114547200</available>
  <source>
  </source>
  <target>
    <path>/data</path>
    <permissions>
      <mode>0755</mode>
      <owner>0</owner>
      <group>0</group>
      <label>unconfined_u:object_r:default_t:s0</label>
    </permissions>
  </target>
</pool>
```

⑧关闭存储池。

```
[root@ centos ~]# virsh pool - destroy data
Pool data destroyed
```

⑨删除存储池。

```
[root@ centos ~]# virsh pool - delete data
Pool data deleted
```

⑩取消存储池定义。

```
[root@ centos ~]# virsh pool - undefine data
Pool data has been undefined
```

2）卷管理

（1）卷管理操作命令。

①列出存储池下的卷。

```
virsh vol - list <pool - name >[ -- details]
```

-- details 选项会显示卷的详细信息。

②创建卷。

```
virsh vol - create[ -- pool] <pool - name >[ -- name] <vol - name >[ -- capacity]
<vol - capacity >
```

参数说明如下：

-- pool：指定存储池的名字。

-- name：指定卷的名字。

-- capacity：指定卷的存储容量。

③查询卷的 XML 信息。

```
virsh vol - dumpxml <vol >[ -- pool <pool - name >]
```

④删除卷。

```
virsh vol - delete <vol >[ -- pool <pool - name >][ -- delete - snapshots]
```

添加 -- delete - snapshots 会将快照也一并删除。

（2）卷管理操作实例。

①查看 data 存储池下的卷列表。

```
[root@ centos ~]# virsh vol - list data -- details
Name    Path            Type     Capacity  Allocation
-----------------------------------------
vol1   /data/vol1   file   1.00 GiB        1.00 GiB
```

②data 存储池下创建 vol2 卷，设置容量为 1 GB。

```
[root@ centos ~]# virsh vol - create - as -- pool data -- name vol2 -- capacity 1G
Vol vol2 created
```

③查看 vol2 的 XML 配置文件。

```
[root@ centos ~]# virsh vol-dumpxml vol2 --pool data
<volume type='file'>
  <name>vol2</name>
  <key>/data/vol2</key>
  <source>
  </source>
  <capacity unit='bytes'>1073741824</capacity>
  <allocation unit='bytes'>1073741824</allocation>
  <physical unit='bytes'>1073741824</physical>
  <target>
    <path>/data/vol2</path>
    <format type='raw'/>
    <permissions>
      <mode>0600</mode>
      <owner>0</owner>
      <group>0</group>
      <label>system_u:object_r:default_t:s0</label>
    </permissions>
    <timestamps>
      <atime>1664343990.011405825</atime>
      <mtime>1664343990.010405825</mtime>
      <ctime>1664343990.010405825</ctime>
    </timestamps>
  </target>
</volume>
```

④删除 data 存储池下的 vol2 卷。

```
[root@ centos ~]# virsh vol-delete vol2 --pool data
Vol vol2 deleted
```

6. 虚拟网络管理

qemu 的虚拟网络分为三种模式：

nat：虚拟机通过 NAT 与外界通信。

routed：虚拟机通过路由方式与外界通信。

isolated：虚拟机不与外界通信。

前两种模式需要修改 Linux 内核，将内核的 net. ipv4. ip_forward 值设置为 1，开启包转发功能。

```
[root@ centos ~]# vi/etc/sysctl.conf
net.ipv4.ip_forward=1
[root@ centos ~]# sysctl -p
net.ipv4.ip_forward=1
```

每个虚拟网络都有一个 XML 配置文件，文件存放在/etc/libvirt/qemu/networks 目录下。

1）虚拟网络管理命令

（1）列出虚拟网络。

```
virsh net-list[--all]
```

（2）创建虚拟网络。

```
virsh net-create <xmlfile>
```

（3）启动虚拟网络。

```
virsh net-start <net-name>
```

（4）停止虚拟网络。

```
virsh net-destroy <net-name>
```

（5）设置虚拟网络自动启动。

```
virsh net-autostart <net-name>
```

（6）取消虚拟网络自动启动。

```
virsh net-autostart <net-name> --disable
```

（7）查看网络的配置文件。

```
virsh net-dumpxml <net-name>
```

（8）删除网络。

```
virsh net-undefine <net-name>
```

2）虚拟网络管理实例

（1）查看虚拟网络列表。

```
[root@ centos ~]# virsh net-list --all
Name                          State      Autostart     Persistent
----------------------------------------------
default                     active      yes             yes
```

（2）创建虚拟网络。

- NAT 模式网络

在/etc/libvirt/qemu/network 目录下创建 net1.xml 文件，以 NAT 模式创建虚拟网络，NAT 模式的虚拟网络配置文件案例如下：

```
<network>
  <name>net1</name>
  <uuid>48390cba-9ddc-4d5a-a702-74f8f9f6e47d</uuid>
  <forward mode='nat'/>
  <bridge name='br1' stp='on' delay='0'/>
  <mac address='ea:68:05:79:6a:80'/>
  <ip address='192.168.1.1' netmask='255.255.255.0'>
```

```
    <dhcp>
      <range start='192.168.1.2' end='192.168.1.254'/>
    </dhcp>
  </ip>
</network>
[root@ centos ~]# virsh net-create/etc/libvirt/qemu/networks/net1.xml
Network net1 created from/etc/libvirt/qemu/networks/net1.xml
[root@ centos ~]# virsh net-list --all
Name                    State       Autostart      Persistent
------------------------------------------
default                 active      yes            yes
net1                    active      no             no
```

此时创建的 net1 不是持久性的网络，停止网络后会自动删除，想要创建持久性的网络，需要使用 virsh net-define 命令。

```
[root@ centos ~]# virsh net-define/etc/libvirt/qemu/networks/net1.xml
Network net1 defined from/etc/libvirt/qemu/networks/net1.xml
[root@ centos ~]# virsh net-list --all
Name                    State       Autostart      Persistent
------------------------------------------
default                 active      yes            yes
net1                    inactive    no             yes
```

- Routed 模式网络

在/etc/libvirt/qemu/network 目录下创建 net2.xml 文件，以 Routed 模式创建虚拟网络，Routed 模式的虚拟网络配置文件案例如下：

```
<network>
  <name>net2</name>
  <uuid>fa738dc5-9ddc-4d5a-a702-74f8f9f6e47d</uuid>
  <forward dev='ens33' mode='route'>
    <interface dev='ens33'/>
  </forward>
  <bridge name='br2' stp='on' delay='0'/>
  <mac address='3a:68:05:79:6a:80'/>
  <domain name='net2'/>
  <ip address='192.168.2.1' netmask='255.255.255.0'>
  </ip>
  <route family='ipv4' address='192.168.2.0' prefix='24' gateway=
'192.168.2.1'/>
</network>
[root@ centos ~]# virsh net-create/etc/libvirt/qemu/networks/net2.xml
Network net2 created from/etc/libvirt/qemu/networks/net2.xml
```

- Isolated 模式网络

在/etc/libvirt/qemu/network 目录下创建 net3. xml 文件，以 Isolated 模式创建虚拟网络，Isolated 模式的虚拟网络配置文件案例如下：

```
<network>
  <name>net3</name>
  <uuid>69238dc5-9ddc-4d5a-a702-74f8f9f6e47d</uuid>
  <bridge name='br3' stp='on' delay='0'/>
  <mac address='3a:68:05:79:6a:a3'/>
  <domain name='net3'/>
  <ip address='192.168.3.1' netmask='255.255.255.0'>
  </ip>
</network>
[root@ centos ~]# virsh net-create/etc/libvirt/qemu/networks/net3.xml
Network net3 created from/etc/libvirt/qemu/networks/net3.xml
```

（3）启动 net1 虚拟网络。

```
[root@ centos ~]# virsh net-start net1
Network net1 started
```

（4）停止 net1 虚拟网络。

```
[root@ centos ~]# virsh net-destroy net1
Network net1 destroyed
```

（5）设置 net1 虚拟网络自动启动。

```
[root@ centos ~]# virsh net-autostart net1
Network net1 marked as autostarted
```

（6）取消 net1 虚拟网络自动启动。

```
[root@ centos ~]# virsh net-autostart net1 --disable
Network net1 unmarked as autostarted
```

（7）查看 net1 虚拟网络的配置文件。

```
[root@ centos ~]# virsh net-dumpxml net1
<network>
  <name>net1</name>
  <uuid>48390cba-9ddc-4d5a-a702-74f8f9f6e47d</uuid>
  <forward mode='nat'/>
  <bridge name='br1' stp='on' delay='0'/>
  <mac address='ea:68:05:79:6a:80'/>
  <ip address='192.168.1.1' netmask='255.255.255.0'>
      <dhcp>
      <range start='192.168.1.2' end='192.168.1.254'/>
    </dhcp>
  </ip>
</network>
```

（8）删除 net1 虚拟网络。

```
[root@ centos ~]# virsh net-undefine net1
Network net1 has been undefined
```

（9）配置虚拟机使用网络。

先关闭虚拟机，然后通过修改虚拟机配置文件的方式配置虚拟机使用虚拟网络。

```
[root@ centos ~]# virsh destroy vm001
Domain vm001 destroyed
[root@ centos ~]# virsh edit vm001
```

修改配置文件网络设置内容如下：

```
<interface type='network'>
    <mac address='52:54:00:44:f1:0c'/>
    <source network='default'/>
    <model type='virtio'/>
    <address type='pci' domain='0x0000' bus='0x00' slot='0x03' function='
0x0'/>
</interface>
```

配置文件信息说明如下：

<interface type = 'network'>：使用虚拟网络。

<source network = 'default'/>：设置虚拟网络名称。

<model type = 'virtio'/>：网络硬件类型。

【任务工单】

工单号：2－2

<table>
<tr><td colspan="2">项目名称：虚拟化技术</td><td colspan="2">任务名称：使用 KVM</td></tr>
<tr><td colspan="2">班级：</td><td>学号：</td><td>姓名：</td></tr>
<tr><td>任务安排</td><td colspan="3">□安装 KVM 软件包
□利用 virt－install 命令的三种方式创建 KVM 虚拟机
□完成虚拟机管理的基本操作
□完成虚拟机快照管理的基本操作
□完成虚拟机存储管理的基本操作
□完成虚拟机网络管理的基本操作</td></tr>
<tr><td>成果交付形式</td><td colspan="3">操作过程截图整理形成操作文档，上传至教学平台</td></tr>
<tr><td>任务实施总结</td><td colspan="3">任务自评（0～10 分）：
项目收获：

改进点：</td></tr>
<tr><td>成果验收</td><td colspan="3">□完全满足任务要求
□基本满足任务要求
要求全部完成，能够使用命令行对 KVM 虚拟机的各项资源进行管理，但操作不熟练：

□不能满足需求
要求未全部完成，对管理 KVM 虚拟机各项资源的指令存在问题：</td></tr>
</table>

【知识巩固】

1. KVM 虚拟机可以通过图形界面化创建，也可以通过命令行创建。(　　)

A. 对　　　　B. 错

2. KVM 虚拟化架构中包含 kvm、qemu 和 libvirt 组件，可以构成一套虚拟化解决方案的两个组件是（　　）。

A. kvm　　　　B. qemu　　　　C. libvirt

【小李的反思】

差之毫厘，谬以千里。

源自《礼记·经解》，意思是说，事情开始时有很小的差错，若不及时纠正，最后就会造成大的错误。同样，本任务中 KVM 的执行指令有一点错误，就会导致运行失败，或者达不到对虚拟机的预期管理效果，所以，在技术学习和科学研究过程中需保持谨慎、精益求精、一丝不苟。

2019 年 2 月 1 日，美国“哥伦比亚”航天飞机着陆前发生爆炸，7 名航天员全部遇难，全世界为之震惊。事后调查结果显示，造成此次灾难的原因竟是一块隔热瓦脱落。“哥伦比亚”飞机有 2 000 多块隔热瓦，能抵御 3 000 ℃高温，避免航天飞机返回大气层时外壳被熔化。航天飞机是高科技产品，许多标准是非常严格的，但就一块脱落的隔热瓦，0.05% 的差错葬送了价值连城的航天飞机，还有无法用价值衡量的 7 条生命。

1% 的错误导致 100% 的失败，对于产品品质而言，不是 100 分，就是 0 分，任何一个生产工序的问题都可能使所有的努力白费。在做技术和搞科研的时候，我们也要时刻保持一丝不苟、精益求精的品质，不因毫厘之差而造成千里之谬。

项目评价

<table>
<tr><td colspan="6">项目名称：虚拟化技术</td></tr>
<tr><td colspan="2">班级：</td><td>学号：</td><td colspan="3">姓名：</td></tr>
<tr><td colspan="2">评价指标</td><td>评价等级及分值</td><td>学生自评</td><td>组内互评</td><td>教师评分</td></tr>
<tr><td rowspan="2">素质目标达成情况（30%）</td><td>精益求精的工匠精神（20%）</td><td>A（20 分）：虚拟机安装和使用过程中能不断修正问题，精益求精，追求完美
B（15 分）：能够完成虚拟机安装和使用的任务，但对过程中存在的问题修改积极性不高
C（7 分）：对虚拟机安装和使用过程中存在的问题放任不管</td><td></td><td></td><td></td></tr>
<tr><td>自我学习热情（10%）</td><td>A（10 分）：自我学习热情高涨，积极和同学探讨学习问题
B（7 分）：自我学习热情较好，能够自主完成学习
C（3 分）：自我学习热情一般，学习积极性不高</td><td></td><td></td><td></td></tr>
<tr><td rowspan="3">知识目标达成情况（40%）</td><td>任务实施完成情况（20%）</td><td>A（20 分）：任务实施全部完成，且基本满足任务要求
B（16 分）：任务实施大部分完成，存在一些小的问题
C（10 分）：任务实施部分完成，存在一些影响任务实施结果的问题</td><td></td><td></td><td></td></tr>
<tr><td>测验作业完成情况（10%）</td><td>A（10 分）：测验作业全部完成，知识理解透彻
B（7 分）：测验作业大部分完成，能基本完成知识的理解
C（3 分）：测验作业部分完成，对知识的理解较为片面</td><td></td><td></td><td></td></tr>
<tr><td>课上活动（10%）</td><td>A（10 分）：积极参与课上抢答、提问、主题讨论等
B（7 分）：能够参与课上抢答、提问、主题讨论等
C（3 分）：部分课上抢答、提问、主题讨论等</td><td></td><td></td><td></td></tr>
</table>

续表

评价指标		评价等级及分值	学生自评	组内互评	教师评分
能力目标达成情况（30%）	任务实施完成质量（20%）	A（20 分）：任务实施完成质量优秀 B（16 分）：任务实施完成质量良好 C（10 分）：任务实施完成质量一般			
	超凡脱俗（10%）	A（10 分）：能够帮助同组同学解决虚拟机安装过程中存在的问题，并能整理问题解决手册 B（7 分）：能够帮助同组同学解决虚拟机安装过程中存在的问题 C（3 分）：能够规定时间内完成学习任务			

项目总结

虚拟化是实现云计算最重要的技术基础，虚拟化技术实现了物理资源的逻辑抽象和统一表示。通过虚拟化技术可以提高资源的利用率，并能够根据用户业务需求的变化，快速、灵活地进行资源部署。本项目通过探索虚拟化技术和使用 KVM 两个任务讲述了虚拟化的一些关键技术，在探索虚拟化技术任务中，主要讲述了虚拟化的概念、特点、服务器虚拟化方式、虚拟化体系架构等，并通过在 VMware Workstation Pro 上部署 CentOS 7 虚拟机，加强对虚拟化技术的理解；在使用 KVM 任务中，由于在搭建 OpenStack 私有云时会更多用到基于 Linux 内核的 KVM 虚拟机管理器，在该任务中主要讲述了 KVM 的安装部署、使用 KVM 虚拟机管理器对虚拟机的基本管理、虚拟机的快照管理、虚拟机的存储管理和虚拟机的网络管理等。

项目三

OpenStack 云平台部署

项目导入

小李已经在前面的准备工作中了解了 OpenStack 和虚拟化技术，小李觉得自己可以开始大干一场着手 OpenStack 私有云平台部署了，从前面 OpenStack 的学习中，小李知道 OpenStack 是一个云操作系统框架，本身并不提供计算、存储、网络等服务，而是通过在 OpenStack 框架内集成各种组件提供相应的功能，小李作为初学者，计划部署 Keystone、Nova、Glance、Neutron、Horizon、Cinder 这些核心组件支撑云平台的基本功能，保障云平台能够平稳运行起来。

本项目将随着小李的学习脚步，进行 OpenStack 云平台的部署：首先，进行 OpenStack 部署的环境准备，包括网络、NTP、数据库、消息队列等；其次，进行 OpenStack 核心组件的部署和验证，包括 Keystone 组件、Glance 组件、Nova 组件、Neutron 组件、Horizon 组件和 Cinder 组件。

项目目标

【素质目标】

- 培养学生精益求精的工匠精神
- 培养学生乐于探索的学习精神
- 培养学生云计算工程师的职业素养

【知识目标】

- 了解 OpenStack 部署的实验环境规划
- 掌握 CentOS 7 服务器上网络、NTP、数据库、消息队列等准备环境的配置
- 掌握认证服务 Keystone 组件的配置和验证过程
- 掌握镜像服务 Glance 组件的配置和验证过程
- 掌握计算服务 Nova 组件的配置和验证过程
- 掌握网络服务 Neutron 组件的配置和验证过程
- 掌握 Web 服务 Horizon 组件的配置和验证过程
- 掌握块存储 Cinder 组件的配置和验证过程

【能力目标】

- 能够完成 CentOS 7 服务器的网络配置，完成 NTP、数据库、消息队列等的配置
- 能够自主完成云平台的搭建，并能够对云平台各功能组件部署正确性进行验证

任务 1 OpenStack 部署环境准备

【任务描述】

小李对云平台的部署工作还在学习过程中，为保障生产环境的安全性，不能直接在生产环境中部署云平台，需要先在实验环境部署正确之后，再迁移到生产环境，所以小李在 VMware 里创建了两台 CentOS 7 的服务器，作为控制节点和计算节点，以该双节点作为实验环境部署 OpenStack 云平台。开始部署前需要进行环境准备工作，小李通过查看参考官方文档（https://docs. openstack. org/install - guide/environment. html）得知，环境准备工作包括安全配置、网络配置、NTP 配置、数据库配置和消息队列配置等。

【知识要点】

1. SELinux

SELinux 是一种基于域 - 类型模型（domain - type）的强制访问控制安全系统，它由 NSA 编写并设计成内核模块包含到内核中，相应地，某些安全相关的应用也被打了 SELinux 的补丁，最后还有一个相应的安全策略。

SELinux 的配置文件为/etc/selinux/config，SELinux 有 disabled、permissive、enforcing 3 种选择：

- enforcing 是强制模式系统，它受 SELinux 保护，违反了策略就无法继续操作下去。
- permissive 是提示模式系统不会受到 SELinux 保护，只是收到警告信息。permissive 就是 SELinux 有效，即使违反了策略，它也会继续操作，但是会把违反的内容记录下来（警告信息）。
- disabled 禁用 SELinux。

2. NTP

NTP（Network Time Protocol）是用来使计算机时间同步化的一种协议，它可以使计算机对其服务器或时钟源（如石英钟、GPS 等）做同步化，它可以提供高精准度的时间校正，并且可以加密确认的方式来防止恶毒的协议攻击。

NTP 提供准确时间，首先要有准确的时间来源，这一时间应该是国际标准时间 UTC。时间按 NTP 服务器的等级传播。按照离外部 UTC 源的远近，将所有服务器归入不同的 Stratum（层）中。Stratum - 1 在顶层，有外部 UTC 接入，而 Stratum - 2 则从 Stratum - 1 获取时间，Stratum - 3 从 Stratum - 2 获取时间，依此类推。Stratum 层的总数限制在 15 以内。所有这些服务器在逻辑上形成阶梯式的架构相互连接，而 Stratum - 1 的时间服务器是整个系统的

基础。

计算机主机一般与多个时间服务器连接，利用统计学的算法过滤来自不同服务器的时间，以选择最佳的路径和来源来校正主机时间。即使主机在长时间无法与某一时间服务器相联系，NTP 服务也依然有效运转。

为防止对时间服务器的恶意破坏，NTP 使用了识别（Authentication）机制，检测来对时的信息是否真正来自所声称的服务器，并检测资料的返回路径，以提供对抗干扰的保护机制。

3. 部署环境要求

OpenStack 项目是一个支持各种云环境的开源云计算平台，该项目旨在实现简单的实现、巨大的可扩展性和丰富的功能集，OpenStack 通过各种服务的组合，提供完整的 IaaS 服务解决方案。OpenStack 物理上由一个服务器集群组成，不同的服务组件一般安装在不同的物理机上，一般来说，一个 OpenStack 平台包含以下功能节点：

- 控制节点：管理 OpenStack，其上运行的服务有 Keystone、Glance、Horizon 以及 Nova 和 Neutron 中管理相关的组件。控制节点也运行支持 OpenStack 的服务，例如 SQL 数据库（通常是 MySQL）、消息队列（通常是 RabbitMQ）和网络时间服务 NTP。
- 网络节点：其上运行的服务为 Neutron，为 OpenStack 提供 L2 和 L3 网络，包括虚拟机网络、DHCP、路由、NAT 等。
- 存储节点：提供块存储（Cinder）或对象存储（Swift）服务。
- 计算节点：其上运行 Hypervisor（默认使用 KVM）。同时，运行 Neutron 服务的 ent，为虚拟机提供网络支持。

实验中，受学习计算机硬件限制，同时又需要功能完备，在计算机上安装 VMware Workstation Pro 虚拟机管理器。VMware 中创建两台虚拟机来部署 OpenStack Rocky 版本，两台虚拟机规格见表 3－1。

表 3－1　两台虚拟机规格

节点名称	功能角色	硬件	网络规划
controller	控制节点 网络节点	内存：8 GB 硬盘：40 GB CPU：四核，开启虚拟化功能 网卡 1：用作管理网络 网卡 2：用作外网	网卡 1：192. 168. 16. 100 网卡 2：不配置 IP 地址
compute	计算节点 存储节点	内存：4 GB 硬盘：40 GB CPU：四核，开启虚拟化功能 网卡 1：用作管理网络 网卡 2：用作外网	网卡 1：192. 168. 16. 200 网卡 2：不配置 IP 地址

虚拟机 1：控制节点（网络节点），CentOS 7. 6 1810（minal），40 GB 硬盘，8 GB 内存。

虚拟机 2：计算节点（存储节点），CentOS 7. 6 1810（minal），40 GB 硬盘，4 GB 内存。

【任务实施】

1. 网络配置

OpenStack 的各个节点间需要保持良好的网络通信，实验中需要使用两个网络，一个用作管理网络，另一个用作外部网络，本实验中将计算节点作为存储节点使用，所以拓扑图中存储节点的 IP 地址和计算节点的一致。如果设备数量足够，可以使用单独的存储节点，此时存储节点的 IP 可设置为 192.168.16.0/24 网段内的其他 IP 地址。网络拓扑如图 3－1 所示。

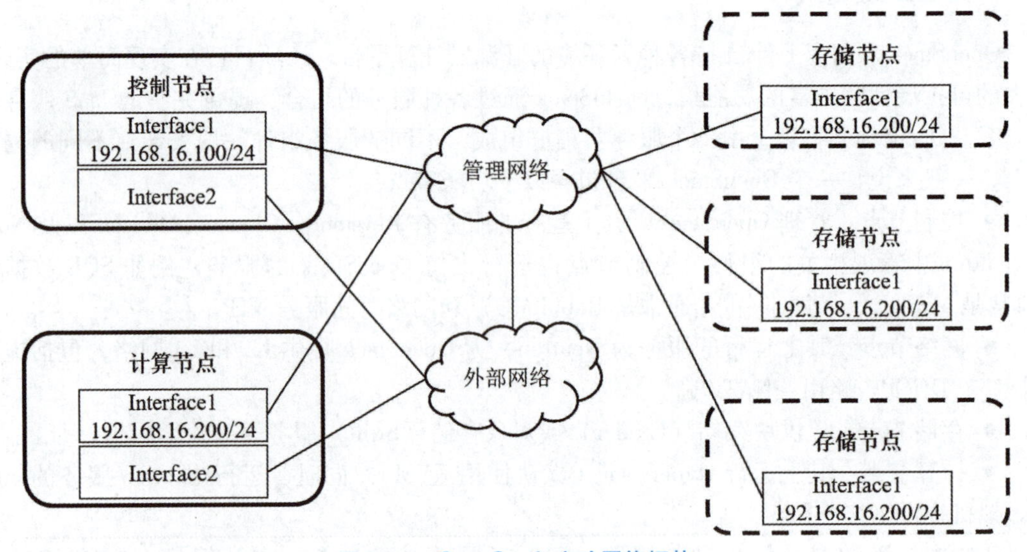

图 3－1　OpenStack 实验网络拓扑

1）VMware 网络配置

（1）开启两台主机的 CPU 虚拟化功能。

在 VMware Workstation 中选择要设置的虚拟机，单击“编辑虚拟机设置”，在“虚拟机设置”的“硬件”选项卡下选择“处理器”选项，勾选“虚拟化 Intel VT－x/EPT 或 AMD－V/RVI(V)”复选框，如图 3－2 所示。

（2）同时分别给 controller 和 compute 两个节点添加一个网卡。

在 VMware Workstation 中选择要设置的虚拟机，单击“编辑虚拟机设置”，在“虚拟机设置”对话框的“硬件”选项卡下单击“添加”按钮，在“添加硬件向导”页面选择“网络适配器”并单击“完成”按钮，如图 3－3 所示。

（3）VMnet8 的 DHCP 配置。

在 VMware Workstation 的菜单栏中单击“编辑”按钮，在编辑菜单栏中选择“虚拟网络编辑器”，打开“虚拟网络编辑器”对话框。在网卡列表选择“VMnet8”，在 DHCP 设置下的子网 IP 文本框中输入 IP 网段（192.168.16.0），在子网掩码文本框中输入子网掩码（255.255.255.0），如图 3－4 所示。

（4）NAT 配置。

单击“NAT 设置”按钮，打开“NAT 设置”对话框，检查网关设置是否正确（192.168.16.2），如图 3－5 所示。

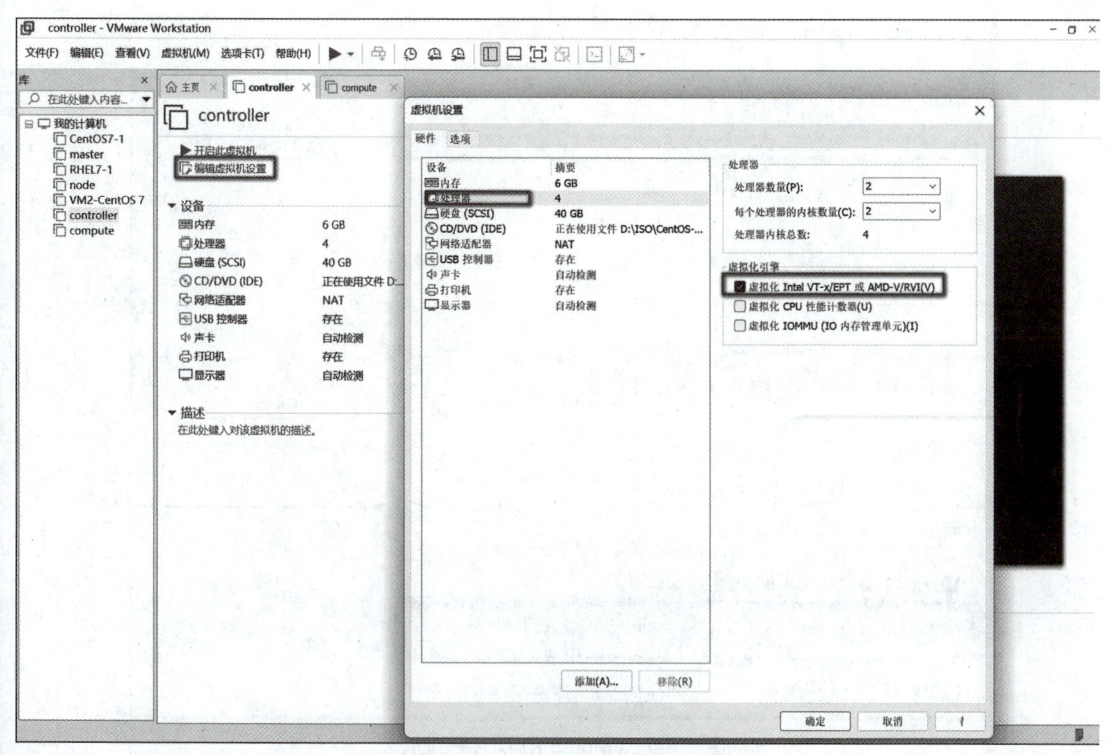

图 3－2　开启处理器虚拟化

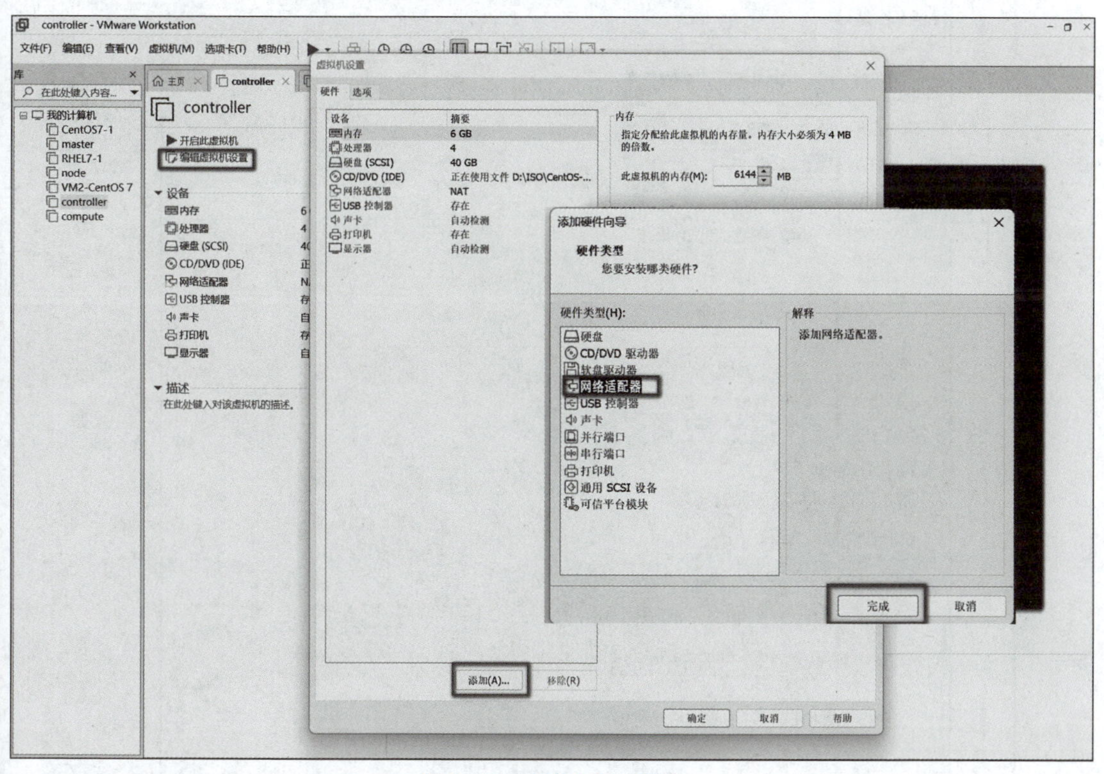

图 3－3　虚拟机添加网卡

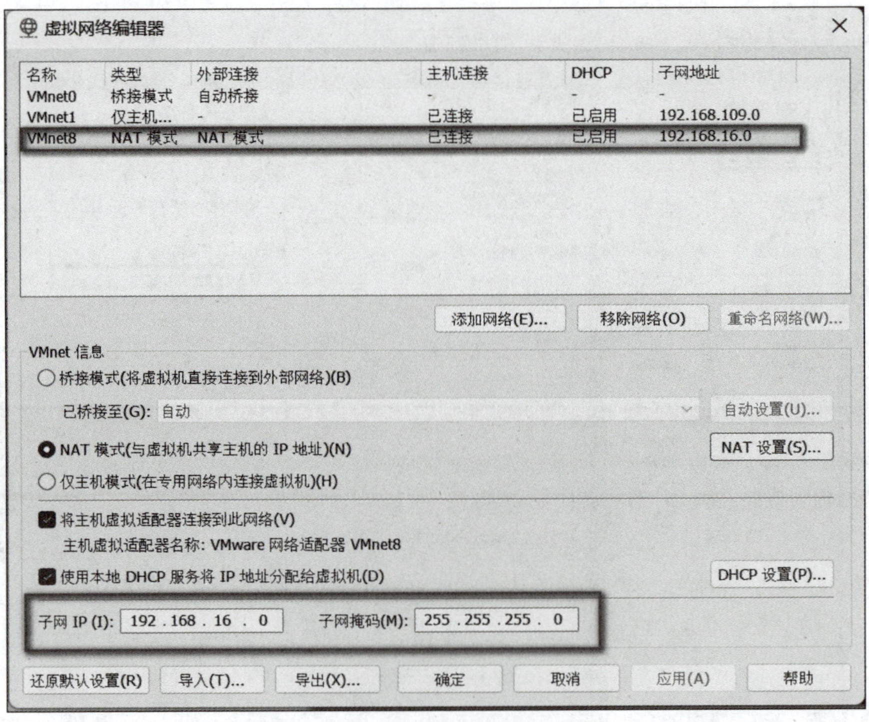

图 3-4 VMnet8 的 DHCP 配置

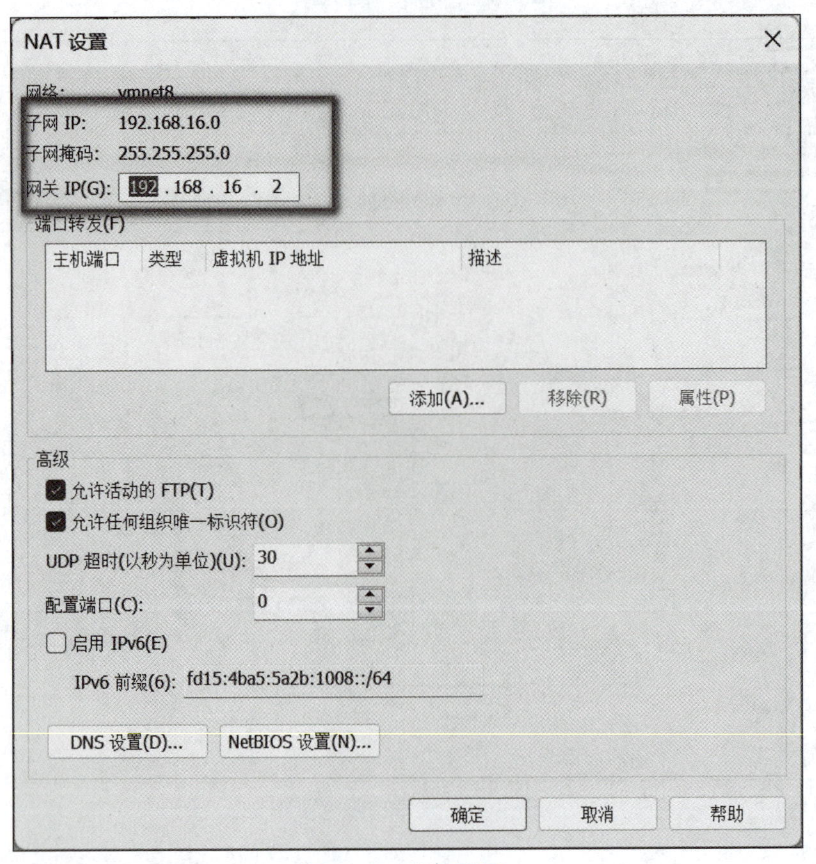

图 3-5 NAT 配置

2）节点主机网卡设置

（1）使用 ip addr 命令查看当前网络配置（ens33 和 ens36 两块网卡），如图 3－6 所示。

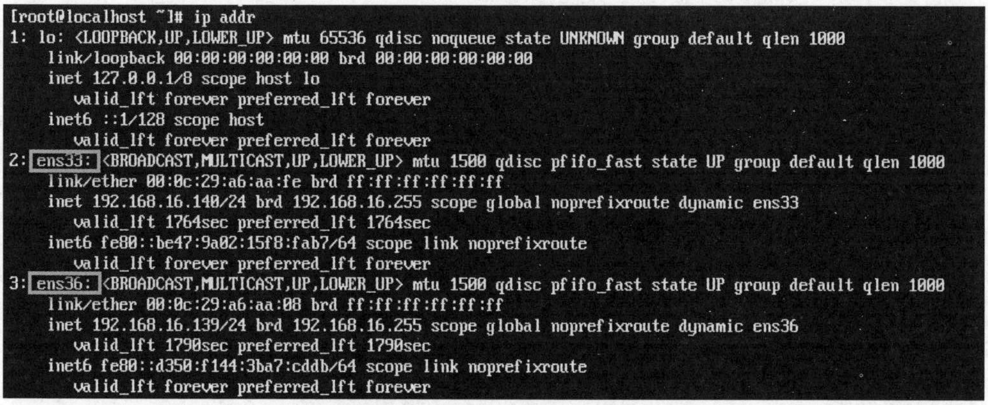

```
[root@localhost ~]# ip addr
1: lo: <LOOPBACK,UP,LOWER_UP> mtu 65536 qdisc noqueue state UNKNOWN group default qlen 1000
    link/loopback 00:00:00:00:00:00 brd 00:00:00:00:00:00
    inet 127.0.0.1/8 scope host lo
       valid_lft forever preferred_lft forever
    inet6 ::1/128 scope host
       valid_lft forever preferred_lft forever
2: ens33: <BROADCAST,MULTICAST,UP,LOWER_UP> mtu 1500 qdisc pfifo_fast state UP group default qlen 1000
    link/ether 00:0c:29:a6:aa:fe brd ff:ff:ff:ff:ff:ff
    inet 192.168.16.140/24 brd 192.168.16.255 scope global noprefixroute dynamic ens33
       valid_lft 1764sec preferred_lft 1764sec
    inet6 fe80::be47:9a02:15f8:fab7/64 scope link noprefixroute
       valid_lft forever preferred_lft forever
3: ens36: <BROADCAST,MULTICAST,UP,LOWER_UP> mtu 1500 qdisc pfifo_fast state UP group default qlen 1000
    link/ether 00:0c:29:a6:aa:08 brd ff:ff:ff:ff:ff:ff
    inet 192.168.16.139/24 brd 192.168.16.255 scope global noprefixroute dynamic ens36
       valid_lft 1790sec preferred_lft 1790sec
    inet6 fe80::d350:f144:3ba7:cddb/64 scope link noprefixroute
       valid_lft forever preferred_lft forever
```

图 3－6 controller 节点网络状态

ens33 网卡：作为管理网络使用，配置 IP 为 192. 168. 16. 140/24。

ens36 网卡：作为外部网络使用，不配置 IP。

（2）将 ens33 的网卡文件复制并改名为 ens36。

```
[root@ localhost ~]# cp/etc/sysconfig/network-scripts/ifcfg-ens33/etc/sysconfig/network-scripts/ifcfg-ens36
```

（3）ens33 网卡配置：vi/etc/sysconfig/network－scripts/ifcfg－ens33，配置文件内容如下：

```
TYPE=Ethernet
PROXY_METHOD=none
BROWSER_ONLY=no
BOOTPROTO=none
DEFROUTE=yes
IPV4_FAILURE_FATAL=no
IPV6INIT=yes
IPV6_AUTOCONF=yes
IPV6_DEFROUTE=yes
IPV6_FAILURE_FATAL=no
IPV6_ADDR_GEN_MODE=stable-privacy
NAME=ens33
UUID=2e7beaf5-4e9e-4a27-83eb-568d35390872
DEVICE=ens33
ONBOOT=yes
IPADDR=192.168.16.100
NETMASK=255.255.255.0
GATEWAY=192.168.16.2
```

（4）ens36 网卡配置：vi/etc/sysconfig/network－scripts/ifcfg－ens36，配置文件内容

如下：

```
TYPE = Ethernet
PROXY_METHOD = none
BROWSER_ONLY = no
BOOTPROTO = dhcp
DEFROUTE = yes
IPV4_FAILURE_FATAL = no
IPV6INIT = yes
IPV6_AUTOCONF = yes
IPV6_DEFROUTE = yes
IPV6_FAILURE_FATAL = no
IPV6_ADDR_GEN_MODE = stable - privacy
NAME = ens36
UUID = 2e083153 - 2da2 - 308e - 8f34 - e1fad0cc3eb8
DEVICE = ens36
ONBOOT = yes
```

（5）ens36 网卡的 UUID 通过 nmcli con show 命令查看所有连接显示，如图 3 – 7 所示。

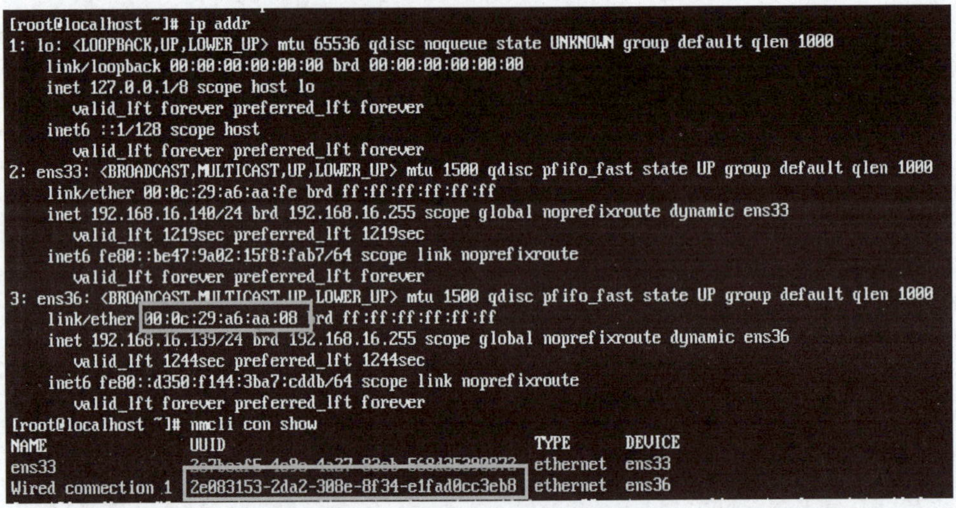

```
[root@localhost ~]# ip addr
1: lo: <LOOPBACK,UP,LOWER_UP> mtu 65536 qdisc noqueue state UNKNOWN group default qlen 1000
    link/loopback 00:00:00:00:00:00 brd 00:00:00:00:00:00
    inet 127.0.0.1/8 scope host lo
       valid_lft forever preferred_lft forever
    inet6 ::1/128 scope host
       valid_lft forever preferred_lft forever
2: ens33: <BROADCAST,MULTICAST,UP,LOWER_UP> mtu 1500 qdisc pfifo_fast state UP group default qlen 1000
    link/ether 00:0c:29:a6:aa:fe brd ff:ff:ff:ff:ff:ff
    inet 192.168.16.140/24 brd 192.168.16.255 scope global noprefixroute dynamic ens33
       valid_lft 1219sec preferred_lft 1219sec
    inet6 fe80::be47:9a02:15f8:fab7/64 scope link noprefixroute
       valid_lft forever preferred_lft forever
3: ens36: <BROADCAST,MULTICAST,UP,LOWER_UP> mtu 1500 qdisc pfifo_fast state UP group default qlen 1000
    link/ether 00:0c:29:a6:aa:08 brd ff:ff:ff:ff:ff:ff
    inet 192.168.16.139/24 brd 192.168.16.255 scope global noprefixroute dynamic ens36
       valid_lft 1244sec preferred_lft 1244sec
    inet6 fe80::d350:f144:3ba7:cddb/64 scope link noprefixroute
       valid_lft forever preferred_lft forever
[root@localhost ~]# nmcli con show
NAME                UUID                                  TYPE      DEVICE
ens33               [illegible]                           ethernet  ens33
Wired connection 1  2e083153-2da2-308e-8f34-e1fad0cc3eb8  ethernet  ens36
```

图 3 – 7　controller 节点所有网络连接

（6）重启网络服务。

```
[root@ localhost ~]# systemctl restart network
```

• compute 主机网卡配置过程同 controller 主机网卡，具体参数根据实际情况进行配置。

3）修改主机名

（1）将 controller 主机的主机名修改为 controller。

```
[root@ localhost ~]# hostnamectl set - hostname controller
```

（2）将 compute 主机的主机名修改为 compute。

```
[root@ localhost ~]# hostnamectl set-hostname compute
```

- 配置完成后重启虚拟机，新的主机名即可生效。

4）配置 IP 地址与主机名映射

（1）分别编辑 controller 节点和 compute 节点的/etc/hosts 文件，在文件中添加 controller 和 compute 的 IP 地址与域名映射（192.168.16.100 controller 和 192.168.16.200 compute）。

```
[root@ controller ~]# vi/etc/hosts
127.0.0.1    localhost localhost.localdomain localhost4 localhost4.localdomain4
::1          localhost localhost.localdomain localhost6 localhost6.localdomain6
192.168.16.100 controller
192.168.16.200 compute
```

- compute 节点配置过程与 controller 节点配置过程一致。

（2）节点间网络连通测试：

①controller 节点 ping compute 节点，可以正常 ping 通，如图 3-8 所示。

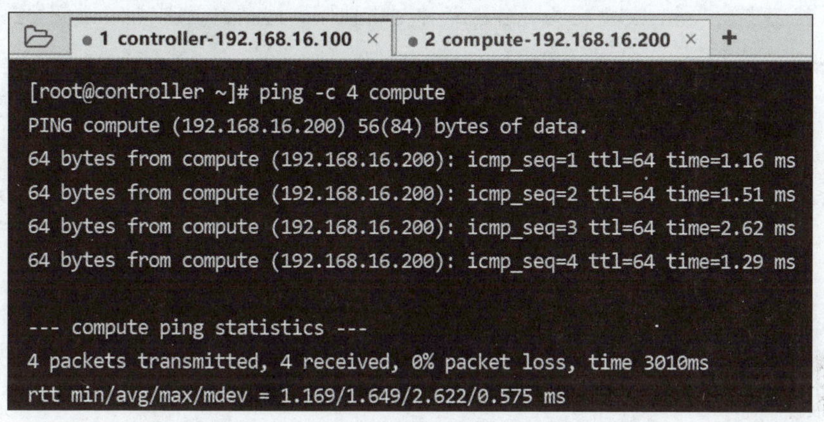

图 3-8 controller 节点 ping compute 节点

②compute 节点 ping controller 节点，可以正常 ping 通，如图 3-9 所示。

```
1 controller-192.168.16.100    2 compute-192.168.16.200
[root@compute ~]# ping -c 4 controller
PING controller (192.168.16.100) 56(84) bytes of data.
64 bytes from controller (192.168.16.100): icmp_seq=1 ttl=64 time=0.635 ms
64 bytes from controller (192.168.16.100): icmp_seq=2 ttl=64 time=1.70 ms
64 bytes from controller (192.168.16.100): icmp_seq=3 ttl=64 time=4.49 ms
64 bytes from controller (192.168.16.100): icmp_seq=4 ttl=64 time=0.710 ms

--- controller ping statistics ---
4 packets transmitted, 4 received, 0% packet loss, time 3013ms
rtt min/avg/max/mdev = 0.635/1.886/4.497/1.565 ms
```

图 3-9 compute 节点 ping controller 节点

2. 安全配置

1）防火墙配置

（1）通过 systemctl status firewalld. service 查看防火墙的状态，防火墙状态 active（running）表示防火墙开启（CentOS 7 默认开启状态）。

```
[root@ controller ~]# systemctl status firewalld. service
● firewalld. service - firewalld - dynamic firewall daemon
   Loaded: loaded(/usr/lib/systemd/system/firewalld. service; enabled; vendor preset:enabled)
   Active:active(running)since 四 2022 -10 -20 12:58:34 CST;5min ago
     Docs:man:firewalld(1)
 Main PID:8873(firewalld)
   CGroup:/system. slice/firewalld. service
           └─8873/usr/bin/python -Es/usr/sbin/firewalld --nofork --nopid

10 月 20 12:58:33 controller systemd[1]:Starting firewalld - dynamic firewall daemon...
10 月 20 12:58:34 controller systemd[1]:Started firewalld - dynamic firewall daemon.
```

（2）关闭防火墙。在 OpenStack 部署过程中，有些步骤会因为防火墙而失败，在初学过程中，建议暂时关闭防火墙。

```
[root@ controller ~]# systemctl stop firewalld. service
[root@ controller ~]# systemctl disable firewalld. service
Removed symlink/etc/systemd/system/multi - user. target. wants/firewalld. service.
Removed symlink/etc/systemd/system/dbus - org. fedoraproject. FirewallD1. service.
```

2）SELinux 配置

关闭 SELinux 防火墙并将 SELinux 的值修改为 disabled。

```
[root@ controller ~]# setenforce 0
[root@ controller ~]# vi/etc/selinux/config
# This file controls the state of SELinux on the system.
# SELINUX = can take one of these three values:
#       enforcing - SELinux security policy is enforced.
#       permissive - SELinux prints warnings instead of enforcing.
#       disabled - No SELinux policy is loaded.
SELINUX = disabled
# SELINUXTYPE = can take one of three values:
#       targeted - Targeted processes are protected,
#       minimum - Modification of targeted policy. Only selected processes are protected.
#       mls - Multi Level Security protection.
SELINUXTYPE = targeted
```

3. NTP 配置

为了实现各节点间的时间同步，需要配置 NTP 服务，在配置过程中，controller 节点引用更精确的官方机构网络时间服务器的时间，其他节点应用 controller 节点的时间。

1）controller 节点安装

（1）安装 chrony 包。

```
[root@ controller ~]# yum install chrony
```

（2）设置 controller 节点的 chrony 服务器可以被哪些网段访问，在/etc/chrony. conf 文件中添加 allow 192. 168. 16. 0/24 代码，以实现 conpute 节点（192. 168. 16. 200）可以正常同步 controller 的时间。

```
[root@ compute ~]# vi/etc/chrony. conf
allow 192. 168. 16. 0/24
```

（3）重启 NTP 服务。

```
[root@ controller ~]# systemctl enable chronyd. service
[root@ controller ~]# systemctl start chronyd. service
```

2）compute 节点安装

（1）安装 chrony 包。

```
[root@ compute ~]# yum install chrony
```

（2）在/etc/chrony. conf 配置文件中添加 server controller iburst 设置 compute 节点的时间同步于 controller 节点的 NTP 服务器，并将其他 NTP 服务器注释掉。

```
# Use public servers from the pool. ntp. org project.
# Please consider joining the pool(http://www. pool. ntp. org/join. html).
server controller iburst
#server 0. centos. pool. ntp. org iburst
#server 1. centos. pool. ntp. org iburst
#server 2. centos. pool. ntp. org iburst
#server 3. centos. pool. ntp. org iburst
```

（3）重启 NTP 服务。

```
[root@ compute ~]# systemctl enable chronyd. service
[root@ compute ~]# systemctl start chronyd. service
```

3）实验验证

（1）controller 节点查看 NTP 时间同步源，包含 4 个同步源。

```
[root@ controller ~]# chronyc sources
210 Number of sources =4
MS Name/IP address          Stratum Poll Reach LastRx Last sample
==========================================================================
```

```
^+electrode.felixc.at          3  6  77   24   -80ms[   -80ms]+/-   111ms
^*  pingless.com               2  6  377  25   +1290us[   +13ms]+/-   240ms
^? time.cloudflare.com         3  7  40   354  -91ms[   -39ms]+/-   175ms
^-makaki.miuku.net             2  6  37   26   +55ms[   +66ms]+/-   216ms
```

（2）controller 节点查看 NTP 时间同步源，只有 controller 节点一个。

```
[root@ compute ~]# chronyc sources
210 Number of sources =1
MS Name/IP address           Stratum Poll Reach LastRx Last sample
=========================================================================
^*  controller                 3  6  77  47   -320us[   -35ms]+/-   205ms
```

4. OpenStack 包安装

（1）双节点安装 OpenStack 的 Rocky 版本软件包。

```
[root@ controller ~]# yum install centos-release-openstack-rocky
```

（2）双节点更新软件包。

```
[root@ controller ~]# yum upgrade
```

（3）重启系统后更新生效。

（4）安装 OpenStack 客户端。

```
[root@ controller ~]# yum install python-openstackclient
```

（5）安装 openstack-selinux 软件包，以自动管理 OpenStack 服务的安全策略。

```
[root@ controller ~]# yum install openstack-selinux -y
```

5. 数据库安装

大多数的 OpenStack 服务组件都需要数据库存储信息，根据版本的不同，可以是 MariaDB 或 MySQL 等。本实验安装 MariaDB 数据库，数据库运行在 controller 节点。

（1）安装数据库软件包。

```
[root@ controller ~]# yum install mariadb mariadb-server python2-PyMySQL
```

（2）创建并编辑配置文件/etc/my.cnf.d/openstack.cnf，在该文件中创建［mysqld］部分，设置 bind-address 值为 controller 节点管理网络的 IP 地址，以便其他节点可以通过管理网络访问数据库。具体配置文件如下：

```
[mysqld]
bind-address =192.168.16.100            #controller 节点的管理网络 IP 地址
default-storage-engine = innodb         #设置默认存储引擎
innodb_file_per_table = on              #每个表的数据单独保存
max_connections =4096                   #设置最大连接数
collation-server = utf8_general_ci      #设置字符集
character-set-server = utf8
```

（3）启动数据库并设置随系统启动。

```
[root@ controller ~]# systemctl enable mariadb.service
[root@ controller ~]# systemctl start mariadb.service
```

（4）运行 mysql_secure_installation 脚本来保护数据库服务。为 root 账户设置密码。

```
[root@ controller ~]# mysql_secure_installation
```

脚本执行过程中，会有相关操作的提示和密码设置，具体提示及设置信息如下：

```
NOTE:RUNNING ALL PARTS OF THIS SCRIPT IS RECOMMENDED FOR ALL MariaDB
SERVERS IN PRODUCTION USE! PLEASE READ EACH STEP CAREFULLY!
In order to log into MariaDB to secure it,we'll need the current
password for the root user. If you've just installed MariaDB,and
you haven't set the root password yet,the password will be blank,
so you should just press enter here.
Enter current password for root(enter for none):  #直接按 Enter 键
OK,successfully used password,moving on...
Setting the root password ensures that nobody can log into the MariaDB
root user without the proper authorisation.
Set root password? [Y/n]y  #是否设置 root 密码
New password:  #输入两次 root 密码 admin123
Re-enter new password:
Password updated successfully!
Reloading privilege tables..
... Success!
By default,a MariaDB installation has an anonymous user,allowing anyone
to log into MariaDB without having to have a user account created for
them. This is intended only for testing,and to make the installation
go a bit smoother. You should remove them before moving into a
production environment.
Remove anonymous users? [Y/n]y  #是否删除匿名用户
... Success!
Normally,root should only be allowed to connect from 'localhost'. This
ensures that someone cannot guess at the root password from the network.
Disallow root login remotely? [Y/n]y  #是否禁止 root 远程登录
... Success!
By default,MariaDB comes with a database named 'test' that anyone can
access. This is also intended only for testing,and should be removed
before moving into a production environment.
Remove test database and access to it? [Y/n]y  #是否删除 test 库
 - Dropping test database...
... Success!
 - Removing privileges on test database...
... Success!
```

```
Reloading the privilege tables will ensure that all changes made so far
will take effect immediately.
Reload privilege tables now? [Y/n]y  #加载权限表
... Success!
Cleaning up...
All done! If you've completed all of the above steps,your MariaDB
```

6. 消息队列配置

OpenStack 使用消息队列协调组件间的操作和状态信息，支持多种消息队列，包括 RabbitMQ、Qpid 和 ZeroMQ 等，本实验采用 RabbitMQ。消息队列服务运行在 controller 节点。

（1）安装 RabbitMQ 软件包。

```
[root@ controller ~]# yum install rabbitmq - server - y
```

（2）启动 RabbitMQ 并设置为随系统启动。

```
[root@ controller ~]# systemctl start rabbitmq - server. service
[root@ controller ~]# systemctl enable rabbitmq - server. service
Created symlink from/etc/systemd/system/multi - user. target. wants/rabbitmq - server. service to/usr/lib/systemd/system/rabbitmq - server. service.
```

（3）添加 OpenStack 用户并设置密码，用户名为 openstack，密码为 openstack。

```
[root@ controller ~]# rabbitmqctl add_user openstack openstack
```

（4）为 OpenStack 用户设置/目录的配置、写和读权限。

```
[root@ controller ~]# rabbitmqctl set_permissions openstack ".*" ".*" ".*"
```

7. Memcached 配置

服务的身份认证机制会使用 Memcached 缓存令牌，缓存服务 Mencached 一般运行在 controller 节点。

（1）安装 Memcached 软件包。

```
[root@ controller ~]# yum install memcached python - memcached
```

（2）编辑/etc/sysconfig/memcached 配置文件，配置服务使用控制节点的管理 IP 地址，同时也可以使其他节点通过管理网络访问 Memcached 服务。配置文件修改内容如下：

```
OPTIONS = "- l127.0.0.1,::1,controller"
```

（3）启动 Memcached 服务。

```
[root@ controller ~]# systemctl start memcached. service
[root@ controller ~]# systemctl enable memcached. service
Created symlink from/etc/systemd/system/multi - user. target. wants/memcached. service to/usr/lib/systemd/system/memcached. service.
```

（4）查看端口信息，验证 Memcached 是否正常启用。

```
[root@ controller ~]# netstat -lntp
Active Internet connections(only servers)
Proto Recv-Q Send-Q Local Address      Foreign Address  State   PID/Program name
tcp        0      0 0.0.0.0:25672          0.0.0.0:*     LISTEN  45865/beam.smp
tcp        0      0 192.168.16.100:3306    0.0.0.0:*     LISTEN  20102/mysqld
tcp        0      0 192.168.16.100:11211   0.0.0.0:*     LISTEN  54408/memcached
tcp        0      0 127.0.0.1:11211        0.0.0.0:*     LISTEN  54408/memcached
tcp        0      0 0.0.0.0:4369           0.0.0.0:*     LISTEN  1/systemd
tcp        0      0 0.0.0.0:22             0.0.0.0:*     LISTEN  1011/sshd
tcp        0      0 127.0.0.1:25           0.0.0.0:*     LISTEN  1421/master
tcp6       0      0 :::5672                :::*          LISTEN  45865/beam.smp
tcp6       0      0 ::1:11211              :::*          LISTEN  54408/memcached
tcp6       0      0 :::22                  :::*          LISTEN  1011/sshd
tcp6       0      0 ::1:25                 :::*          LISTEN  1421/master
```

8. Etcd 配置

OpenStack 服务使用 Etcd（一种分布式可靠的密钥值存储）进行分布式密钥锁定、存储配置、服务实时追踪及一些其他场景。

（1）安装 Etcd 软件包。

```
[root@ controller ~]# yum install etcd
```

（2）编辑/etc/etcd/etcd.conf 配置文件，配置文件内容如下：

```
#[Member]
ETCD_DATA_DIR="/var/lib/etcd/default.etcd"
ETCD_LISTEN_PEER_URLS=http://192.168.16.100:2380
ETCD_LISTEN_CLIENT_URLS=http://192.168.16.100:2379
ETCD_NAME="controller"
#[Clustering]
ETCD_INITIAL_ADVERTISE_PEER_URLS=http://192.168.16.100:2380
ETCD_ADVERTISE_CLIENT_URLS="http://192.168.16.100:2379"
ETCD_INITIAL_CLUSTER="controller=http://192.168.16.100:2380"
ETCD_INITIAL_CLUSTER_TOKEN="etcd-cluster-01"
ETCD_INITIAL_CLUSTER_STATE="new"
```

（3）启动 Etcd 服务。

```
[root@ controller ~]# systemctl start etcd.service
[root@ controller ~]# systemctl enable etcd.service
Created symlink from/etc/systemd/system/multi-user.target.wants/etcd.service
to/usr/lib/systemd/system/etcd.service.
```

【任务工单】

工单号：3－1

项目名称：OpenStack 云平台部署		任务名称：OpenStack 部署环境准备
班级：	学号：	姓名：
任务安排	□完成 controller 节点和 conpute 节点的网络配置 □关闭双节点防火墙和 SELinux □完成网络时间服务 NTP 的配置 □安装 OpenStack 的 Rocky 版本软件包 □controller 节点部署 MariaDB 数据库 □controller 节点部署 RabbitMQ 消息队列服务 □controller 节点部署 Memcached 和 Etcd	
成果交付形式	将命令操作验证结果或配置内容截图，上传至教学平台	
任务实施总结	任务自评（0～10 分）： 任务收获： 改进点：	
成果验收	□完全满足任务要求 □基本满足任务要求 要求基本完成，存在一些需要微调的小问题： □不能满足需求 要求未全部完成，虚拟机环境存在较大问题，需要修改的问题较多：	

【知识巩固】

1. /etc/selinux/config 配置文件中，SELinux 取值为 enforcing 时，表示的是（　　）。

A. 强制模式　　B. 提示模式　　C. 禁用 SELinux　　D. 无任何含义

2. CentOS 7 虚拟机的网卡配置文件存放路径是__________。

3. 在 RabbitMQ 中如何添加用户名为 user，密码为 password 的用户？

【小李的反思】

不积跬步，无以至千里；不积小流，无以成江海。

源自荀子《劝学》，意思是不积累一步半步的行程，就没有办法达到千里之远；不积累细小的流水，就没办法汇成江河大海。比喻学习必须要慢慢积累，任何事情的成功都是从小到大逐渐积累的，任何事情都要从一点一滴的小事开始做起。同学们在部署 OpenStack 私有云平台时候，需要部署很多功能组件，每一步都可能会遇到各种技术问题，但不能因为事情难而退步，勇敢地迈出第一步，就已经成功一半，在实践中一点一滴完成每个组件的部署，逐渐积累，最终一定可以完成整个平台的部署。

杰克·伦敦是美国出名的故事家，他的学识全靠自修得来，他常常把词典和书里的词句抄在小小的纸片上，然后把这些纸片挂在窗帘上、衣架上、橱窗上、床帐上，甚至塞在镜子缝里，以便在刮脸、穿衣、睡觉前后能看一看。他把部分纸片放在衣兜里，外出参加音乐会、拜访亲友或散步时，抽出空闲时间念一念。由于他不断记忆，最后掌握了大量的词汇，写起文章来得心应手。

没有人可以随便成就大业，任何事情的成功都源于点滴的积累，不积跬步，无以至千里；不积小流，无以成江海，从小事做起，不断积累，由量变引起质变，才能获得真正的成功。

任务 2　认证服务 Keystone 部署

【任务描述】

小李已经完成了 OpenStack 部署的环境准备工作，现想部署第一个组件：认证服务 Keystone。通过查阅资料，小李知道该组件需要部署在 controller 节点上，现参考官方文档（https://docs.openstack.org/keystone/rocky/install/index-rdo.html）进行第一个组件认证服务 Keystone 的部署工作，包括 Keystone 数据库创建，安装并配置 Keystone 组件，配置 Apache HTTP 服务，创建域、项目、用户、角色，创建 OpenStack 客户端环境脚本等。

【知识要点】

1. Keystone 组件作用

Keystone（OpenStack identity service）是 OpenStack 中的一个独立的身份认证服务模块，主要负责 OpenStack 用户的身份认证、令牌管理、提供访问资源的服务目录，以及基于用户角色的访问控制。

其他服务通过 Keystone 来注册其服务的 Endpoint（服务访问的 URL），任何服务之间相互调用，需要经过 Keystone 的身份验证，获得目标服务的 Endpoint 来找到目标服务。它主要功能包括：

- 身份验证（authentication）：令牌的发放和验证。
- 授权（authorization）：基于角色 role 的权限管理。
- 服务目录（catalog of services）：提供服务目录（ServiceCatalog，包括 service 和 endpoint）服务，用户（无论是 Dashboard 还是 APIClient）都需要访问 Keystone 获取服务列表，以及每个服务的地址（OpenStack 中称为 Endpoint）。
- 用户管理：管理用户账户。

2. Keystone 中的概念

1）认证（Authentication）

认证是确认一个用户身份的过程。要确认传入的请求是否是合法的，OpenStack 的 Keystone 服务会通过验证用户提供的一组证书来确认用户的身份。

起初，这些凭据是用户名和密码，或者是用户名和 API 密钥。当 identity 验证用户证书后，会给用户提供一个认证 Token 供用户后续请求所用。

2）证书（Credentials）

证书是确认用户身份的数据，是一个可以证明身份的身份证，包含以下三种：

- 用户名和密码。
- 用户名和 API key。
- Keystone 提供的认证 Token。

3）端点（Endpoint）

Endpoint 就是访问一个服务的地址，即 URL。

Keystone 中包含一个 endpoint 模板，模板中包含一个 URL 列表。列表中的每个 URL 对应一个服务实例的访问地址，具有 public、internal、admin 三种权限。

- public：全局可访问。
- internal：内部局域网访问。
- admin：管理员访问。

4）项目（Project）

Project 在 OpenStack 的早期版本称为租户（Tenant），它是各服务可访问资源的集合，不同 Project 之间的资源是彼此隔离的。每个 Project 中包含有多个 User，每个 User 会根据权限划分来使用 Project 中的资源。User 访问 Project 的资源前，必须要与该 Project 关联，并且指定 User 在 Project 下的 Role，一个 assignment（关联）即 Project - User - Role。

5）角色（Role）

Role 就是一组用户权限，通过给 User 指定 Role，使 User 获得 Role 对应的操作权限，确定一个用户可以做哪些操作、不可以做哪些操作。

Keystone 返回给 User 的 Token 包含了 Role 列表，被访问的 Services 会判断访问它的 User 和 User 提供的 Token 中所包含的 Role，以及每个 Role 访问资源或者进行操作的权限。系统默认使用管理角色 Admin 和成员角色 User（过去的普通用户角色是_member_）。

6）服务（Service）

OpenStack 中的各个服务组件：Nova、Glance、Cinder、Swift、Neutron 等。

7）令牌（Token）

一个 User 对于某个目标（Project 或者 Domain）的一个有限时间段内的身份令牌。

8）域（Domain）

域是项目和用户的集合，定义了身份认证实体的边界，一个域可以表示个人、公司或运营商。用户可以被授予域的管理员角色。域管理员可以在域中创建项目、用户和组，并为域中的用户和组分配角色。

9）组（Group）

组是域内用户的集合，为一个组授予域或项目的权限时，会对组内的所有用户生效，向组内添加或删除用户，也会相应地随着是否在组内而增加或减少对域或项目的权限。

3. Keystone 认证服务流程

用户请求云主机的流程涉及认证服务（Keystone）、计算服务（Nova）、镜像服务（Glance）和网络服务（Neutron）等，在整个过程中，每一个服务的请求都需要进行身份验证，令牌作为凭证在服务间传递。Keystone 的认证流程如图 3-10 所示。

下面以用户 Alice 创建云主机实例为例，说明 Keystone 的认证流程，用户想要创建虚拟机，向 Keystone 发送账户和密码进行验证，Keystone 验证通过之后，会给用户返回 Token 令牌和服务访问的 Endpoint；用户根据获取到的 Endpoint 向 Nova 发送令牌和创建云主机的请求，Nova 通过 Keystone 对令牌进行验证，验证成功后，向 Glance 请求创建云主机需要的镜像，Glance 通过 Keystone 验证 Token 是否合法，验证通过后，向 Nova 返回云主机镜像，此

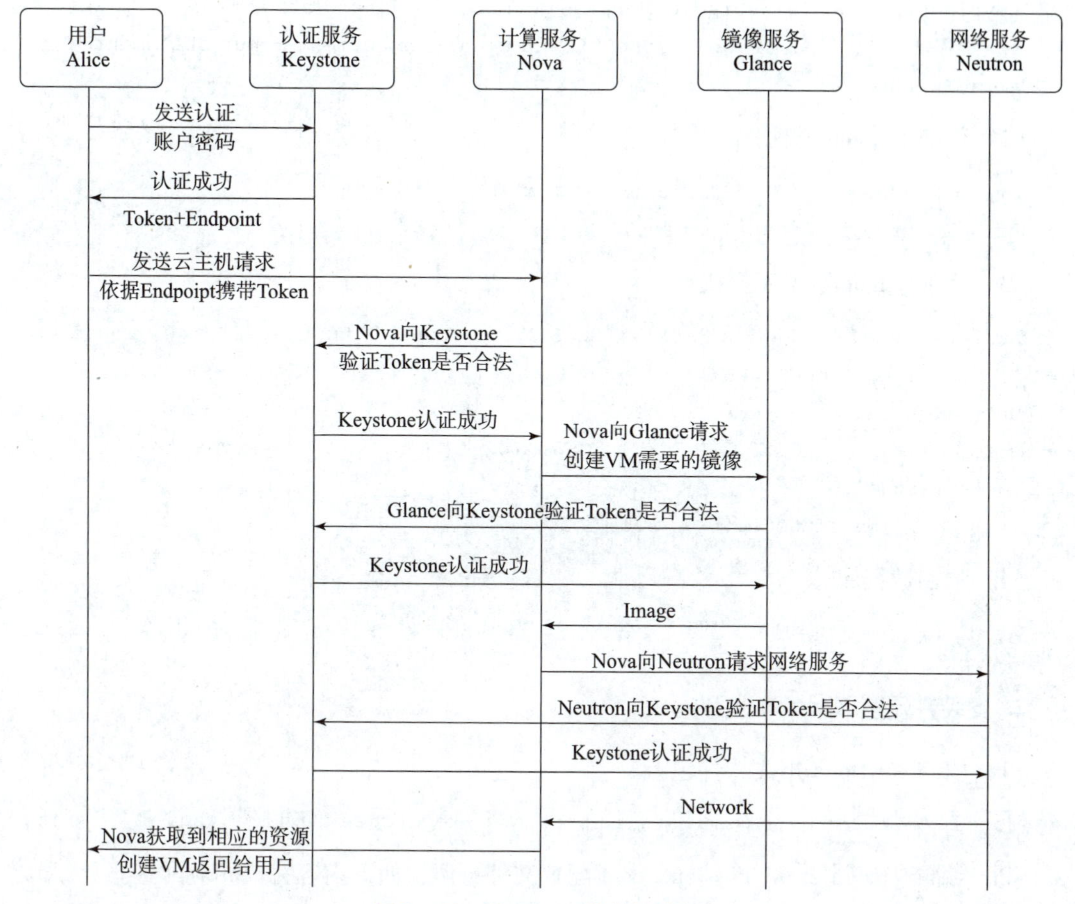

图 3－10　Keystone 的认证流程

时 Nova 已经获取到了云主机的镜像，但还需要申请网络资源，Nova 继续向 Neutron 请求网络服务，Neutron 通过 Keystone 验证 Token 后，向 Nova 返回网络资源，Nova 获取到镜像和网络资源后，创建云主机并返回给用户，此时，用户已经得到了所需要的云主机。

【任务实施】

Keystone 组件为 OpenStack 提供身份认证服务，部署在 controller 节点。

1. Keystone 数据库创建及授权

在安装配置 Keystone 前，需要创建数据库，并对数据库进行授权。

（1）以 root 用户身份连接到数据库服务器，连接时，输入环境准备阶段安装 MariaDB 时的密码。

```
[root@ controller ~]# mysql -u root -p
Enter password:
Welcome to the MariaDB monitor. Commands end with;or \g.
Your MariaDB connection id is 2
Server version:10.1.20 -MariaDB MariaDB Server
```

```
Copyright(c)2000,2016,Oracle,MariaDB Corporation Ab and others.
Type 'help;' or '\h' for help. Type '\c' to clear the current input statement.
MariaDB[(none)]>
```

（2）创建 Keystone 数据库。

```
MariaDB[(none)]>CREATE DATABASE keystone;
Query OK,1 row affected(0.00 sec)
```

（3）为 Keystone 数据库授权。

```
MariaDB[(none)]>GRANT ALL PRIVILEGES ON keystone.* TO 'keystone'@'localhost' IDENTIFIED BY 'keystone';
Query OK,0 rows affected(0.00 sec)
MariaDB[(none)]>GRANT ALL PRIVILEGES ON keystone.* TO 'keystone'@'%' IDENTIFIED BY 'keystone';
Query OK,0 rows affected(0.00 sec)
```

（4）退出数据库。

```
MariaDB[(none)]>exit
```

2. 安装并配置 Keystone 组件

（1）安装 Keystone 相关软件包。

```
[root@controller ~]# yum install openstack-keystone httpd mod_wsgi
```

（2）编辑/etc/keystone/keystone.conf 配置文件，添加如下内容完成相应配置。

在［database］部分，配置数据库访问路径：

```
[database]
#...
connection=mysql+pymysql://keystone:keystone@controller/keystone
```

在［token］部分，配置 Fernet 认证的令牌提供者：

```
[token]
#...
provider=fernet
```

（3）同步认证服务数据库。

```
[root@controller ~]# su -s/bin/sh -c "keystone-manage db_sync" keystone
```

（4）验证数据库表。

```
[root@controller ~]# mysql -ukeystone -pkeystone -e "use keystone;show tables;"
+-----------------------------------+
| Tables_in_keystone|
+-----------------------------------+
| application_credential_role|
```

```
| assignment |
|       …… |
| user_option |
| whitelisted_config |
+----------------------------------+
```

（5）初始化 Fernet 密钥存储库。

```
[root@ controller ~]# keystone-manage fernet_setup --keystone-user keystone --keystone-group keystone
[root@ controller ~]# keystone-manage credential_setup --keystone-user keystone --keystone-group keystone
```

（6）引导认证服务（此处的 admin 为管理账户密码）。

```
keystone-manage bootstrap --bootstrap-password admin \
--bootstrap-admin-url http://controller:5000/v3/ \
--bootstrap-internal-url http://controller:5000/v3/ \
--bootstrap-public-url http://controller:5000/v3/ \
--bootstrap-region-id RegionOne
```

3. 配置 Apache HTTP 服务

（1）编辑/etc/httpd/conf/httpd. conf，配置 ServerName 为 controller 节点。

```
[root@ controller ~]# vi/etc/httpd/conf/httpd.conf
ServerName controller
```

（2）在/etc/httpd/conf. d/目录下创建/usr/share/keystone/wsgi - keystone. conf 文件的链接。

```
[root@ controller ~]# ln -s/usr/share/keystone/wsgi-keystone.conf/etc/httpd/conf.d/
```

（3）启动 HTTP 服务并配置为随系统启动。

```
[root@ controller ~]# systemctl start httpd.service
[root@ controller ~]# systemctl enable httpd.service
Created symlink from/etc/systemd/system/multi-user.target.wants/httpd.service to/usr/lib/systemd/system/httpd.service.
```

（4）配置管理账户环境变量（ADMIN_PASS 为 keystone - manage bootstrap 命令所配置的密码，本书使用的是 admin）。

```
export OS_USERNAME=admin
export OS_PASSWORD=ADMIN_PASS
export OS_PROJECT_NAME=admin
export OS_USER_DOMAIN_NAME=Default
export OS_PROJECT_DOMAIN_NAME=Default
```

```
export OS_AUTH_URL=http://controller:5000/v3
export OS_IDENTITY_API_VERSION=3
```

4. 创建域、项目、用户、角色

OpenStack 通过 Keystone 服务为各组件提供认证服务，认证服务需要使用域、项目、用户和角色等。

（1）查看域列表，当前系统中已经存在名为 Default 的域。

```
[root@controller ~]# openstack domain list
+---------+---------+---------+--------------------+
| ID      | Name    | Enabled | Description        |
+---------+---------+---------+--------------------+
| default | Default | True    | The default domain |
+---------+---------+---------+--------------------+
```

（2）创建 example 域。

```
[root@controller ~]# openstack domain create --description "An example domain" example
+-------------+----------------------------------+
| Field       | Value                            |
+-------------+----------------------------------+
| description | An example domain                |
| enabled     | True                             |
| id          | 7e4a2925e3ee47c1af8aa85f76db8c64 |
| name        | example                          |
| tags        | []                               |
+-------------+----------------------------------+
```

（3）在默认的 Default 域内创建一个名为 service 的项目。

```
[root@controller ~]# openstack project create --domain default --description "service project" service
+-------------+----------------------------------+
| Field       | Value                            |
+-------------+----------------------------------+
| description | service project                  |
| domain_id   | default                          |
| enabled     | True                             |
| id          | 836801414b99461b91e31be9f1c94b97 |
| is_domain   | False                            |
| name        | service                          |
| parent_id   | default                          |
| tags        | []                               |
+-------------+----------------------------------+
```

常规任务需要非特权账户，需要创建 myproject 项目和 myuser 用户并赋予 myrole 角色。

（4）实际工作中不能所有的操作都在 admin 账户下执行，需要一些非特权账户，此处创建 myproject 项目和 myuser 用户并赋予 myrole 角色。

- 创建 myproject 项目。

```
[root@ controller ~]# openstack project create --domain default --description "
demo project" myproject
+-------------+----------------------------------+
| Field       | Value                            |
+-------------+----------------------------------+
| description | demo project                     |
| domain_id   | default                          |
| enabled     | True                             |
| id          | e598a75ab1434d4ea2cfa4e0da5ae9ee |
| is_domain   | False                            |
| name        | myproject                        |
| parent_id   | default                          |
| tags        | []                               |
+-------------+----------------------------------+
```

- 创建 myuser 用户（设置密码为 myuser）。

```
[root@ controller ~] # openstack user create --domain default --password -
prompt myuser
User Password:
Repeat User Password:
+---------------------+----------------------------------+
| Field               | Value                            |
+---------------------+----------------------------------+
| domain_id           | default                          |
| enabled             | True                             |
| id                  | 1ecd571c56b34ff78dbae47cd7b20c89 |
| name                | myuser                           |
| options             | {}                               |
| password_expires_at | None                             |
+---------------------+----------------------------------+
```

- 创建 myrole 角色。

```
[root@ controller ~]# openstack role create myrole
+-----------+----------------------------------+
| Field     | Value                            |
+-----------+----------------------------------+
| domain_id | None                             |
| id        | 0694847e73ff48a4b06a819494c0008f |
| name      | myrole                           |
```

```
+-----------+-------------------------------------+
```

- 为 myuser 用户赋予 myrole 角色。

```
[root@ controller ~]# openstack role add --project myproject --user myuser myrole
```

5. Keystone 安装验证

（1）取消 OS_AUTH_URL 和 OS_PASSWORD 两个临时变量的设置。

```
[root@ controller ~]# unset OS_AUTH_URL OS_PASSWORD
```

（2）请求 admin 用户的身份验证令牌。

执行命令后，需要输入 admin 用户密码（此处密码为 Keystone 服务的管理账户密码 admin）。

```
[root@ controller ~]# openstack --os-auth-url http://controller:5000/v3\
>--os-project-domain-name Default --os-user-domain-name Default\
>--os-project-name admin --os-username admin token issue
Password:
+------------+-----------------------------------------------------+
| Field      | Value                                               |
+------------+-----------------------------------------------------+
| expires    | 2022-10-20T15:17:54+0000                            |
| id         | gAAAAABjUViSr4lBgbiTTrId1BXIZFTyPwFjFTWNpsA…2V6TspahRKPM|
| project_id | 62b92970ed2a42cb9ea0e8f013004062                    |
| user_id    | 974ca43e23234f0b91847c162982961d                    |
+------------+-----------------------------------------------------+
```

从显示结果可看出，admin 用户的令牌能够被正常请求到。

（3）请求 myuser 用户的身份验证令牌。

执行后，需要输入 myuser 用户密码（创建 myuser 用户时，设置密码为 myuser）。

```
[root@ controller ~]# openstack --os-auth-url http://controller:5000/v3\
>    --os-project-domain-name Default --os-user-domain-name Default\
>    --os-project-name myproject --os-username myuser token issue
Password:
+------------+-----------------------------------------------------+
| Field      | Value                                               |
+------------+-----------------------------------------------------+
| expires    | 2022-10-20T15:25:36+0000                            |
| id         | gAAAAABjUVpgKO11sYtlTgjpC3DBLRfudIOOhFUbg…IaM0TxOgUkxg958 |
| project_id | e598a75ab1434d4ea2cfa4e0da5ae9ee                    |
| user_id    | 1ecd571c56b34ff78dbae47cd7b20c89                    |
+------------+-----------------------------------------------------+
```

admin 用户和 myuser 用户都可以正常获取到身份令牌，说明 Keystone 服务组件部署已完成。

6. 创建用户环境变量脚本

在 OpenStack 中，以不同用户操作时，需要设置不同的环境变量（就像以一个账户登录之后才能执行该账户的操作一样），每次切换用户时都通过命令进行若干环境参数的修改是非常麻烦的，因此，可以为不同用户设置不同的环境变量脚本，切换环境时，只需执行相应的环境变量配置脚本即可，此处为 admin 用户和 myuser 用户配置环境变量脚本。

（1）创建 admin 用户的环境变量脚本 admin - openstack. sh，脚本内容如下：

```
export OS_USERNAME=admin
export OS_PASSWORD=admin #前面设置的 admin 的密码
export OS_PROJECT_NAME=admin
export OS_USER_DOMAIN_NAME=Default
export OS_PROJECT_DOMAIN_NAME=Default
export OS_AUTH_URL=http://controller:5000/v3
export OS_IDENTITY_API_VERSION=3
export OS_IMAGE_API_VERSION=2
```

（2）创建 myuser 用户的环境变量脚本 demo - openstack. sh，脚本内容如下：

```
export OS_USER_DOMAIN_NAME=Default
export OS_PROJECT_NAME=myproject
export OS_USERNAME=myuser
export OS_PASSWORD=myuser #前面设置的 myuser 的密码
export OS_AUTH_URL=http://controller:5000/v3
export OS_IDENTITY_API_VERSION=3
export OS_IMAGE_API_VERSION=2
```

（3）执行 admin - openstack. sh 脚本，同步 admin 用户的环境变量，并获取 admin 用户的 Token。

```
[root@ controller ~]# source admin-openstack.sh
[root@ controller ~]# openstack token issue
+------------+-----------------------------------------------------+
| Field      | Value                                               |
+------------+-----------------------------------------------------+
| expires    | 2022-10-20T15:31:56+0000                            |
| id         | gAAAAABjUVvcUR5lEXL-KHP4YZ02_TvvWD7sPL…J9Zt7kVwSeUraz6Fv9BuIU |
| project_id | 62b92970ed2a42cb9ea0e8f013004062                    |
| user_id    | 974ca43e23234f0b91847c162982961d                    |
+------------+-----------------------------------------------------+
```

【任务工单】

工单号：3－2

<table>
<tr><td colspan="3">项目名称：OpenStack 云平台部署</td><td colspan="2">任务名称：认证服务 Keystone 部署</td></tr>
<tr><td colspan="2">班级：</td><td colspan="2">学号：</td><td>姓名：</td></tr>
<tr><td>任务安排</td><td colspan="4">□Keystone 数据库的创建与授权
□安装 Keystone 组件并完成 Keystone 配置文件的配置
□配置 Apache HTTP 服务
□创建 example 域和 service 项目
□创建非特权账户（包括 myproject 项目和 myuser 用户并赋予 myrole 角色）
□请求 admin 和 myuser 用户的 Token 验证 Keystone 部署的正确性
□配置 admin 和 myuser 用户的环境变量脚本</td></tr>
<tr><td>成果交付形式</td><td colspan="4">获取 admin 和 myuser 用户的 Token，同学间互相验证 Keystone 部署的正确性并打分</td></tr>
<tr><td>任务实施总结</td><td colspan="4">任务自评（0～10 分）：
任务收获：____________________
____
____
改进点：____________________
____
____________________</td></tr>
<tr><td>成果验收</td><td colspan="4">□完全满足任务要求
□基本满足任务要求
Keystone 部署正确，能够正常获取到用户的 Token，但存在一些不影响 Keystone 使用的小问题：
____
____
□不能满足需求
Keystone 部署完成，但功能存在问题，获取用户 Token 失败，需要核对部署过程修改：
____
____
____________________</td></tr>
</table>

【知识巩固】

1. Endpoint 具有三种权限，分别是（　　）。

A. internal：内部局域网访问　　B. admin：管理员访问

C. public：全局可访问　　D. external：外部可访问

2. Keystone 服务默认监听端口号是（　　）。

A. 9292　　B. 5000　　C. 8778　　D. 8774

【小李的反思】

独学而无友，则孤陋而寡闻。

源自《礼记·学记》，意思是说，如果只是独自一个人学习而没有朋友一起讨论，就会孤陋寡闻。学习要吸取他人长处，朋友间共同学习、集思广益、取长补短、彼此帮助、彼此讨论/切磋，才能弥补自身的缺憾，并获得更多的知识。同学们在完成本任务时，是以团队的形式完成的，对于不理解的知识点，团队之间互相讨论，互帮互学，促进知识的理解、技术的掌握。

玄奘是唐代的名僧，由于强烈的求知欲，二十岁时就被称为佛门的“千里驹”，但当时佛教传入中国都是通过翻译的，由于理解不一，产生了各种教派系。玄奘深知，要彻底理解佛经的教义，唯一的办法就是去佛教的发源地天竺（今印度）求法交流学习，以弄清佛教的真谛。公元630年，玄奘历经千难万险，经过大小二十多个国家，终于到达天竺（印度）。在天竺他遍游佛教圣迹，遍访名师，虚心求教，最终在佛学上取得巨大成功，同时也促进了东西方的文化交流。

在学习的过程中，不能闭门造车沉浸在自己的世界里，一味地沉浸在自己的世界里就容易变成井底之蛙，要能够多进行学术的交流，在交流的过程中取长补短，互相学习，像蜜蜂一样，不断吸取群芳精华，经过反复加工，酿造知识精华。

任务 3　镜像服务 Glance 部署

【任务描述】

小李已经完成了 OpenStack 第一个组件认证服务 Keystone 的部署并通过了验证，对云平台的部署项目充满了信心和工作热情。他想到云平台需要对学生下发虚拟机，虚拟机需要有操作系统，那么就应该有一个组件对虚拟机操作系统镜像进行管理，通过查阅资料，小李知道 Glance 组件可以完成此功能，该组件需要部署在 controller 节点上，现参考官方文档（https://docs.openstack.org/glance/rocky/install/install-rdo.html）进行镜像服务 Glance 组件的部署工作，包括创建 Glance 数据库，创建 Glance 用户、服务实体和服务 API 端点，安装并配置 Glance 组件等。

【知识要点】

1. Glance 组件基础

1）Glance 组件的作用

Glance（OpenStack image service）：OpenStack 中的镜像服务，主要功能是使用户能发现、注册、检索虚拟机镜像，对外提供一个能够查询虚拟机镜像元数据和检索真实镜像的 REST API，其主要功能包括：

- 查询和获取镜像的元数据和镜像本身。
- 注册和上传虚拟机镜像，包括镜像的创建、上传、下载和管理。
- 维护镜像信息，包括元数据和镜像本身。
- 支持多种方式存储镜像，包括普通的文件系统、Swift、Amazon S3 等。
- 对虚拟机实例执行创建快照命令来创建新的镜像，或者备份虚拟机的状态。

Glance 本身并不实现对镜像的存储功能，Glance 只是一个代理，充当镜像存储服务与 OpenStack 的其他组件之间的纽带。镜像可以被保存在 Glance 节点的文件系统中，也可以对接后端存储。具体使用哪种，可以在/etc/glance/glance-api.conf 文件中的［glance_store］模块下配置。前者机制相对比较简单，但由于没有备份机制，当文件系统损伤时，所有的镜像将不可用；后者可以对接 Cinder 的块存储服务或 Swift 和 Ceph 的对象存储服务等。由于后端存储一般具有非常健壮的备份还原机制，可以减少因为文件系统损伤而造成的镜像不可用情况。

2）Glance 组件 API 版本

Glance 服务的 REST API 有两个版本：API v1 和 API v2，这两个版本对镜像存储的支持相同，v1 版本从 Newton 版本开始已经过时。

v1 版本只提供基本的镜像和成员操作功能，包括镜像创建、删除、下载、列表、详细信息查询、更新，以及镜像租户成员的创建、删除和列表。

v1 版本的实现如图 3-11 所示，Glance 服务的 v1 版本 API 具有 glance-api 和 glance-registry 两个 WSGI 服务，二者都提供 REST API，使用者通过 REST API 执行关于镜像的各种操作。但两个 API 的使用者是不同的，glance - api 的使用者是 OpenStack 命令行工具、Horizon 或 Nova 服务，而 glance - registry 提供的 REST API 仅供 glance - api 使用。

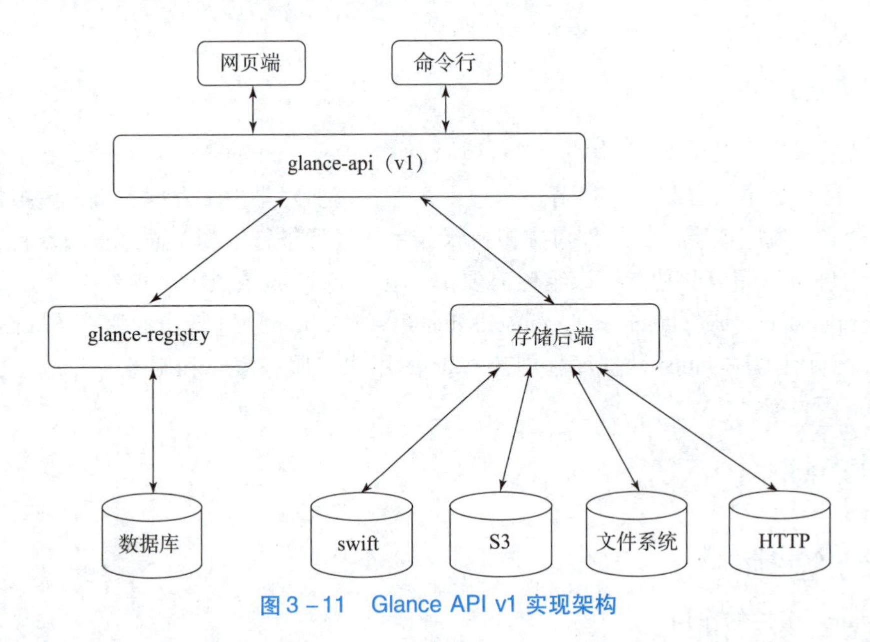

图 3 - 11　Glance API v1 实现架构

glance - api 是系统后台运行的服务进程，是进入 Glance 服务的入口，对外提供 REST API，负责接收用户的 REST API 请求，响应镜像查询、获取和存储的调用。glance - api 在处理 API 请求时，如果是与镜像本身相关的操作，则将请求转发给存储后端，通过后端存储系统提供相应镜像操作

glance - registry 是系统后台运行的 Glance 注册服务进程，负责处理与镜像元数据相关的 REST API 请求，元数据包括镜像大小、类型等信息。glance - api 接收的请求如果是与镜像的元数据相关的操作，glance - api 会把请求转发给 glance - registry，由 glance - registry 解析请求内容，并与数据库交互，存储、处理、检索镜像的元数据。glance - api 对外提供 API，而 glance - registry 的 API 只由 glance - api 使用。

v2 版本除了支持 v1 的所有功能外，主要增加了镜像位置的添加、删除、修改，元数据、名称空间操作以及镜像标记操作。

v2 版本的实现如图 3 - 12 所示，将 glance - registry 集成到了 glance - api 内部，glance - api 在接收到与镜像元数据相关的操作时，会直接操作，无须将请求转给 glance - registry 处理，这么做的好处是减少了一个中间的处理环节。

3）镜像的访问权限

镜像的访问权限包含 4 种，分别是：

- Public（公共的）：可以被所有的项目使用。
- Private（私有的）：只有被镜像所有者所在的项目使用。
- Shared（共享的）：一个非共有的镜像，可以共享给其他项目。
- Protected（受保护的）：镜像受保护，任何角色不能删除该镜像。

4）虚拟机镜像磁盘格式

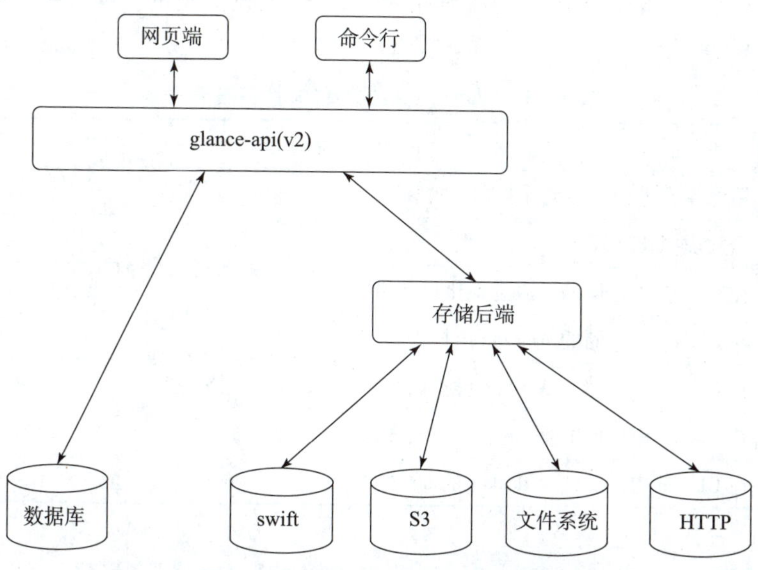

图 3-12　Glance API v2 实现架构

在 OpenStack 中添加镜像到 Glance 时，需要指定虚拟机镜像的磁盘格式和容器格式，Glance 支持的虚拟机镜像文件磁盘格式见表 3-2。

表 3-2　Glance 支持的虚拟机镜像文件磁盘格式

磁盘格式	说明
raw	非结构化的磁盘格式
qcow2	由 QEMU 仿真支持，可动态扩展，支持写时复制（Copy on Write）的磁盘格式，是 KVM 虚拟机默认使用的磁盘文件格式
vhd	该格式通用于 VMware、Xen、VirtualBox 以及其他虚拟机管理程序
vhdx	VHD 格式的增强版本，支持更大的磁盘尺寸
vmdk	一种比较通用的虚拟机磁盘格式
vdi	由 VirtualBox 虚拟机监控程序和 QEMU 仿真器支持的磁盘格式
iso	用于光盘（CD-ROM）数据内容的档案格式
ploop	由 Virtuozzo 支持，用于运行 OS 容器的磁盘格式
aki	在 Glance 中存储的 Amazon 内核格式
ari	在 Glance 中存储的 Amazon 虚拟内存盘（Ramdisk）格式
ami	在 Glance 中存储的 Amazon 机器格式

5）虚拟机镜像容器格式

从文件角度，Glance 中的容器格式是指虚拟镜像的文件格式，Glance 对镜像文件进行管理，往往把镜像元数据装载于一个容器中。在这个容器中，包含了虚拟机的元数据（metadata）和其他相关信息等数据。在创建虚拟镜像文件的时候，需要管理员指定镜像的容器格式。但需要注意的是，容器格式字符串在当前并没有被 Glance 或其他 OpenStack 组件使用，所以，如果不确定容器格式，简单地将其指定 bare 是比较安全的。Glance 支持的镜像文件容器格式见表 3-3。

表 3-3　Glance 支持的镜像文件容器格式

容器格式	说明
bare	镜像没有进行容器或元数据的封装
ovf	开放虚拟化格式
ova	在 Glance 中存储的开放虚拟化设备格式
aki	在 Glance 中存储的 Amazon 内核格式
ari	在 Glance 中存储的 Amazon 虚拟内存盘（Ramdisk）格式
ami	在 Glance 中存储的是 Amazon 机器图像
docker	在 Glance 中存储的是 docker 的 tar 文件

2. 虚拟机状态

镜像状态是 Glance 管理镜像的重要内容，Glance 可以通过虚拟机镜像的状态感知某一镜像的使用情况。镜像从创建到上传成功通过异步任务方式一步一步完成，整个过程会经历镜像的多个状态。OpenStack 中的 Glance 服务镜像状态见表 3-4。

表 3-4　Glance 服务镜像状态

镜像状态	说明
queued	queued 是镜像初始化状态，表示镜像文件刚刚被创建，在 Glance 数据库中已经保存了镜像元数据，但还没有上传至 Glance 中
saving	表示镜像数据正在上传到 Glance
uploading	表明已经进行了镜像上传。在此状态下，不允许调用 PUT/file（请注意，对一个已经在队列中的镜像调用 PUT/files 会使得镜像变成 saving 状态。当镜像处于 saving 状态时，不允许执行 PUT/stage 操作。因此，不能对相同的镜像使用这两种上传方法）
importing	表示已经进行了导入操作，但镜像还没有准备好使用
active	active 是当镜像上传成功以后的一种状态，表明镜像在 Glance 中可用
deactivated	表示不允许任何非管理员用户访问镜像数据，禁止下载镜像、镜像导出和镜像克隆等操作
killed	表示在上载镜像数据期间发生错误，并且镜像不可读
deleted	Glance 保留了关于镜像的信息，但不再可用，此状态下的图像将在以后自动删除
pending_delete	Glance 尚未删除镜像数据，处于此状态的图像无法恢复

Glance 对镜像进行全生命周期的管理，镜像状态的转换过程如图 3-13 所示。Glance 在处理镜像过程中，镜像一般会经历 queued、saving、active 和 deleted 等几个状态，其他状态只有在特殊情况下才会出现。

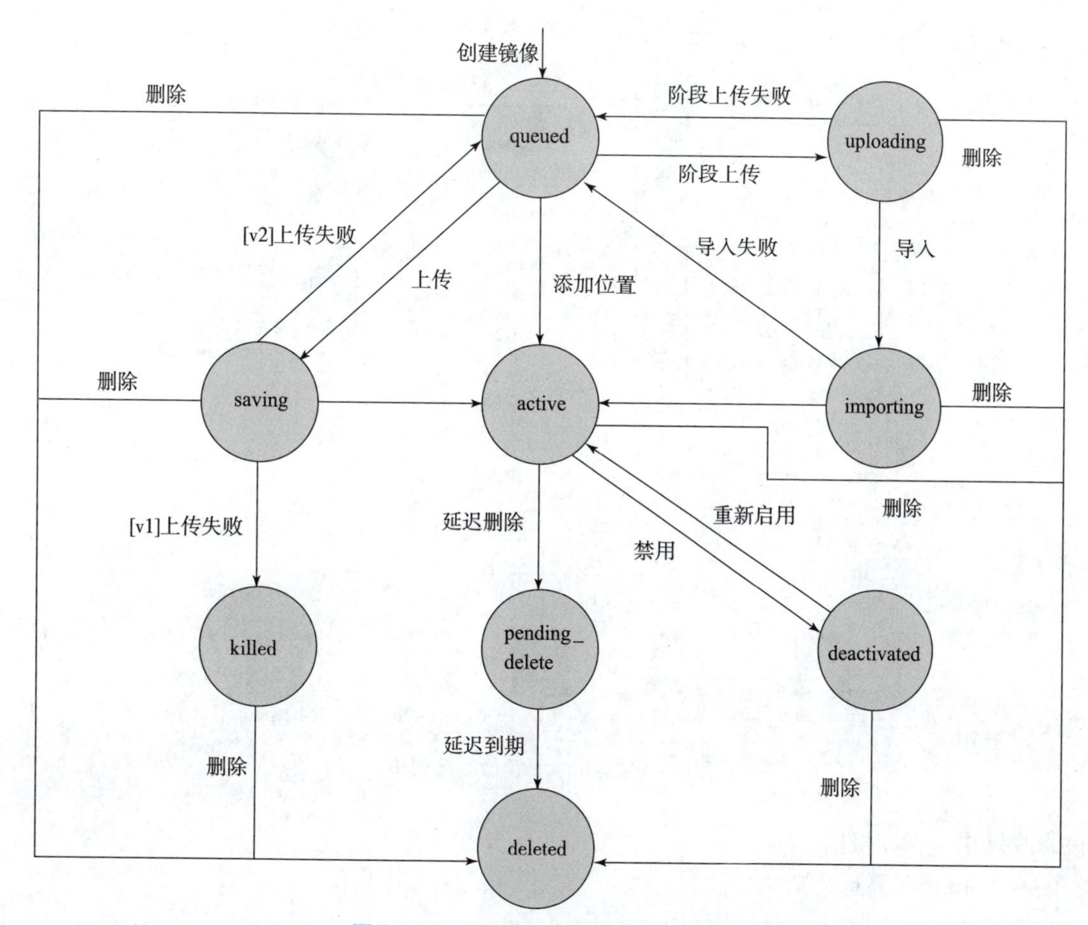

图 3－13 Glance 镜像状态的转换过程

3. 镜像服务基础架构

OpenStack 的 Glance 服务采用 C/S 架构，对外提供 REST API 来完成服务请求，通过域控制器管理内部服务操作，把各个服务分发到各层，各服务在各自特定的层完成相应操作。所有的文件操作通过 glance_store 库与外部存储后端及本地文件交互，并提供统一接口访问后端存储。Glance 在所有组件之间共享一个基于 SQL 的中央数据库 Glance DB，存储、处理、检索镜像的元数据，如图 3－14 所示。

A client（客户端）：使用 Glance 服务的应用程序。

REST API（应用程序接口）：对外提供的应用操作接口，使用者通过 REST API 执行镜像的各种操作。

Glance Domain Controller（Glance 域控制器）：实现主要功能的中间件，例如：认证、通知、策略、数据库连接等。相当于是一个调度员，将 Glance 内部服务的操作分发到各层（Auth 授权、Notifier 通知、Policy 策略、Quota 配额、Location 定位、DB 数据库连接），具体任务由每个层实现。

- Auth（授权）

验证镜像自己或者它的属性是否可以被修改，只有管理员和该镜像的拥有者才可以执行

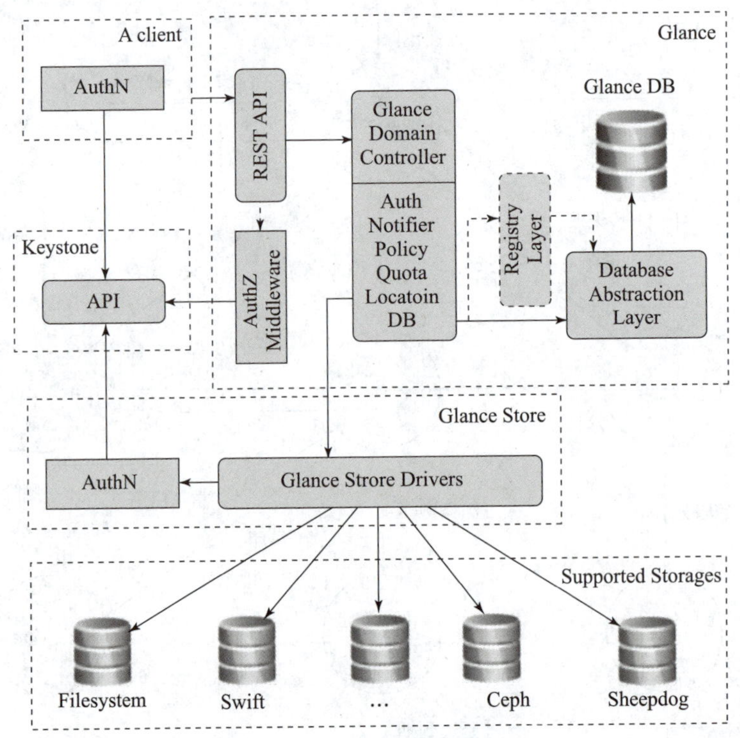

图 3－14　Glance 镜像基础架构

该修改操作，否则报错。

- Notifier（消息通知）

把下列信息添加到 queue 队列中：

①关于所有镜像修改的通知。

②在使用过程中发生的所有异常和警告。

- Policy（规则定义）

①定义操作镜像的访问规则 rules，这些规则都定义在/etc/policy. json 文件中。

②监控 rules 的执行。

- Quota（配额）

如果针对一个用户，管理员为其规定好他能够上传的所有镜像的大小配额，此处的 Quota 层就是用来检测用户上传是否超出配额限制：

①如果没有超出配额限制，那么添加镜像的操作成功。

②如果超出了配额，那么添加镜像的操作失败并且报错。

- Location（定位）

与 Glance Store 交互，如上传、下载等。由于可以有多个存储后端，不同的镜像存放的位置都被该组件管理。

①当一个新的镜像位置被添加时，检测该 URI 是否正确。

②当一个镜像位置被改变时，负责从存储中删除该镜像。

③阻止镜像位置的重复。

- DB（数据库）

①实现了与数据库 API 的交互。

②将镜像转换为相应的格式以记录在数据库中。并且从数据库接收的信息转换为可操作的镜像对象。

Database Abstraction Layer（DAL，数据库抽象层）：Glance 和数据库之间通信的一个 API 接口。

Registry Layer（注册层）：可选层，通过一个单独的服务，组织 Glance Domain Controller 和 Glance DB 之间安全通信。

Glance Store（Glance 存储）：用来组织处理 Glance 和各种存储后端的交互。

所有的镜像文件操作都是通过调用 Glance Store 库执行的，它负责与外部存储端和（或）本地文件系统的交互。Glance Store 提供了一个统一的接口来访问后端的存储。

【任务实施】

Glance 服务为 OpenStack 提供镜像管理服务，用户可以通过 Glance 发现、检索和注册虚拟机镜像。Glance 提供 REST API，用户可以通过它来查询虚拟机镜像元数据并检索实际镜像，通过镜像服务可以将虚拟机存储在文件系统、对象存储系统等多种存储位置。

1. Galnce 数据库创建及授权

在安装配置 Glance 前，需要创建数据库，并对数据库进行授权。

（1）以 root 用户身份连接到数据库服务器，连接时，输入环境准备阶段安装 MariaDB 时的密码。

```
[root@ controller ~]# mysql -u root -p
Enter password:
Welcome to the MariaDB monitor. Commands end with;or \g.
Your MariaDB connection id is 3
Server version:10.1.20 -MariaDB MariaDB Server
Copyright(c)2000,2016,Oracle,MariaDB Corporation Ab and others.
Type 'help;' or '\h' for help. Type '\c' to clear the current input statement.
MariaDB[(none)]>
```

（2）创建 Glance 数据库。

```
MariaDB[(none)]>CREATE DATABASE glance;
Query OK,1 row affected(0.00 sec)
```

（3）为 Glance 数据库授权。

```
MariaDB[(none)]>GRANT ALL PRIVILEGES ON glance.* TO 'glance'@ 'localhost' IDENTIFIED BY 'glance';
Query OK,0 rows affected(0.00 sec)
MariaDB[(none)]>GRANT ALL PRIVILEGES ON glance.* TO 'glance'@ '%' IDENTIFIED BY 'glance';
Query OK,0 rows affected(0.00 sec)
```

（4）退出数据库。

```
MariaDB[(none)]>exit
```

2. 创建 Glance 服务的凭证

（1）执行 admin-openstack. sh 脚本，获取管理员的命令行执行权限。

```
[root@ controller ~]# source. /admin-openstack. sh
```

（2）创建 Glance 用户。

需要输入 Glance 用户密码，此处设置为 glance。

```
[root@ controller ~]# openstack user create --domain default --password-prompt glance
User Password:
Repeat User Password:
+---------------------+-----------------------------------+
| Field               | Value                             |
+---------------------+-----------------------------------+
| domain_id           | default                           |
| enabled             | True                              |
| id                  | fbf37de9362643ddb8f9f401b92fedd0  |
| name                | glance                            |
| options             | {}                                |
| password_expires_at | None                              |
+---------------------+-----------------------------------+
```

（3）为 Glance 用户和 Service 项目添加管理员（admin）角色。

```
[root@ controller ~]# openstack role add --project service --user glance admin
```

（4）创建 Glance 服务实体。

```
[root@ controller ~]# openstack service create --name glance \
--description "OpenStack Image" image
+------------+----------------------------------+
| Field      | Value                            |
+------------+----------------------------------+
| description| OpenStack Image                  |
| enabled    | True                             |
| id         | 19b0cf1e7b4d4a54a6320ed8851bcf53 |
| name       | glance                           |
| type       | image                            |
+------------+----------------------------------+
```

（5）创建镜像服务的 API 端点。

Endpoint 是一个可以通过网络来访问和定位某个 OpenStack 服务的地址，通常是一个 URL。

OpenStack 的 Endpoint 有三种类型：admin、internal、public。

admin：给 admin 用户使用。

internal：内部使用，OpenStack 内部服务使用它来跟别的服务通信。

public：互联网用户可以访问的地址。

①创建公共端点。

```
[root@ controller ~]# openstack endpoint create --region RegionOne \
image public http://controller:9292
+--------------+----------------------------------+
| Field        | Value                            |
+--------------+----------------------------------+
| enabled      | True                             |
| id           | 8d082cc510ba4200859390a24fae9895 |
| interface    | public                           |
| region       | RegionOne                        |
| region_id    | RegionOne                        |
| service_id   | 19b0cf1e7b4d4a54a6320ed8851bcf53 |
| service_name | glance                           |
| service_type | image                            |
| url          | http://controller:9292           |
+--------------+----------------------------------+
```

②创建内部端点。

```
[root@ controller ~]# openstack endpoint create --region RegionOne \
image internal http://controller:9292
+--------------+----------------------------------+
| Field        | Value                            |
+--------------+----------------------------------+
| enabled      | True                             |
| id           | fde09cb8174d428dabb798b50d28d3b3 |
| interface    | internal                         |
| region       | RegionOne                        |
| region_id    | RegionOne                        |
| service_id   | 19b0cf1e7b4d4a54a6320ed8851bcf53 |
| service_name | glance                           |
| service_type | image                            |
| url          | http://controller:9292           |
+--------------+----------------------------------+
```

③创建管理端点。

```
[root@ controller ~]# openstack endpoint create --region RegionOne \
image admin http://controller:9292
+--------------+----------------------------------+
| Field        | Value                            |
+--------------+----------------------------------+
```

```
| enabled          | True                               |
| id               | 75bdbc117d754676a192f1f259491af0   |
| interface        | admin                              |
| region           | RegionOne                          |
| region_id        | RegionOne                          |
| service_id       | 19b0cf1e7b4d4a54a6320ed8851bcf53   |
| service_name     | glance                             |
| service_type     | image                              |
| url              | http://controller:9292             |
+------------------+------------------------------------+
```

3. 安装并配置 Glance 组件

（1）安装 Glance 软件包。

```
[root@ controller ~]# yum install openstack-glance
```

（2）编辑/etc/glance/glance-api.conf 配置文件，进行如下配置。

①在［database］部分，配置数据库访问地址。

#链接中第二个 glance 为 Glance 数据库的密码

```
[database]
connection=mysql+pymysql://glance:glance@ controller/glance
```

②在［keystone_authtoken］和［paste_deploy］部分，配置身份认证访问信息。

```
[keystone_authtoken]
www_authenticate_uri=http://controller:5000
auth_url=http://controller:5000
memcached_servers=controller:11211
auth_type=password
project_domain_name=Default
user_domain_name=Default
project_name=service
username=glance
password=glance

[paste_deploy]
flavor=keystone
```

③在［glance_store］部分，配置本地文件系统存储和映像文件的位置。

```
[glance_store]
stores=file,http
default_store=file
filesystem_store_datadir=/var/lib/glance/images/
```

（3）编辑/etc/glance/glance-registry.conf 配置文件，配置内容如下。

①在［database］部分，配置数据库访问地址。

#链接中第二个 glance 为 Glance 数据库的密码

```
[database]
#...
connection = mysql + pymysql://glance:glance@ controller/glance
```

②在［keystone_authtoken］和［paste_deploy］部分，配置身份认证访问信息。

```
[keystone_authtoken]
#...
www_authenticate_uri = http://controller:5000
auth_url = http://controller:5000
memcached_servers = controller:11211
auth_type = password
project_domain_name = Default
user_domain_name = Default
project_name = service
username = glance
password = glance

[paste_deploy]
#...
flavor = keystone
```

（4）同步 Glance 数据库。

```
[root@ controller ~]# su - s/bin/sh - c "glance - manage db_sync" glance
/usr/lib/python2.7/site - packages/oslo_db/sqlalchemy/enginefacade.py:1352:OsloDBDeprecationWarning:EngineFacade is deprecated;please use oslo_db.sqlalchemy.enginefacade
......
Database is synced successfully.
```

（5）验证 Glance 是否可以正常查询 Glance 数据库中数据表。

```
[root@ controller ~]# mysql - uglance - pglance - e "use glance;show tables;"
+------------------------------------+
| Tables_in_glance                   |
+------------------------------------+
| alembic_version                    |
| image_locations                    |
| image_members                      |
| image_properties                   |
| image_tags                         |
| images                             |
| metadef_namespace_resource_types   |
```

```
| metadef_namespaces                    |
| metadef_objects                       |
| metadef_properties                    |
| metadef_resource_types                |
| metadef_tags                          |
| migrate_version                       |
| task_info                             |
| tasks                                 |
+---------------------------------------+
```

（6）启动镜像服务并配置为随系统启动。

```
[root@ controller ~]# systemctl start openstack-glance-api.service \
openstack-glance-registry.service
[root@ controller ~]# systemctl enable openstack-glance-api.service \
openstack-glance-registry.service
Created symlink from/etc/systemd/system/multi-user.target.wants/openstack-glance-api.service to/usr/lib/systemd/system/openstack-glance-api.service.
Created symlink from/etc/systemd/system/multi-user.target.wants/openstack-glance-registry.service to/usr/lib/systemd/system/openstack-glance-registry.service.
```

4. 验证 Glance 服务

（1）执行 admin-openstack.sh 脚本使能 admin 权限。

```
[root@ controller ~]# source./admin-openstack.sh
```

（2）下载测试用镜像。

```
[root@ controller ~]# wget http://download.cirros-cloud.net/0.4.0/cirros-0.4.0-x86_64-disk.img
```

（3）使用 QCOW2 磁盘格式，bare 容器格式将镜像上载到 Glance 服务，可见性设置为 public，以便所有项目都可以访问它。

```
[root@ controller ~]# openstack image create "cirros" \
--file cirros-0.4.0-x86_64-disk.img \
--disk-format qcow2 --container-format bare \
--public
+------------------+--------------------------------------------+
| Field            | Value                                      |
+------------------+--------------------------------------------+
| checksum         | 443b7623e27ecf03dc9e01ee93f67afe           |
| container_format | bare                                       |
| created_at       | 2022-10-21T13:53:33Z                       |
| disk_format      | qcow2                                      |
```

```
| file             | /v2/images/a58a2e2f-fd7e-4ef3-b8c1-bd4446f3d33d/
                                                                  file|
| id               | a58a2e2f-fd7e-4ef3-b8c1-bd4446f3d33d                |
| min_disk         | 0                                                   |
| min_ram          | 0                                                   |
| name             | cirros                                              |
| owner            | 62b92970ed2a42cb9ea0e8f013004062                    |
| properties       | os_hash_algo='sha512',os_hash_value='6513f2…350f78',
                     os_hidden='False'                                   |
| protected        | False                                               |
| schema           | /v2/schemas/image                                   |
| size             | 12716032                                            |
| status           | active                                              |
| tags             |                                                     |
| updated_at       | 2022-10-21T13:53:33Z                                |
| virtual_size     | None                                                |
| visibility       | public                                              |
+------------------+-----------------------------------------------------+
```

（4）确认镜像已上传成功，并且状态为 active。

```
[root@ controller ~]# openstack image list
+--------------------------------------+--------+--------+
| ID                                   | Name   | Status |
+--------------------------------------+--------+--------+
| a58a2e2f-fd7e-4ef3-b8c1-bd4446f3d33d | cirros | active |
+--------------------------------------+--------+--------+
```

【任务工单】

工单号：3－3

<table>
<tr><td colspan="2">项目名称：OpenStack 云平台部署</td><td colspan="2">任务名称：镜像服务 Glance 部署</td></tr>
<tr><td colspan="2">班级：</td><td>学号：</td><td>姓名：</td></tr>
<tr><td>任务安排</td><td colspan="3">□Glance 数据库的创建与授权
□创建 Glance 服务的凭证，包括创建 Glance 用户并授权和创建 Glance 服务端点
□安装 Glance 组件，并配置 Glance 的配置文件完成部署
□上传镜像到 Glance，验证 Glance 服务部署的正确性</td></tr>
<tr><td>成果交付形式</td><td colspan="3">创建镜像并上传至 Glance 服务，通过 openstack image list 命令查看镜像是否上传成功</td></tr>
<tr><td>任务实施总结</td><td colspan="3">任务自评（0～10 分）：
任务收获：

改进点：

</td></tr>
<tr><td>成果验收</td><td colspan="3">□完全满足任务要求
□基本满足任务要求
Glance 部署正确，能够正常上传镜像到 Glance 服务，但存在一些不影响 Glance 使用的小问题：

□不能满足需求
Glance 部署完成，但镜像上传后存在问题无法使用，需要核对部署过程修改：

</td></tr>
</table>

【知识巩固】

Glance 服务默认监听端口号是（　　）。

A. 9292　　B. 5000　　C. 8778　　D. 8774

【小李的反思】

锲而舍之，朽木不折；锲而不舍，金石可镂。

源自《荀子·劝学》，意思是说，雕刻一会儿就中断的话，连朽木都不能折断；不停地雕刻，金石也可雕刻出美丽的图纹。学习是一个漫长的、艰苦的过程，也是一个长期积累的过程，必须心无旁骛，专心致志，“锲而不舍”，才能学有所成；反之，“锲而舍之”，就将半途而废，一事无成。

欧洲文艺复兴时期的著名画家达·芬奇，从小爱好绘画。父亲送他到当时意大利的名城佛罗伦萨，拜名画家佛罗基奥为师。老师要他从画蛋入手。他画了一个又一个，足足画了十多天。老师见他有些不耐烦了，便对他说：“不要以为画蛋容易，要知道，1 000 个蛋中从来没有两个是完全相同的；即使是同一个蛋，只要变换一下角度去看，形状也就不同了，蛋的椭圆形轮廓就会有差异。所以，要在画纸上把它完美地表现出来，非得下番苦功不可。”从此，达·芬奇用心学习素描，经过长时期勤奋艰苦的艺术实践，终于创作出许多不朽的名画。

如果达·芬奇画了几个蛋之后就放弃了，那么就没有这一著名画家了，正是他的锲而不舍，铸就了这一时代的伟大画家。同学们在学习过程中会遇到各种困难，遇到困难首先想到的是如何去解决困难迎难而上，而不是恐惧困难中途放弃，锲而不舍坚持做下去，终会有所收获。“锲而不舍”不仅是走向成功的必要途径，也是一种精神，它已经融入我们民族的血液和心灵，成为我们民族重要的美德之一。

任务 4 计算服务 Nova 部署

【任务描述】

小李已经完成了镜像服务 Glance 的部署并经过了验证，他想到计算机都有 CPU 这样的计算部件，那么 OpenStack 云平台是不是也应该需要一个像 CPU 这样的计算组件呢？带着这个疑问，小李通过网络查询资料得知，OpenStack 中的 Nova 组件为其提供计算功能，现在小李参考官方文档（https://docs.openstack.org/nova/rocky/install/）进行计算服务 Nova 组件的部署工作，包括 controller 节点部署 Nova、compute 节点部署 Nova 及部署结果验证等。

【知识要点】

1. Nova 作用

Nova 是 OpenStack 最核心的服务模块，负责管理和维护云计算环境的计算资源，负责整个云环境虚拟机生命周期的管理。

Nova 自身并没有提供任何虚拟化能力，它提供计算服务，使用不同的虚拟化驱动来与底层支持的 Hypervisor（虚拟机管理器）进行交互。所有的计算实例（虚拟服务器）由 Nova 进行生命周期的调度管理（启动、挂起、停止、删除等），Nova 还需要 Keystone、Glance、Neutron、Cinder 和 Swift 等其他服务的支持，能与这些服务集成，实现如加密磁盘、裸金属计算实例等。

2. Nova 架构

Nova 由多个提供不同功能的服务进程组成，对外通过 REST API 进行通信，对内 Nova 子进程间通过 RPC（Remote Procedure Call，远程过程调用）消息传递机制进行通信。API 处理 REST 请求，主要包括数据库读/写，向其他 Nova 服务发送 RPC 消息，生成对 REST 调用的响应等。RPC 消息传递是通过 oslo.messaging 库实现的，oslo.messaging 是在消息队列上的一个抽象。大多数 Nova 组件可以在多个服务器上运行，并有一个侦听 RPC 消息的管理器，但 nova-compute 是一个例外，单个进程运行在它所管理的 Hypervisor 上（使用 VMware 或 Ironic 驱动程序时除外）。Nova 组件的架构如图 3-15 所示。

用于接收 HTTP 请求、转换命令、通过消息队列或 HTTP 与其他组件通信。

Conductor：处理需要协调的请求（构建虚拟机或调整虚拟机大小），充当数据库代理或者处理对象转换。

Scheduler：用于决定哪台计算节点承载计算实例。

Compute：管理虚拟机管理器与虚拟机之间通信。

DB：用于数据存储的 SQL 数据库，通常是 MariaDB。

Placement：用于跟踪资源提供者清单和使用情况，该组件在 Newton 版本集成于 Nova 组件，在 Stein 版本从 Nova 组件中独立出去成为单独的服务组件。

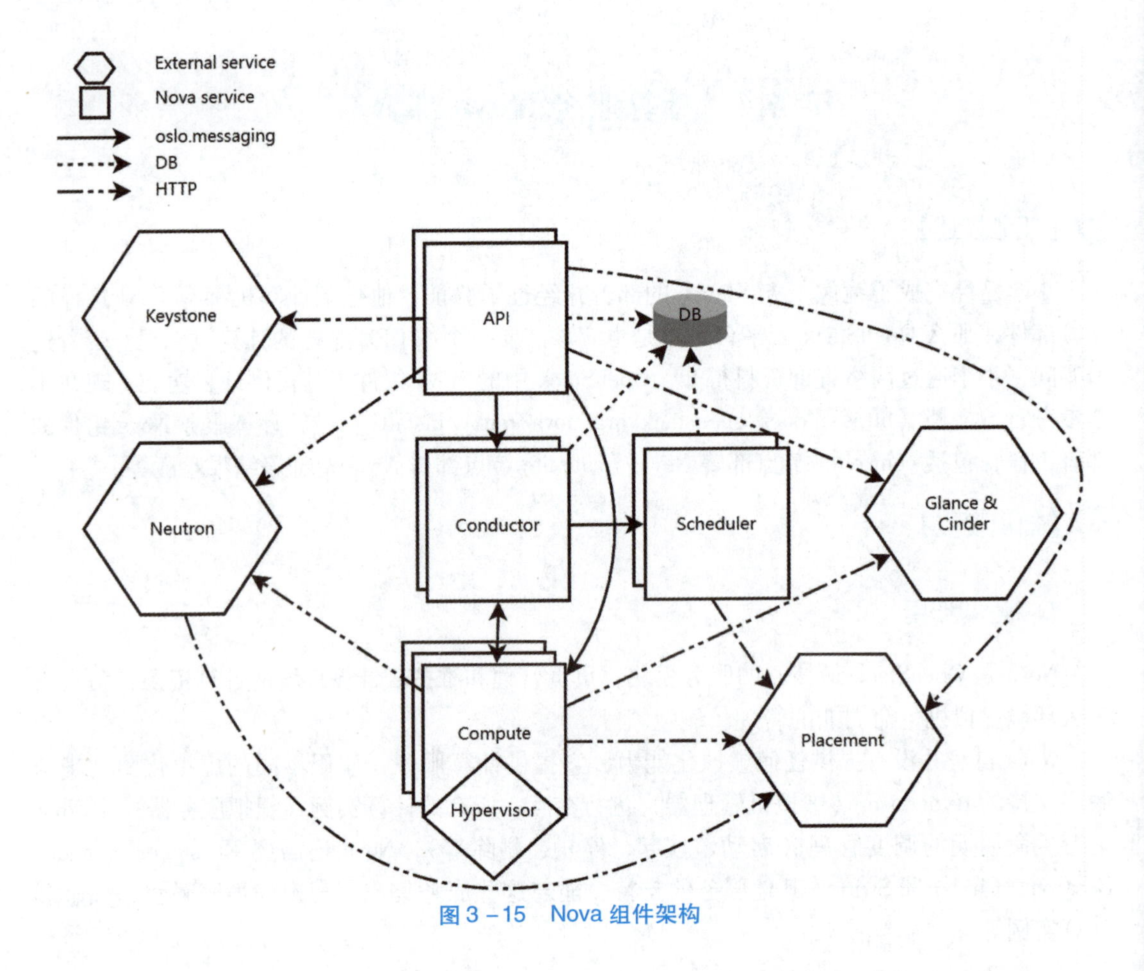

图 3－15 Nova 组件架构

3. Nova 组件

1）API 组件

API 是整个 Nova 组件的门户，是外部访问 Nova 的唯一途径，由 nova－api 服务实现。nova－api 服务接收来自 OpenStack 命令行、Dashbord 和其他需要跟 Nova 交换的外部组件的请求，通过消息队列将请求发送给内部其他的服务并响应外部的请求。nova－api 对接收到的 API 请求做以下处理：

- 检查客户端传入的参数是否合法有效。
- 调用 Nova 其他服务来处理客户端 HTTP 请求。
- 格式化 Nova 其他子服务返回结果并返回给客户端。

2）Conductor 组件

Conductor 组件是计算节点访问数据库的中间件，由 nova－conductor 模块实现，旨在为数据库的访问提供一层安全保障。引入 nova－conductor 的好处主要包括以下两点：

- 实现更高的系统安全性

在 OpenStack 的早期版本中，nova－compute 可以直接访问数据库，但这样存在非常大的安全隐患。因为 nova－compute 这个服务是部署在计算节点上的，为了能够访问控制节点上的数据库，就必须在计算节点的/etc/nova/nova. conf 中配置访问数据库的连接信息。如果任意一个计算节点被黑客入侵，都会导致部署在控制节点上的数据库面临极大风险。为了解决这个问题，从 G 版本开始，Nova 引入了一个新服务 nova - conductor，将 nova - compute 访问数据库的全部操作都放到 nova - conductor 中，而且 nova - conductor 是部署在控制节点上的，这样就避免了 nova - compute 直接访问数据库，增加了系统的安全性。

- 实现更好的系统伸缩性

nova - compute 与 nova - conductor 是通过消息中间件交互的。这种松散的架构允许配置多个 nova - conductor 实例。在一个大规模的 OpenStack 部署环境里，管理员可以通过增加 nova - conductor 的数量来应对日益增长的计算节点对数据库的访问。

3）Scheduler 组件

Scheduler 组件是虚机调度服务，由 nova - scheduler 服务实现，主要解决的是如何选择在哪个计算节点上启动实例的问题，它应用多种规则，考虑内存使用率、CPU 负载率、CPU 架构（Intel/AMD）等多种因素，根据一定的算法，确定虚拟机实例能够运行在哪一台计算服务器上。创建实例时，用户会提出资源需求，例如 CPU、内存、磁盘各需要多少。OpenStack 将这些需求定义在 flavor 中，用户只需要指定用哪个 flavor 就可以了，nova - scheduler 服务会从队列中接收一个虚拟机实例的请求，通过读取数据库的内容，从可用资源池中选择最合适的计算节点来创建新的虚拟机实例。

nova - scheduler 服务决策一个虚拟机应该调度到某物理节点，需要分为两个步骤：

①过滤（filter）：过滤出可以创建虚拟机的主机。

②计算权值（weight）：根据权重大小进行分配，默认根据资源可用空间进行权重排序。

（1）Nova 调度器的类型。

①随机调度器（chance scheduler）：从所有正常运行 nova - compute 服务的节点中随机选择。

②过滤器调度器（filter scheduler）：根据指定的过滤条件以及权重选择最佳的计算节点。

③缓存调度器（caching scheduler）：可看作随机调度器的特殊类型，在随机调度的基础上，将主机资源信息缓存在本地内存中，然后通过后台的定时任务定时从数据库中获取最新的主机资源信息。

调度器可在/etc/nova/nova. conf 文件中的 scheduler_driver 选项配置。

（2）调度器调度过程。

当过滤调度器需要执行调度操作时，会让过滤器对计算节点进行判断，返回 True 或 False，表示是否可以在其上创建实例。经过过滤器的过滤，nova - scheduler 选出了能够部署实例的计算节点，如果有多个计算节点通过了过滤，那么最终选择哪个节点呢？nova - scheduler 会对每个计算节点依据权重值打分，得分最高的获胜。

调度器调度过程主要分两个阶段：

①通过指定的过滤器选择满足条件的计算节点，比如内存使用率小于 50%，可以使用多个过滤器依次进行过滤。

②对过滤之后的主机列表进行权重计算并排序，选择最优的计算节点来创建虚拟机实例。

调度器调度过程如图 3 - 16 所示。在服务器集群中有 6 台计算节点，通过多个过滤器过滤之后，主机 2 和主机 4 不符合过滤条件，被过滤器过滤，剩下 4 台计算节点都可以用来放置实例，在这剩下的 4 台计算节点中，会按照权重值进行排序，按照优先级从高到低依次是主机 5、主机 6、主机 1、主机 3。根据排序结果，主机 5 优先级最高，最终入选。

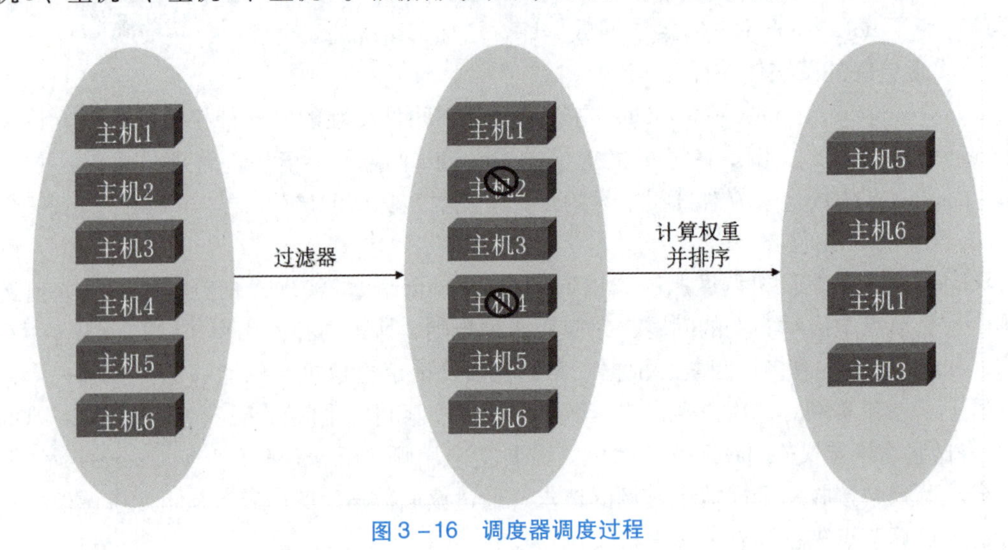

图 3－16　调度器调度过程

（3）权重（weight）。

nova－scheduler 服务可以使用多个过滤器依次进行过滤，过滤之后的节点再通过计算权重选出能够部署实例的节点。所有权重位于 nova/scheduler/weights 目录下，目前默认实现的是根据计算节点空闲的内存量计算权重值，空闲越多，权重越大，实例将被部署到当前空闲内存最多的计算节点上。

4）compute 组件

nova－compute 在计算节点上运行，负责管理节点上的实例，OpenStack 对实例的操作最后都是交给 nova－compute 来完成。nova－compute 与 Hypervisor 一起实现 OpenStack 对实例生命周期的管理，如图 3－17 所示。

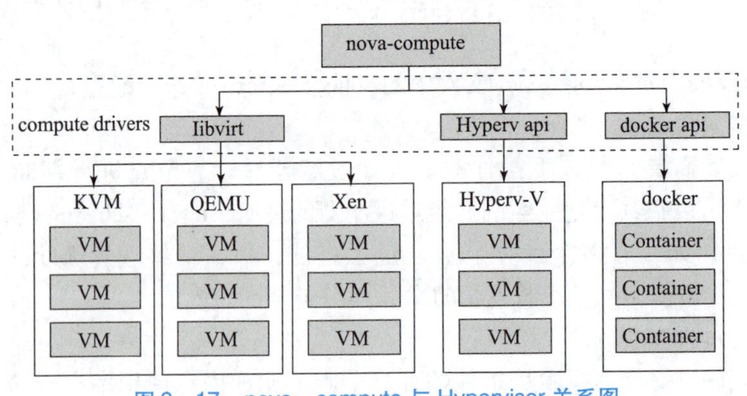

图 3－17　nova－compute 与 Hypervisor 关系图

nova－compute 的功能可以分为两类：

（1）定时向 OpenStack 报告计算节点的状态。

每隔一段时间，nova－compute 就会报告当前计算节点的资源使用情况和 nova－compute

服务状态，这样 OpenStack 就能得知每个计算节点的 vcpu、ram、disk 等信息。nova - scheduler 的很多过滤器才能根据计算节点的资源使用情况进行过滤，选择符合规格要求的计算节点。

（2）实现实例生命周期的管理。

OpenStack 对实例最主要的操作都是通过 nova - compute 实现的，包括实例的启动、关闭、重启、暂停、恢复、删除、调整实例大小、迁移、创建快照等。

5）placement 组件

placement 服务跟踪资源（比如计算节点、存储资源池、网络资源池等）的使用情况，最早在 Newton 版本被引入 OpenStack/Nova，以 API 的形式进行孵化，所以也经常被称呼为 placement API。它参与到 nova - scheduler 选择目标主机的调度流程中，负责跟踪记录资源提供者的库存和使用情况，并对资源进行分类和特征标记。

- 对于使用共享存储解决方案的用户，希望 Nova 和 Horizon 能够正确报告共享存储磁盘资源的总量和使用量信息。
- 对于 Neutron，会使用外部的第三方路由网络功能，希望 Nova 能够掌握和使用特定的网络端口与特定的子网池相关联，确保虚拟机能够在该子网池上启动。
- 作为高级的 Cinder，希望当 nova boot 命令中指定了 cinder volume - id 后，Nova 能够知道哪一些计算节点与请求卷所在的 Cinder 存储池相关联。

所以，当资源类型和提供者变得多样时，自然就需求一种高度抽象且简单统一的管理方法，让用户和代码能够便捷地使用、管理、监控整个 OpenStack 的系统资源，这就是 Placement（布局）。

4. Cell 架构

当 OpenStack 的 Nova 集群规模变大时，数据库和消息队列服务就会出现“瓶颈”问题。Nova 为提高水平扩展及分布式、大规模的部署能力，同时又不增加数据库和消息中间件的复杂度，从 Grzzly 版本引入了 Cell 概念。

Cell 可译为单元。为支持更大规模的部署，OpenStack 将大的 Nova 集群分成小的单元，每个单元都有自己的消息队列和数据库，可以解决规模增大时引起的“瓶颈”问题。在 Cell 中，Keystone、Neutron、Cinder、Glance 等资源是共享的。

Cell v2 的架构如图 3 - 18 所示。所有的 Cell 形成一个扁平架构，API 与 Cell 节点之间存在边界。API 节点只需要数据库，不需要消息队列。nova - api 依赖 nova - api 和 nova - cello 两个数据库。API 节点上部署 nova - scheduler 服务，在调度的时候，只需要在数据库中查出对应的 Cell 信息，就能直接连接过去，从而出现一次调度就可以确定具体在哪个 Cell 的哪台机器上启动。在 Cell 节点中，只需要安装 nova - compute 和 nova - conductor 服务，以及它所依赖的消息队列和数据库。API 上的服务会直接连接 Cell 的消息队列和数据库，Cell 下的计算节点只需要注册到所在的 Cell 节点下就可以了。

API 节点包含 nova - api 和 nova - cell0 两个数据库，其中，nova_api 数据库中存放全局信息，这些全局数据表是从 Nova 库迁过来的，如 flavor（实例模型）、instance groups（实例组）、quota（配额）。nova_cell0 数据库的模式与 Nova 的一样，主要用途就是当实例调度失败时，实例的信息不属于任何一个 Cell，因而存放到 nova_cell0 数据库中。

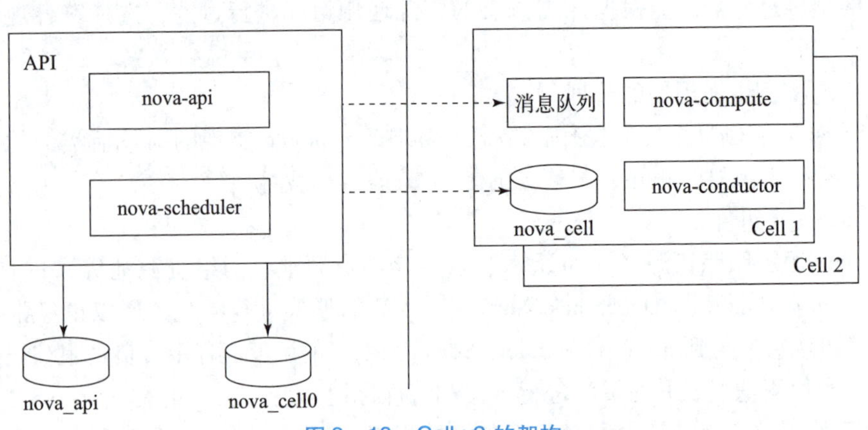

图 3－18　Cell v2 的架构

5. Nova 创建虚拟机的流程

Nova 创建虚拟机的详细流程如图 3－19 所示。

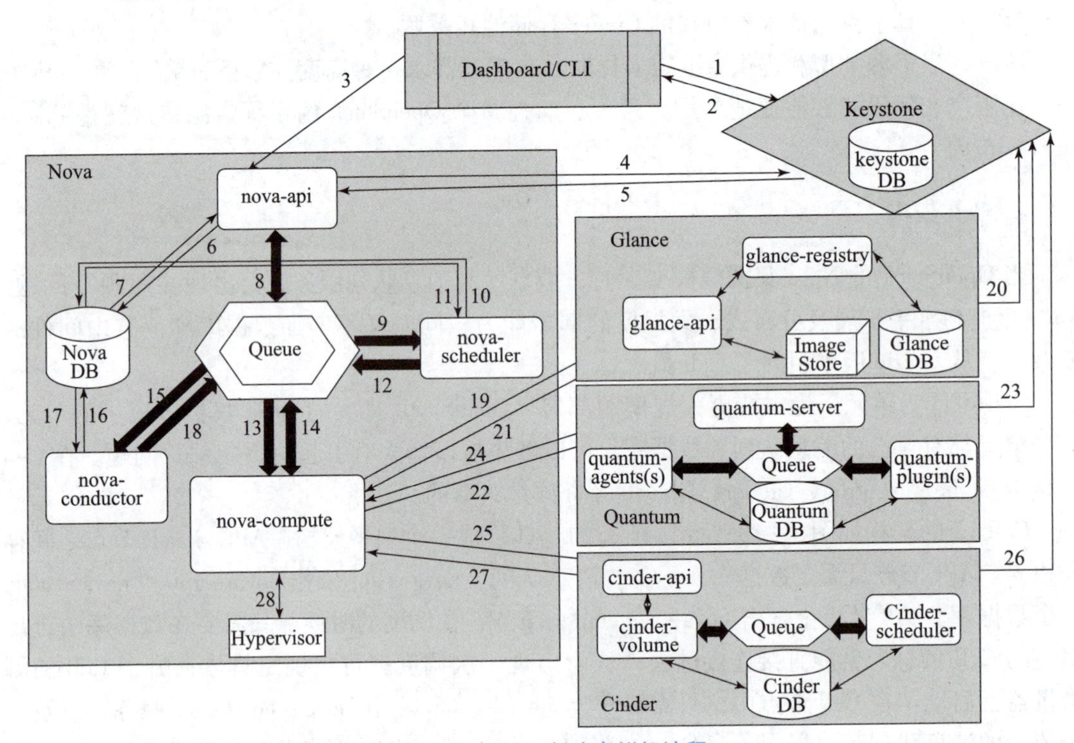

图 3－19　Nova 创建虚拟机流程

（1）界面或命令行通过 RESTful API 向 Keystone 获取认证信息。

（2）Keystone 通过用户请求认证信息，正确后生成 Token 返回给对应的认证请求。

（3）界面或命令行通过 RESTful API 向 nova－api 发送一个创建虚拟机的请求（携带 Token）。

（4）nova-api 接受请求后，向 Keystone 发送认证请求，查看 Token 是否为有效用户。

（5）Keystone 验证 Token 是否有效，如有效，则返回有效的认证和对应的角色（注：有些操作需要有角色权限才能操作）。

（6）通过认证后，nova-api 检查创建虚拟机参数是否有效合法后，和数据库通信。

（7）当所有的参数有效后，初始化新建虚拟机的数据库记录。

（8）nova-api 通过 rpccall 向 nova-scheduler 请求是否有创建虚拟机的资源（Host ID）。

（9）nova-scheduler 进程侦听消息队列，获取 nova-api 的请求。

（10）nova-scheduler 通过查询 nova 数据库中计算资源的情况，并通过调度算法计算符合虚拟机创建需要的主机。

（11）对于有符合虚拟机创建需要的主机，nova-scheduler 更新数据库中虚拟机对应的物理主机信息。

（12）nova-scheduler 通过 rpccast 向 nova-compute 发送对应的创建虚拟机请求的消息。

（13）nova-compute 会从对应的消息队列中获取创建虚拟机请求的消息。

（14）nova-compute 通过 rpccall 向 nova-conductor 请求获取虚拟机消息。

（15）nova-conductor 从消息队列中拿到 nova-compute 请求消息。

（16）nova-conductor 根据消息查询虚拟机对应的信息。

（17）nova-conductor 从数据库中获得虚拟机对应信息。

（18）nova-conductor 把虚拟机信息通过消息的方式发送到消息队列中。

（19）nova-compute 从对应的消息队列中获取虚拟机信息消息。

（20）nova-compute 通过 Keystone 的 RESTfull API 拿到认证的 Token，并通过 HTTP 请求 glance-api 获取创建虚拟机所需的镜像。

（21）glance-api 向 Keystone 认证 Token 是否有效，并返回验证结果。

（22）Token 验证通过，nova-compute 获得虚拟机镜像信息（URL）。

（23）nova-compute 通过 Keystone 的 RESTfull API 拿到认证 k 的 Token，并通过 HTTP 请求 neutron-server 获取创建虚拟机所需的网络信息。

（24）neutron-server 向 Keystone 认证 Token 是否有效，并返回验证结果。

（25）Token 验证通过，nova-compute 获得虚拟机网络信息。

（26）nova-compute 通过 Keystone 的 RESTfull API 拿到认证的 Token，并通过 HTTP 请求 cinder-api 获取创建虚拟机所需的持久化存储信息。

（27）cinder-api 向 Keystone 认证 Token 是否有效，并返回验证结果。

（28）Token 验证通过，nova-compute 获得虚拟机持久化存储信息。

（29）nova-compute 根据 instance 的信息调用配置的虚拟化驱动来创建虚拟机。

【任务实施】

Nova 为 OpenStack 提供计算服务，用于为单个用户或使用群组管理虚拟机实例的整个生命周期，根据用户需求来提供虚拟服务。Nova 负责管理整个云的计算资源、网络资源、授权及测度。

Nova 服务需要在 controller 节点和 compute 节点双节点部署。

1. controller 节点部署

1）nova 数据库创建及授权

在安装配置 Nova 前，需要创建数据库，Nova 需要使用 nova_api、nova、nova_cell0 和 placement 四个数据库，并对四个数据库进行授权。

①以 root 用户身份连接到数据库服务器，连接时，输入环境准备阶段安装 MariaDB 时的密码。

```
[root@ controller ~]# mysql -u root -p
Enter password:
Welcome to the MariaDB monitor. Commands end with;or \g.
Your MariaDB connection id is 21
Server version:10.1.20 -MariaDB MariaDB Server
Copyright(c)2000,2016,Oracle,MariaDB Corporation Ab and others.
Type 'help;' or '\h' for help. Type '\c' to clear the current input statement.
MariaDB[(none)] >
```

②创建 nova_api、nova、nova_cell0 和 placement 数据库。

```
MariaDB[(none)] >CREATE DATABASE nova_api;
Query OK,1 row affected(0.00 sec)
MariaDB[(none)] >CREATE DATABASE nova;
Query OK,1 row affected(0.00 sec)
MariaDB[(none)] >CREATE DATABASE nova_cell0;
Query OK,1 row affected(0.00 sec)
MariaDB[(none)] >CREATE DATABASE placement;
Query OK,1 row affected(0.00 sec)
```

③为四个数据库授权（将 nova_api、nova、nova_cell0 数据库密码设置为 nova，placement 数据库密码设置为 placement）。

```
MariaDB[(none)] >GRANT ALL PRIVILEGES ON nova_api.*  TO 'nova'@ 'localhost'\
        IDENTIFIED BY 'nova';
Query OK,0 rows affected(0.00 sec)
MariaDB[(none)] >GRANT ALL PRIVILEGES ON nova_api.*  TO 'nova'@ '%'\
        IDENTIFIED BY 'nova';
Query OK,0 rows affected(0.00 sec)

MariaDB[(none)] >GRANT ALL PRIVILEGES ON nova.*  TO 'nova'@ 'localhost'\
        IDENTIFIED BY 'nova';
Query OK,0 rows affected(0.00 sec)
MariaDB[(none)] >GRANT ALL PRIVILEGES ON nova.*  TO 'nova'@ '%'\
        IDENTIFIED BY 'nova';
Query OK,0 rows affected(0.00 sec)
```

```
placement admin http://controller:8778
+--------------+--------------------------------------+
| Field        | Value                                |
+--------------+--------------------------------------+
| enabled      | True                                 |
| id           | aa8a59edacfc4f2189bd30d75a90f2fa     |
| interface    | admin                                |
| region       | RegionOne                            |
| region_id    | RegionOne                            |
| service_id   | 85a6a0fc51d44256bdfad03c59740354     |
| service_name | placement                            |
| service_type | placement                            |
| url          | http://controller:8778               |
+--------------+--------------------------------------+
```

3）安装并配置 Nova 组件

（1）安装 Nova 服务相关组件。

```
[root@ controller ~]# yum install openstack-nova-api openstack-nova-conduc-
tor\
    openstack-nova-console openstack-nova-novncproxy\
    openstack-nova-scheduler openstack-nova-placement-api
```

（2）编辑/etc/nova/nova.conf 文件，文件配置内容如下：

仅启用计算和元数据 API

```
[DEFAULT]
#...
enabled_apis=osapi_compute,metadata
```

配置数据库访问

```
[api_database]
#...
connection=mysql+pymysql://nova:nova@ controller/nova_api
[database]
#...
connection=mysql+pymysql://nova:nova@ controller/nova
[placement_database]
#...
connection=mysql+pymysql://placement:placement@ controller/placement
```

配置 RabbitMQ 消息队列访问

```
[DEFAULT]
#...
transport_url=rabbit://openstack:openstack@ controller
```

配置身份服务访问

```
[api]
#...
auth_strategy = keystone
[keystone_authtoken]
#...
auth_url = http://controller:5000/v3
memcached_servers = controller:11211
auth_type = password
project_domain_name = Default
user_domain_name = Default
project_name = service
username = nova
password = nova
```

#配置 IP 地址

```
[DEFAULT]
#...
my_ip = 192.168.30.100
```

启用对网络服务的支持

```
[DEFAULT]
#...
use_neutron = true
firewall_driver = nova.virt.firewall.NoopFirewallDriver
```

配置 VNC 代理，以使用控制器节点的管理接口 IP 地址

```
[vnc]
#...
enabled = true
#...
server_listen = $my_ip
server_proxyclient_address = $my_ip
```

配置 Image 服务 API 的位置

```
[glance]
#...
api_servers = http://controller:9292
```

配置锁定路径

```
[oslo_concurrency]
#...
lock_path = /var/lib/nova/tmp
```

配置 placement API

```
[placement]
#...
region_name = RegionOne
project_domain_name = Default
project_name = service
auth_type = password
user_domain_name = Default
auth_url = http://controller:5000/v3
username = placement
password = placement
```

（3）由于一个包的 bug 需要把以下配置添加到/etc/httpd/conf.d/00 - nova - placement - api.conf 来启用对 placement API 的访问。

添加到配置文件最后：

```
<Directory/usr/bin >
  <IfVersion > =2.4 >
    Require all granted
  </IfVersion >
  <IfVersion <2.4 >
    Order allow,deny
    Allow from all
  </IfVersion >
</Directory >
```

（4）重启 httpd 服务。

```
[root@ controller ~]# systemctl restart httpd
```

4）数据库同步

（1）同步 nova - api 和 placement 数据库。

```
[root@ controller ~]# su - s/bin/sh - c "nova - manage api_db sync" nova
```

（2）注册 cell0 数据库。

```
[root@ controller ~]# su - s/bin/sh - c "nova - manage cell_v2 map_cell0" nova
```

（3）创建 cell1 单元。

```
[root@ controller ~]# su - s/bin/sh - c "nova - manage cell_v2 create_cell -- name =
cell1 -- verbose" nova
```

（4）同步 nova 数据库。

```
[root@ controller ~]# su - s/bin/sh - c "nova - manage db sync" nova
```

（5）验证 Nova 的 cell0 和 cell1 已完成注册，如图 3 - 20 所示。

```
[root@ controller ~]# su - s/bin/sh - c "nova - manage cell_v2 list_cells" nova
```

（6）检查各数据库下是否能正常显示数据表，验证数据库同步是否正确。

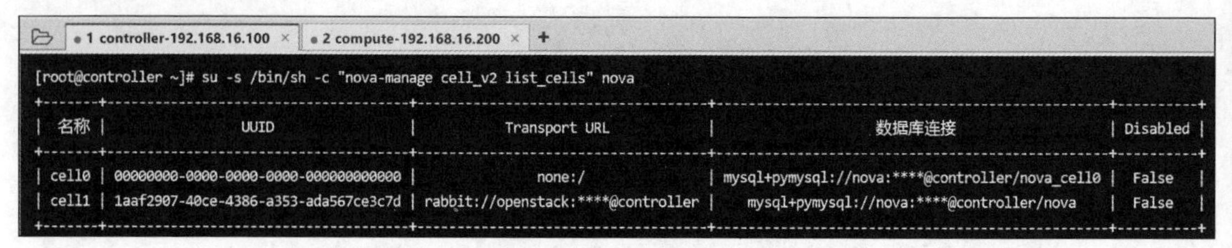

```
[root@controller ~]# su -s /bin/sh -c "nova-manage cell_v2 list_cells" nova
+-------+--------------------------------------+------------------------------------+----------------------------------------------+----------+
| 名称  |                 UUID                 |           Transport URL            |                  数据库连接                  | Disabled |
+-------+--------------------------------------+------------------------------------+----------------------------------------------+----------+
| cell0 | 00000000-0000-0000-0000-000000000000 |               none:/               | mysql+pymysql://nova:****@controller/nova_cell0 |  False   |
| cell1 | 1aaf2907-40ce-4386-a353-ada567ce3c7d | rabbit://openstack:****@controller |   mysql+pymysql://nova:****@controller/nova   |  False   |
+-------+--------------------------------------+------------------------------------+----------------------------------------------+----------+
```

图 3－20　查看 cell_v2 的单元列表

①验证 nova 数据库。

```
[root@ controller ~]# mysql -unova -pnova -e "use nova;show tables;"
+--------------------------------------------+
| Tables_in_nova                             |
+--------------------------------------------+
| agent_builds                               |
| aggregate_hosts                            |
| ……                                         |
| virtual_interfaces                         |
| volume_id_mappings                         |
| volume_usage_cache                         |
+--------------------------------------------+
```

②验证 nova_api 数据库。

```
[root@ controller ~]# mysql -unova -pnova -e "use nova_api;show tables;"
+------------------------------+
| Tables_in_nova_api           |
+------------------------------+
| aggregate_hosts              |
| aggregate_metadata           |
|……                            |
| resource_provider_traits     |
| resource_providers           |
| traits                       |
| users                        |
+------------------------------+
```

③验证 nova_cell0 数据库。

```
[root@ controller ~]# mysql -unova -pnova -e "use nova_cell0;show tables;"
+--------------------------------------------+
| Tables_in_nova_cell0                       |
+--------------------------------------------+
| agent_builds                               |
| aggregate_hosts                            |
| aggregate_metadata                         |
```

```
| ……                                            |
| virtual_interfaces                            |
| volume_id_mappings                            |
| volume_usage_cache                            |
+-----------------------------------------------+
```

④验证 placement 数据库。

```
[root@ controller ~]# mysql - uplacement - pplacement - e "use placement;show tables;"
+------------------------------+
| Tables_in_placement          |
+------------------------------+
| aggregate_hosts              |
| aggregate_metadata           |
| aggregates                   |
| ……                           |
| resource_provider_traits     |
| resource_providers           |
| traits                       |
| users                        |
+------------------------------+
```

5）启动 Nova 控制节点服务

启动 Nova 相关的 openstack - nova - api. service、openstack - nova - consoleauth、openstack - nova - scheduler. service、openstack - nova - conductor. service 和 openstack - nova - novncproxy. service 五个服务并设置为随系统启动。

```
[root@ controller ~]# systemctl start openstack - nova - api. service \
    openstack - nova - consoleauth openstack - nova - scheduler. service \
    openstack - nova - conductor. service openstack - nova - novncproxy. service
[root@ controller ~]# systemctl enable openstack - nova - api. service \
    openstack - nova - consoleauth openstack - nova - scheduler. service \
    openstack - nova - conductor. service openstack - nova - novncproxy. service
Created symlink from/etc/systemd/system/multi - user. target. wants/openstack - nova - api. service to/usr/lib/systemd/system/openstack - nova - api. service.
Created symlink from/etc/systemd/system/multi - user. target. wants/openstack - nova - consoleauth. service to/usr/lib/systemd/system/openstack - nova - consoleauth. service.
Created symlink from/etc/systemd/system/multi - user. target. wants/openstack - nova - scheduler. service to/usr/lib/systemd/system/openstack - nova - scheduler. service.
Created symlink from/etc/systemd/system/multi - user. target. wants/openstack - nova - conductor. service to/usr/lib/systemd/system/openstack - nova - conductor. service.
```

```
Created symlink from/etc/systemd/system/multi-user.target.wants/openstack-nova-novncproxy.service to/usr/lib/systemd/system/openstack-nova-novncproxy.service.
```

2. compute 节点部署

1）安装并配置组件

（1）安装 openstack-nova-compute 软件包。

```
[root@ compute ~]# yum install openstack-nova-compute
```

（2）编辑/etc/nova/nova.conf 配置文件，配置内容如下：

#启用计算和元数据 API

```
[DEFAULT]
#...
enabled_apis=osapi_compute,metadata
```

配置 RabbitMQ 消息队列访问

```
transport_url=rabbit://openstack:openstack@ controller
```

#配置 IP 地址，IP 地址为 compute 节点的管理 IP

```
my_ip=192.168.16.200
```

启用对网络服务的支持

```
use_neutron=true
firewall_driver=nova.virt.firewall.NoopFirewallDriver
```

配置身份服务访问

```
[api]
#...
auth_strategy=keystone
[keystone_authtoken]
auth_url=http://controller:5000/v3
memcached_servers=controller:11211
auth_type=password
project_domain_name=Default
user_domain_name=Default
project_name=service
username=nova
password=nova
```

配置 VNC 代理，以使用控制器节点的管理接口 IP 地址

```
[vnc]
#...
enabled=true
```

```
server_listen = 0.0.0.0
server_proxyclient_address = $my_ip
novncproxy_base_url = http://192.168.16.100:6080/vnc_auto.html
```

配置 Image 服务 API 的位置

```
[glance]
#...
api_servers = http://controller:9292
```

配置锁定路径

```
[oslo_concurrency]
#...
lock_path = /var/lib/nova/tmp
```

配置 placement API

```
[placement]
#...
region_name = RegionOne
project_domain_name = Default
project_name = service
auth_type = password
user_domain_name = Default
auth_url = http://controller:5000/v3
username = placement
password = placement
```

（3）硬件加速配置。

确认 compute 节点是否支持虚拟机的硬件加速：

```
[root@ compute ~]# egrep -c '(vmx|svm)' /proc/cpuinfo
```

如果该命令返回值大于或等于 1，说明 compute 节点支持硬件加速，不需要做任何其他处理；如果该命令返回值为 0，说明 compute 节点不支持硬件加速，此时需要在/etc/nova/nova.conf 配置文件的［libvirt］部分，将 virt_type 参数修改为 qemu。

```
[libvirt]
#...
virt_type = qemu
```

（4）启动计算服务组件及其依赖组件，并设置为随系统启动。

```
[root@ compute ~]# systemctl start libvirtd.service openstack-nova-compute.service
[root@ compute ~]# systemctl enable libvirtd.service openstack-nova-compute.service
Created symlink from/etc/systemd/system/multi-user.target.wants/openstack-nova-compute.service to/usr/lib/systemd/system/openstack-nova-compute.service.
```

2）将计算节点添加到 cell 数据库

以下操作在控制节点执行。

（1）执行 admin-openstack.sh 脚本使能 admin 权限。

```
[root@ controller ~]# source./admin-openstack.sh
```

（2）确认在数据库中包含计算主机。

```
[root@ controller ~]# openstack compute service list --service nova-compute
+----+---------+---------+-----+--------+------+-------+
| ID | Binary      | Host       | Zone    | Status      | State    |Updated At |
+----+---------+---------+-----+--------+------+-------+
| 9  |nova-compute| compute     | nova    | enabled     | up       | 2022-10-
                                                                   22T13:52:
                                                                   29.000000   |
+----+---------+---------+-----+--------+------+-------+
```

（3）发现计算主机。

```
[root@ controller ~]# su-s/bin/sh-c "nova-manage cell_v2 discover_hosts --ver-
bose" nova
Found 2 cell mappings.
Skipping cell0 since it does not contain hosts.
Getting computes from cell 'cell1':013cf7aa-734d-410f-a6ef-2214ba2d6e33
Checking host mapping for compute host 'compute':bf5cd405-4d74-4f30-a149-
fe30f8bcc16c
Creating host mapping for compute host 'compute':bf5cd405-4d74-4f30-a149-
fe30f8bcc16c
Found 1 unmapped computes in cell:013cf7aa-734d-410f-a6ef-2214ba2d6e33
```

（4）添加新计算节点时，必须在控制器节点上运行，以注册这些新计算节点，或者在 /etc/nova/nova.conf 配置文件中设置发现计算节点的时间间隔。

```
[scheduler]
discover_hosts_in_cells_interval=300
```

3. 验证操作

（1）执行 admin-openstack.sh 脚本使能 admin 权限。

```
[root@ controller ~]# source./admin-openstack.sh
```

（2）列出服务组件，以验证每个进程的成功启动和注册：state 为 up 状态，输出结果中会显示有 3 个运行在 controller 节点的服务组件和 1 个运行在 compute 节点的服务组件。

```
[root@ controller ~]# openstack compute service list
+-+--------------+----------+----------+--------+-----+---------+
```

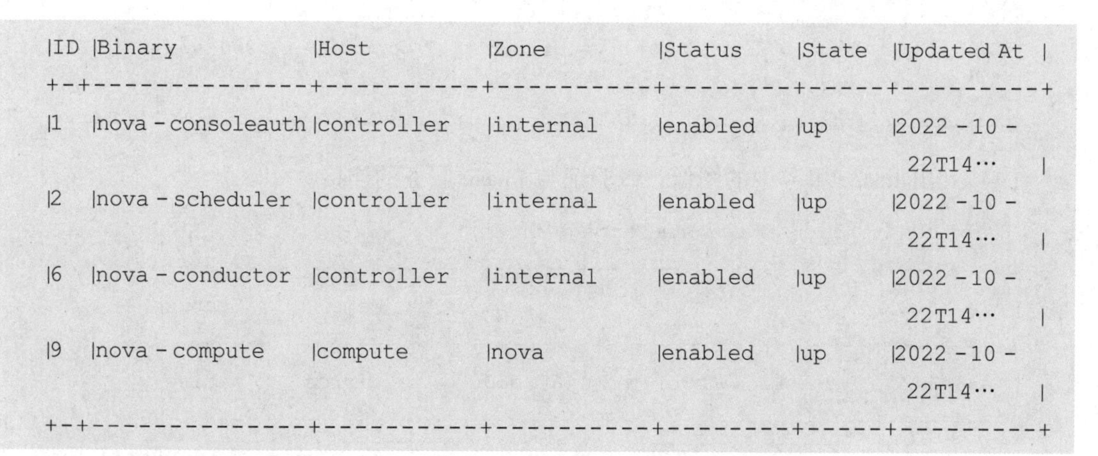

```
|ID |Binary            |Host        |Zone        |Status    |State  |Updated At  |
+-+--------------+----------+----------+--------+-----+---------+
|1  |nova - consoleauth |controller  |internal    |enabled   |up     |2022 -10 -
                                                                     22T14…     |
|2  |nova - scheduler  |controller  |internal    |enabled   |up     |2022 -10 -
                                                                     22T14…     |
|6  |nova - conductor  |controller  |internal    |enabled   |up     |2022 -10 -
                                                                     22T14…     |
|9  |nova - compute    |compute     |nova        |enabled   |up     |2022 -10 -
                                                                     22T14…     |
+-+--------------+----------+----------+--------+-----+---------+
```

（3）列出 Identity 服务中的 API 端点，以验证与 Identity 服务的连接。

```
[root@ controller ~]# openstack catalog list
+-----------+-----------+-------------------------------------+
| Name      | Type      | Endpoints                           |
+-----------+-----------+-------------------------------------+
| glance    | image     | RegionOne                           |
|           |           |   admin:http://controller:9292      |
|           |           | RegionOne                           |
|           |           |   public:http://controller:9292     |
|           |           | RegionOne                           |
|           |           |   internal:http://controller:9292   |
|           |           |                                     |
| nova      | compute   | RegionOne                           |
|           |           |   public:http://controller:8774/v2.1 |
|           |           | RegionOne                           |
|           |           |   internal:http://controller:8774/v2.1 |
|           |           | RegionOne                           |
|           |           |   admin:http://controller:8774/v2.1 |
|           |           |                                     |
| placement | placement | RegionOne                           |
|           |           |   public:http://controller:8778     |
|           |           | RegionOne                           |
|           |           |   internal:http://controller:8778   |
|           |           | RegionOne                           |
|           |           |   admin:http://controller:8778      |
|           |           |                                     |
|keystone   | identity  | RegionOne                           |
|           |           |   internal:http://controller:5000/v3/ |
|           |           | RegionOne                           |
|           |           |   admin:http://controller:5000/v3/  |
|           |           | RegionOne                           |
```

```
|            |            |   public:http://controller:5000/v3/              |
|            |            |                                                   |
+------------+------------+---------------------------------------------------+
```

（4）列出 Image 服务中的图像，以验证与 Image 服务的连接。

```
[root@ controller ~]# openstack image list
+---------------------------------------+--------+--------+
| ID                                    | Name   | Status |
+---------------------------------------+--------+--------+
| a58a2e2f-fd7e-4ef3-b8c1-bd4446f3d33d  | cirros | active |
+---------------------------------------+--------+--------+
```

（5）检查 cells 和 placement API 是否成功运行。

```
[root@ controller ~]# nova-status upgrade check
+----------------------------------+
| 升级检查结果                      |
+----------------------------------+
| 检查:Cells v2                     |
| 结果:成功                         |
| 详情:None                         |
+----------------------------------+
| 检查:Placement API                |
| 结果:成功                         |
| 详情:None                         |
+----------------------------------+
| 检查:Resource Providers           |
| 结果:成功                         |
| 详情:None                         |
+----------------------------------+
| 检查:Ironic Flavor Migration      |
| 结果:成功                         |
| 详情:None                         |
+----------------------------------+
| 检查:API Service Version          |
| 结果:成功                         |
| 详情:None                         |
+----------------------------------+
| 检查:Request Spec Migration       |
| 结果:成功                         |
| 详情:None                         |
+----------------------------------+
| 检查:Console Auths                |
| 结果:成功                         |
```

```
|  详情:None                              |
+----------------------------------------+
```

4. 报错排查案例

执行 su -s /bin/sh -c "nova-manage db sync" nova 同步 nova 数据库时报错，报错提示 Access denied for user'nova@nova'@'controller'using password:NO)，具体信息如图 3-21 所示。

```
[root@controller ~]# su -s /bin/sh -c "nova-manage db sync" nova
ERROR: Could not access cell0.
Has the nova_api database been created?
Has the nova_cell0 database been created?
Has "nova-manage api_db sync" been run?
Has "nova-manage cell_v2 map_cell0" been run?
Is [api_database]/connection set in nova.conf?
Is the cell0 database connection URL correct?
Error: (pymysql.err.OperationalError) (1045, u"Access denied for user 'nova@nova'@'controller' (usin
g password: NO)") (Background on this error at: http://sqlalche.me/e/e3q8)
[root@controller ~]#
```

图 3-21　同步 nova 数据库报错

报错原因：/etc/nova/nova.conf 中的数据库连接的 connection 参数配置出错。

解决方案：

（1）将 connection = mysql + pymysql://nova:nova@nova@172.24.2.10/nova 改成 connection = mysql + pymysql://nova:nova@172.24.2.10/nova，再次执行 su -s /bin/sh -c "nova-manage db sync" nova 同步 nova 数据库，依然报相同错误。

使用 nova-manage cell_v2 list_cells 查看当前注册的 cell，信息如图 3-22 所示。

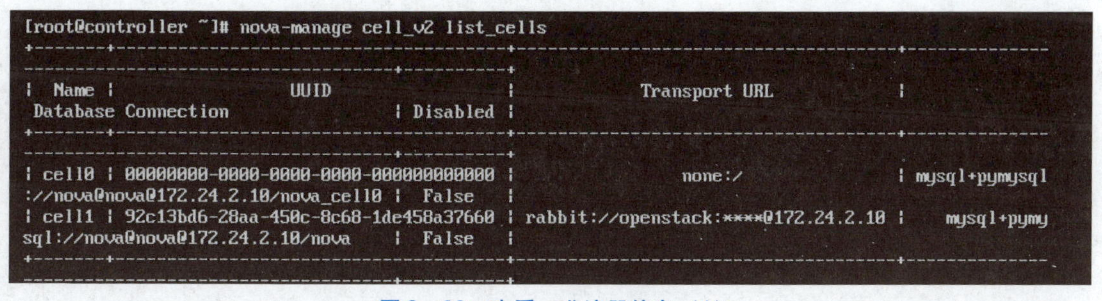

```
[root@controller ~]# nova-manage cell_v2 list_cells
+-------+--------------------------------------+--------------------------------------+-------------
-----------------------------------------+----------+
| Name  |                 UUID                 |             Transport URL            |
 Database Connection                     | Disabled |
+-------+--------------------------------------+--------------------------------------+-------------
-----------------------------------------+----------+
| cell0 | 00000000-0000-0000-0000-000000000000 |                none:/                | mysql+pymysql
://nova@nova@172.24.2.10/nova_cell0 |   False  |
| cell1 | 92c13bd6-28aa-450c-8c68-1de458a37660 | rabbit://openstack:****@172.24.2.10 |   mysql+pymy
sql://nova@nova@172.24.2.10/nova      |   False  |
+-------+--------------------------------------+--------------------------------------+-------------
-----------------------------------------+----------+
```

图 3-22　查看 cell 注册信息（1）

（2）从显示信息可以看出，cell0 和 cell1 已经注册了，Database Connection 没有改过来，删除两个 cell。

```
nova-manage cell_v2 delete_cell --cell_uuid(cell0 的 UUID)
nova-manage cell_v2 delete_cell --cell_uuid(cell1 的 UUID)
```

（3）重新映射 cell0、创建 cell1。

```
su -s/bin/sh -c "nova-manage cell_v2 map_cell0" nova
su -s/bin/sh -c "nova-manage cell_v2 create_cell --name=cell1 --verbose" nova
```

（4）执行 nova - manage cell_v2 list_cells，查看当前注册的 cell，显示正常，如图 3 - 23 所示。

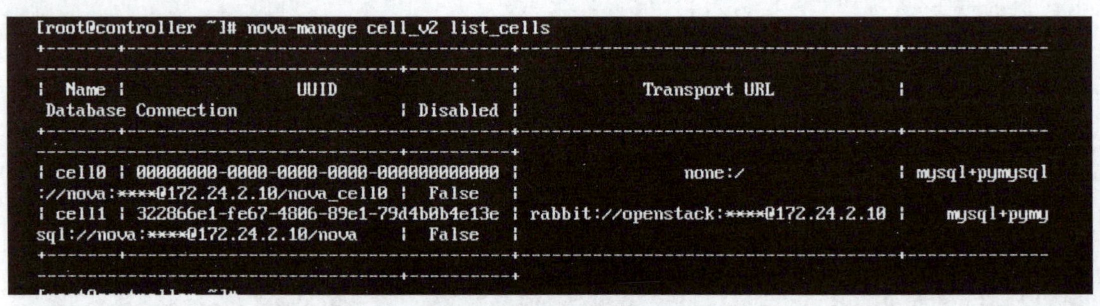

```
[root@controller ~]# nova-manage cell_v2 list_cells
+-------+--------------------------------------+--------------------------------------------+-------------
--------------------------------------+----------+
| Name |                 UUID                 |              Transport URL                 |
 Database Connection                   | Disabled |
+-------+--------------------------------------+--------------------------------------------+-------------
--------------------------------------+----------+
| cell0 | 00000000-0000-0000-0000-000000000000 |                 none:/                     | mysql+pymysql
://nova:****@172.24.2.10/nova_cell0 |  False   |
| cell1 | 322866e1-fe67-4806-89e1-79d4b0b4e13e | rabbit://openstack:****@172.24.2.10 |    mysql+pymy
sql://nova:****@172.24.2.10/nova     |  False   |
+-------+--------------------------------------+--------------------------------------------+-------------
--------------------------------------+----------+
```

图 3 - 23　查看 cell 注册信息（2）

（5）执行 su　- s /bin/sh　- c "nova - manage db sync"。

【任务工单】

工单号：3－4

<table>
<tr><td colspan="2">项目名称：OpenStack 云平台部署</td><td colspan="2">任务名称：计算服务 Nova 部署</td></tr>
<tr><td>班级：</td><td>学号：</td><td colspan="2">姓名：</td></tr>
<tr><td>任务安排</td><td colspan="3">□梳理 Nova 架构及各组件的作用
□Controller 节点部署 Nova 服务
□nova 数据库的创建与授权
□创建计算服务的凭证，包括创建 nova 用户并授权、创建 nova 和 placement 服务端点
□安装 Nova 组件，并配置 Nova 的配置文件完成部署
□compute 节点部署 Nova 服务
□安装 Nova 组件，并配置 Nova 的配置文件完成部署
□计算节点添加到 cell 数据库
□验证 cells 和 placement API 是否成功运行</td></tr>
<tr><td>成果交付形式</td><td colspan="3">执行过程整理成操作文档，上传至教学平台</td></tr>
<tr><td>任务实施总结</td><td colspan="3">任务自评（0～10 分）：
任务收获：____________
____
____
改进点：____________
____
____________</td></tr>
<tr><td>成果验收</td><td colspan="3">□完全满足任务要求
□基本满足任务要求
Nova 部署正确，cells 和 placement API 都能正常运行，但存在一些不影响使用的小问题：
____
____
□不能满足需求
Nova 部署完成，但 cells 或 placement API 运行异常，需要核对部署过程修改：
____
____________</td></tr>
</table>

【知识巩固】

Nova 服务默认监听端口号是（　　）。

A. 9292　　B. 5000　　C. 8778　　D. 8774

【小李的反思】

操千曲而后晓声，观千剑而后识器。

源自刘勰《文心雕龙》，意思是说，练习很多支乐曲之后才能懂得音乐，观察过很多柄剑之后才懂得如何识别剑器。要学会一种技艺，不是容易的事；做个鉴赏家，也要多观察实物，纸上谈兵是不行的。读书要破万卷，下笔才能如有神助。做任何事情，没有一定的经验积累，就不会有很高的造诣。

王羲之 7 岁练习书法，非常刻苦，甚至连吃饭、走路都不放过，到了无时无刻不在练习的地步，他每天坐在池子边练字，没有纸笔，他就在身上画写，送走黄昏，迎来黎明，不知写完了多少墨水、写烂了多少笔头，每天练完字就在池水里洗笔，天长日久，竟将一池水都洗成了墨色，这就是传说中的墨池。正是由于王羲之不断地练习、不断地积累，才使他的书法达到如此高的境界，他的《兰亭序》被誉为“天下第一行书”。

同学们在学习技术时，想要练就高超的技艺和精湛的技能，就需要秉持细致、严谨、负责的态度，奉行精益求精、追求卓越的工作理念，保持对工作的高度认同感、使命感和满足感，在学习中不断打磨技术技能，人一能之己十之，人十能之己百之，在反复的练习中，锻炼自己成为一个 IT 技术的能工巧匠。

任务5 网络服务 Neutron 部署

【任务描述】

小李已经完成了计算服务 Nova 的部署并经过了验证，他想到计算机都需要进行网络配置，那么 OpenStack 云平台是不是也应该需要一个网络管理的组件呢？带着这个疑问，小李通过网络查询资料得知，OpenStack 中的 Neutron 组件为其提供网络管理功能，现小李参考官方文档（https://docs.openstack.org/neutron/rocky/install/install-rdo.html）进行网络服务 Neutron 组件的部署工作，包括 controller 节点部署 Neutron、compute 节点部署 Neutron 及部署结果验证等。

【知识要点】

1. OpenStack 网络

Neutron 是 OpenStack 项目中负责提供网络服务的组件，为虚拟机实例提供网络连接，它基于软件定义网络的思想，实现了网络虚拟化下的资源管理。Neutron 组件在设计上遵循了基于 SDN 实现网络虚拟化的原则，主要功能包括二层交换、三层路由、防火墙、VPN，以及负载均衡等。

Neutron 组件主要功能如下：

（1）二层交换。实例通过虚拟交换机连接到虚拟二层网络，Neutron 支持多种虚拟交换机，一般使用 Linux Bridge 和 Open vSwitch 创建传统的 VLAN 网络，以及基于隧道技术的 Overlay 网络。

（2）三层路由。实例上可以配置不同网段的 IP，Neutron 的虚拟路由器实现实例跨网段通信，同时可以让内网实例通过配置 NAT 的方式访问外网。

（3）负载均衡。LBaaS 支持多种负载均衡产品和方案，不同的实现以插件的形式集成到 Neutron，通过 HAProxy 来实现。

（4）防火墙。Neutron 有两种方式来保障实例和网络的安全性，分别是安全组以及防火墙功能，均可以通过 iptables 来实现，前者是限制进出实例的网络包，后者是进出虚拟路由器的网络包。

2. 网络基础知识

1）Linux 网络虚拟化

传统物理网络中，一系列物理服务器上部署不同的服务和应用，服务器通过网卡连接到交换机形成一个二层网络，实现虚拟化后，网络如图 3-24 所示。在一台物理服务器上可以虚拟多台虚拟机，每台虚拟机上部署不同的服务和应用，虚拟机由虚拟机管理器 Hypervisor 管理，在 Linux 系统中，Hypervisor 通常采用 KVM。在对服务器进行虚拟化的同时，也对网络进行虚拟化。Hypervisor 为虚拟机创建一个或多个虚拟网卡（vNIC），虚拟网卡等同于虚拟机的物理网卡，物理交换机在虚拟网络中被虚拟为虚拟交换机（vSwitch），虚拟机的虚拟网卡连接到虚拟交换机上，虚拟机交换机再通过物理主机的物理网卡连接到外部网络。对于物理网络来说，虚拟化的主要工作是对网卡和交换设备的虚拟化。

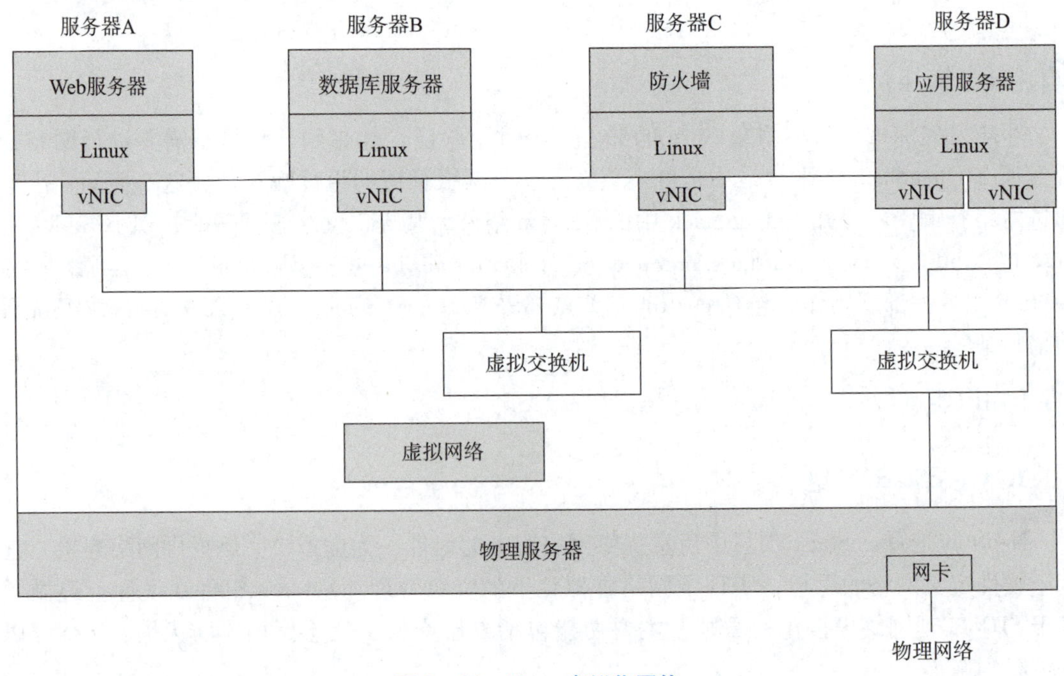

图 3－24　Linux 虚拟化网络

2）Linux 虚拟网桥

虚拟机与物理机不同，没有硬件设备，但是也依然需要与物理机和其他虚拟机进行通信。Linux KVM 提供虚拟网桥设备，像物理交换机具有若干网络接口一样，在网桥上创建多个虚拟的网络接口，每个网络接口再与 KVM 虚拟机的网卡相连。

在 Linux 的 KVM 虚拟系统中，为支持虚拟机的网络通信，网桥接口的名称通常以 vnet 开头，加上从 0 开始顺序编号，如 vnet0、vnet1，在创建虚拟机时，会自动创建这些接口。虚拟网桥 br1 和 br2 分别连接到物理主机的物理网卡 1 和物理网卡 2，物理网卡 1 外连物理网络，如图 3－25 所示。

3）VLAN 技术

虚拟局域网（Virtual Local Area Network，VLAN）是一组逻辑上的设备和用户，这些设备和用户并不受物理位置的限制，可以根据功能、部门及应用等因素将它们组织起来，相互之间的通信就好像它们在同一个网段中一样。当两个主机连接在同一个交换机上时，划分到不同的 VLAN 中，彼此数据流量不可见，OpenStack 也可使用 VLAN 技术隔离不同项目的流量，每个 VLAN 都有一个 ID，编号为 1～4 094。

VLAN 技术可以将一个物理局域网在逻辑上划分成多个广播域。VLAN 技术部署在数据链路层，用于隔离二层流量。同一个 VLAN 内的主机共享同一个广播域，它们之间可以直接进行二层通信。而 VLAN 间的主机属于不同的广播域，不能直接实现二层互通。这样，广播报文就被限制在各个相应的 VLAN 内，同时也提高了网络安全性。

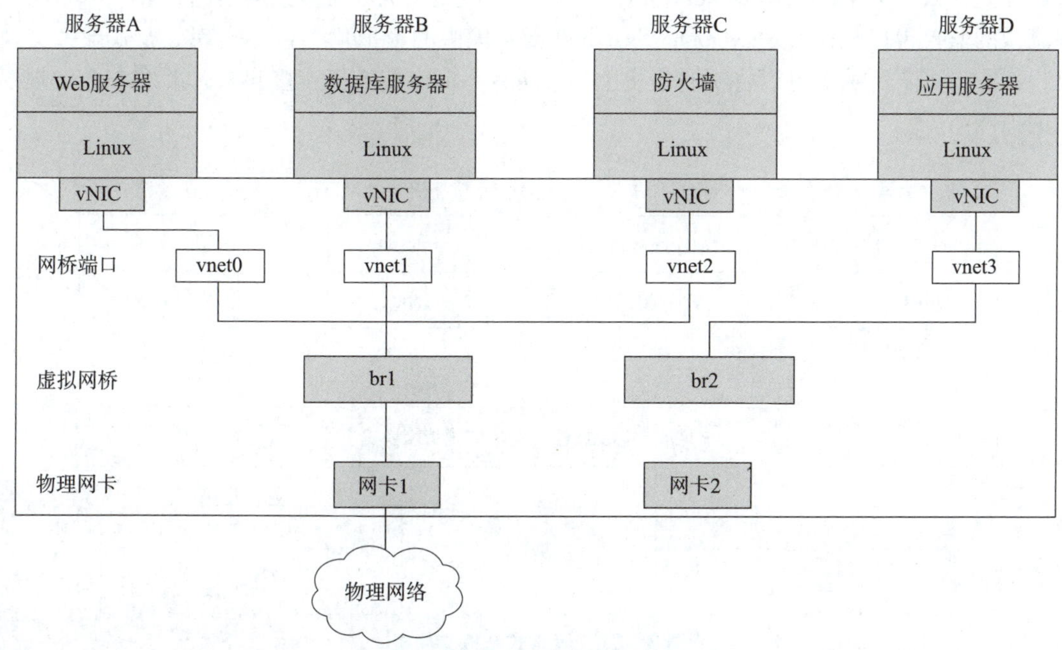

图 3－25　Linux 虚拟网桥示意

如图 3－26 所示，原本属于同一广播域的主机被划分到了两个 VLAN 中，即，VLAN2 和 VLAN3。VLAN 内部的主机可以直接在二层互相通信，VLAN2 和 VLAN3 之间的主机无法直接实现二层通信。

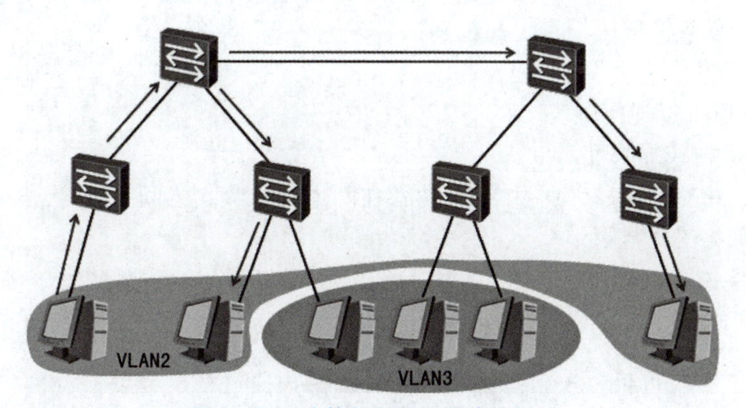

图 3－26　交换机 VLAN 隔离广播域

4）开放虚拟交换机（Open vSwitch）

Open vSwitch 简称 OVS，是一个高质量、多层的虚拟交换软件。它的作用是通过编程扩展支持大规模网络自动化，同时还支持标准的管理接口和协议。OVS 与硬件交换机具备相同特性，可在不同虚拟平台之间移植，具有产品级质量的虚拟交换机，适合在生产环境中部署。

交换设备的虚拟化对虚拟网络来说至关重要。在传统的数据中心，管理员可以对物理交换机进行配置，以控制服务器的网络接入，实现网络隔离、流量监控、QoS 配置、流量优化等目标。在云环境中，采用 Open vSwitch 技术的虚拟交换机可使虚拟网络的管理、网络状态和流量的监控得以轻松实现。Open vSwitch 在云环境中的虚拟化平台上实现分布式虚拟交换机，如图 3－27 所示，可以将不同主机上的 Open vSwitch 交换机连接起来，形成一个大规模的虚拟网络。

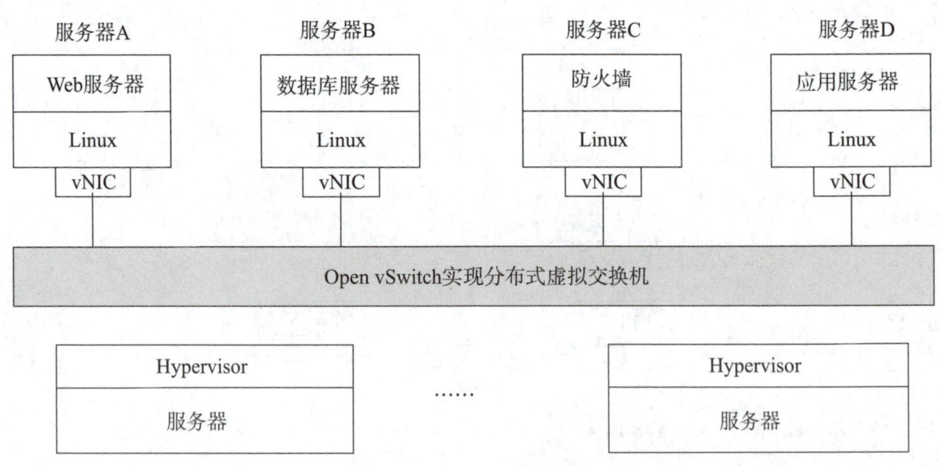

图 3－27　分布式虚拟交换机

5）GRE

通用路由封装（Generic Routing Encapsulation，GRE）：提供了将一种协议报文封装在另外一种协议报文中的机制，是一种隧道封装技术。其根本作用就是要实现隧道功能，通过隧道连接的两个远程网络就如同直连，GRE 在两个远程网络之间模拟出直连链路，从而使网络间达到直连的效果。应用场景主要包括：

- 解决异种网络传输问题。
- IPv4 和 IPv6 互通问题。
- 克服 IGP 协议局限性。

例如，RIP 最大跳数 15，通过 GRE 技术在两个网络节点之间搭建隧道，隐藏它们之间的跳数，扩大网络工作范围。

GRE 需要完成 3 次封装，以图 3－28 为例，解释 GRE 传输数据过程。

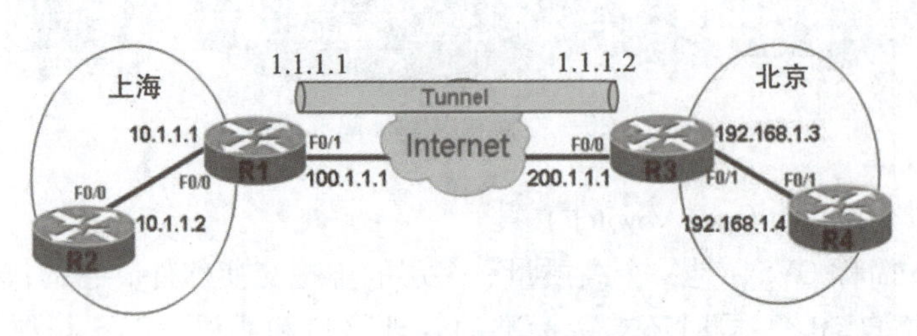

图 3－28　GRE 传输数据过程

GRE 要在远程路由器之间创建虚拟直连链路，也就是隧道（Tunnel），当上海分公司 R2 将数据包 IP 地址封装为 192.168.1.4 发往北京分公司的 R4 时，GRE 操作过程如下：

- 假设 R1 与 R3 的 GRE 虚拟直连链路（隧道）已经建立，隧道链路两端的地址分别

为 1. 1. 1. 1 和 1. 1. 1. 2，隧道两端的起源和终点分别为 100. 1. 1. 1 和 200. 1. 1. 1。

- R1 收到目标 IP 为 192. 168. 1. 4 的数据包后，将原始数据包当作乘客数据包封装进 GRE 协议中，并且添加 GRE 包头，包头中源 IP 为隧道本端地址 1. 1. 1. 1，包头中目标 IP 为隧道对端地址 1. 1. 1. 2，从而完成 GRE 数据包的封装。
- 在封装了 GRE 隧道地址的数据包外面封装 GRE 隧道起源 IP 地址，该 IP 地址为公网地址，即源 IP 为 100. 1. 1. 1，目标 IP 为隧道终点 200. 1. 1. 1，最后将数据包发出去。

封装后的数据包如图 3－29 所示。

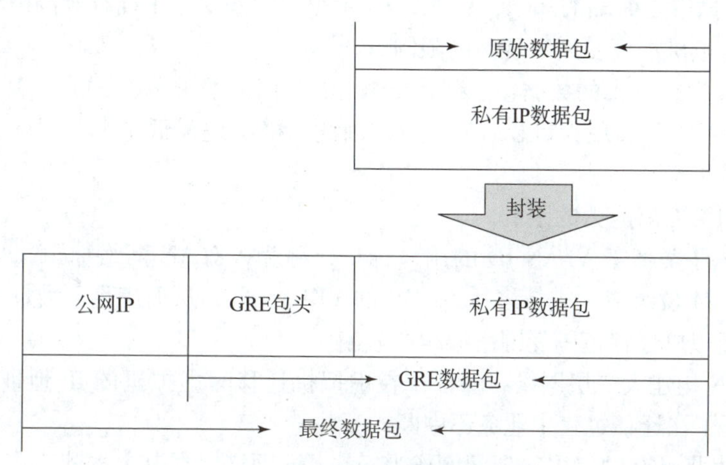

图 3－29　封装后的数据包

数据包被发到 Internet 之后，所有路由器只根据数据包最外面的公网 IP 进行转发，也就是只根据公网目标 IP 地址 200. 1. 1. 1 来转发，直到数据包到达公网 IP 的真正目的地后，即到达 R3（IP：200. 1. 1. 1）之后，公网 IP 包头才会被剥开。当 R3 剥开数据包的公网 IP 包头后，发现 GRE 包头，发现目标 IP 为 1. 1. 1. 2，从而得知自己就是 GRE 隧道的终点，所以继续将 GRE 包头剥开，最后发现目标 IP 地址为 192. 168. 1. 4，然后将数据包发往 192. 168. 1. 4（路由器 R4）。

通过以上 GRE 过程，上海分公司 R2 直接通过私有 IP 地址 192. 168. 1. 4，最终成功与北京分公司 R4 通信。

6）VXLAN

服务器虚拟化技术的广泛部署，极大地增加了数据中心的计算密度；同时，为了实现业务的灵活变更，虚拟机（Virtual Machine，VM）需要能够在网络中不受限制地迁移传统数据中心网络面临的挑战，虚拟机数量的快速增长与虚拟机迁移业务的日趋频繁，给传统的“二层＋三层”数据中心网络带来了新的挑战：

（1）虚拟机规模受网络设备表项规格的限制。

对于同网段主机的通信而言，报文通过查询 MAC 表进行二层转发。服务器虚拟化后，数据中心中 VM 的数量比原有的物理机发生了数量级的增长，伴随而来的便是虚拟机网卡 MAC 地址数量的空前增加。

（2）VLAN 隔离能力不再满足需求。

VLAN 作为当前主流的网络隔离技术，在标准定义中只有 12 位，也就是说，可用的

VLAN 数量只有 4 000 个左右。对于公有云或其他大型虚拟化云计算服务这种动辄上万甚至更多租户的场景而言，VLAN 的隔离能力显然已经力不从心。

（3）虚拟机迁移范围受限。

虚拟机迁移，顾名思义，就是将虚拟机从一个物理机迁移到另一个物理机，但是要求在迁移过程中业务不能中断。要做到这一点，需要保证虚拟机迁移前后，其 IP 地址、MAC 地址等参数维持不变。这就决定了虚拟机迁移必须发生在一个二层域中。而传统数据中心网络的二层域，将虚拟机迁移限制在了一个较小的局部范围内。

针对传统数据中心面临的问题，VXLAN 技术很好地解决了上面提到的限制：

（1）针对虚拟机规模受网络规格的限制。

VXLAN 将虚拟机发出的数据包封装在 UDP 中，并使用物理网络的 IP/MAC 地址作为外层进行封装，对网络只表现为封装后的参数，因此，极大地降低了大二层网络对 MAC 地址规格的需求。

（2）针对网络隔离能力的限制。

VXLAN 引入了类似于 VLAN ID 的用户标识，称为 VXLAN 网络标识（VXLAN Network ID，VNID），由 24 位组成，支持多达 16 MB 的 VXLAN 段，从而满足了大量的用户标识。

（3）针对虚拟机迁移范围受网络架构的限制。

通过 VXLAN 构建大二层网络，保证了在虚拟机迁移时虚拟机的 IP 地址、MAC 地址等保持不变，保证了在迁移过程中业务不中断。

VXLAN 技术将已有的三层物理网络作为 underlay 网络，在其上构建出虚拟的二层网络，即 overlay 网络。overlay 网络通过封装技术，利用 underlay 网络提供的三层转发路径，实现租户二层报文跨越三层网络在不同站点间传递。对租户来说，underlay 网络是透明的，同一租户的不同站点就像工作在一个局域网中。

3. Neutron 网络结构

简化的典型 Neutron 网络结构如图 3－30 所示。由虚拟路由器连接内部网络和外部网络，外部网络负责连接 OpenStack 项目之外的网络环境，又称公共网络，该公共网络是一个虚拟网络，并通过该虚拟网络与外部物理网络连接。内部网络又称私有网络，完全由软件定义，项目用户可以创建自己的内部网络，虚拟机实例连接到内部网络中，默认情况下，项目之间的内部网络是相互隔离的，不能共享。路由器用于将内部网络与外部网络连接起来，因此，要使虚拟机访问外部网络，必须创建一个路由器。

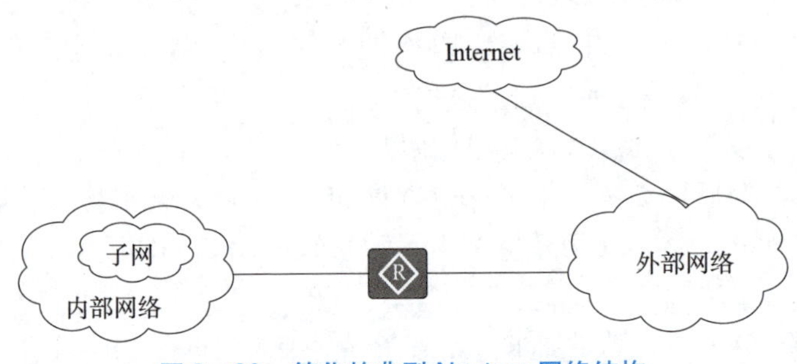

图 3－30　简化的典型 Neutron 网络结构

Neutron 需要实现的主要是内部网络和路由器。内部网络是对二层（L2）网络的抽象，模拟物理网络的二层局域网，对于项目来说，它是私有的。路由器则是对三层（L3）网络的抽象，模拟物理路由器，为用户提供路由、NAT 等服务。

4. 网络拓扑类型

网络虚拟化技术的趋势是在传统单层网络基础上叠加一层逻辑网络，将网络分为两个层次，传统单层网络称为 Underlay（承载网络），叠加其上的逻辑网络称为 Overlay（叠加网络或覆盖网络）。Overlay 网络的节点通过虚拟的或逻辑的连接进行通信，每一个虚拟的或逻辑的连接对应于 Underlay 网络的一条路径，由多个前后衔接的连接组成。Overlay 网络无须对基础网络进行大规模修改，不用关心这些底层实现，是实现云网融合的关键。Neutron 服务支持多种网络拓扑类型。

1）Local

Local 网络与其他网络及节点隔离。该网络中的虚拟机实例只能与位于同一节点上同一网络的虚拟机实例通信，主要用于测试环境。位于同一 Local 网络的实例之间可以通信，位于不同 Local 网络的实例之间无法通信。一个 Local 网络只能位于同一个物理节点上，无法跨节点部署。

2）Flat

Flat 是一种简单的扁平网络拓扑，所有的虚拟机实例都连接在同一网络中，能与位于同一网络的实例进行通信，并且可以跨多个节点。这种网络不使用 VLAN，没有对数据包打 VLAN 标签，无法进行网络隔离。Flat 是基于不使用 VLAN 的物理网络实施的虚拟网络。每个物理网络最多只能实现一个虚拟网络。

3）VLAN

VLAN 是支持 802.1q 协议的虚拟局域网，使用 VLAN 标签标记数据包，实现网络隔离。同一 VLAN 网络中的实例可以通信，不同 VLAN 网络中的实例只能通过路由器来通信。VLAN 网络可以跨节点，是应用最广泛的网络拓扑类型之一。

4）VXLAN

VXLAN（虚拟扩展局域网）可以看作 VLAN 的一种扩展，是基于隧道技术的 overlay 网络，通过唯一的 VNI 区分于其他 VXLAN 网络。相比于 VLAN，它有更大的扩展性和灵活性，是目前支持大规模多租户网络环境的解决方案。由于 VLAN 包头部限长是 12 位，导致 VLAN 的数量限制是 4 096（2^{12}）个，不能满足网络空间日益增长的需求。目前 VXLAN 的封包头部有 24 位用作 VNID 来区分 VXLAN 网段，最多可以支持 16 777 216（2^{24}）个网段。

VLAN 使用 STP 来防止环路，导致一半的网络路径被阻断。VXLAN 的数据包是封装到 UDP 通过三层路由传输和转发的，可以完整地利用三层路由，能克服 VLAN 和物理网络基础设施的限制，更好地利用已有的网络路径。

5）GRE

GRE（通用路由封装）是用一种网络层协议去封装另一种网络层协议的隧道技术。GRE 的隧道由两端的源 IP 地址和目的 IP 地址定义，它允许用户使用 IP 封装 IP 等协议，并支持全部的路由协议。在 OpenStack 环境中，使用 GRE 意味着“IP over IP”，GRE 与 VX-

LAN 的主要区别在于，它是使用 IP 包而非 UDP 进行封装的。

6）GENEVE

GENEVE（通用网络虚拟封装）的目标宣称是仅定义封装数据格式，尽可能实现数据格式的弹性和扩展性。GENEVE 封装的包通过标准的网络设备传送，即通过单播或多播寻址，包从一个隧道端点传送到另一个或多个隧道端点。GENEVE 帧格式由一个封装在 IPv4 或 IPv6 的 UDP 里的简化的隧道头部组成。GENEVE 推出的主要目的是解决封装时添加的元数据信息问题，以适应各种虚拟化场景。

5. Neutron 基本架构

Neutron 采用分布式架构，由多个组件共同对外提供网络服务，其架构如图 3－31 所示。

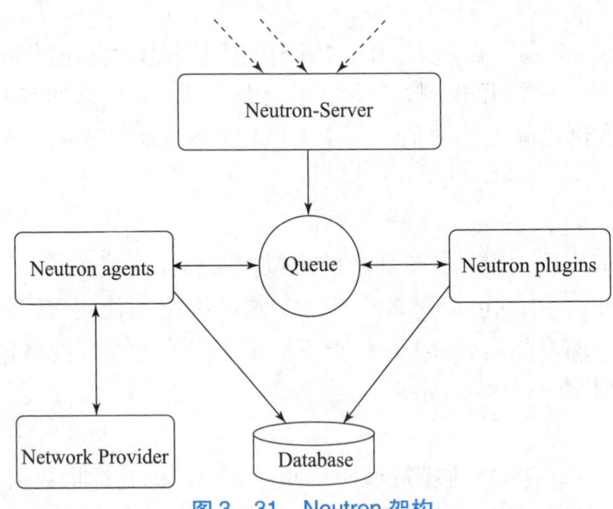

图 3－31　Neutron 架构

Neutron－Server：运行于控制节点，作为 Neutron 的访问入口对外提供 OpenStack 网络 API，接收请求并调用插件处理请求，最终由计算节点和网络节点上的代理完成请求。

Neutron plugins：处理 Neutron－Server 发来的请求，在数据库中维护 OpenStack 逻辑网络的状态，并调用 Agent 处理请求。Plugin 按照功能，分为 core plugin 和 service plugin 两种类型，core plugin 提供基础二层虚拟网络技术，实现网络、子网和端口等核心资源的抽象。service plugin 是指 core plugin 之外的其他插件，提供路由器、防火墙、安全组、负载均衡等服务。

Neutron agents：处理 Plugin 的请求，负责在 Network Provider 上真正实现各种网络功能。

Network Provider：提供网络服务的虚拟或者物理网络设备，比如 Linux Bridge、Open vSwitch 或者其他支持 Neutron 的物理交换机。

Queue：Neutron－Server、Neutron plugins 和 Neutron agents 之间通过消息队列通信和调用。

Database：存放 OpenStack 的网络状态信息，包括 Network、Subnet、Port、Router 等。

6. Neutron 主要插件、代理与服务

1）ML2 插件

Neutron 采用比较开放的架构，通过开发不同的插件和代理来支持不同的网络技术，但随着所支持的网络提供者种类的增加，开发人员发现了两个突出的问题：一个问题是多种网络提供者无法共存，Core Plugin 负责管理和维护 Neutron 二层虚拟网络的状态信息，一个 Neutron 网络只能由一个插件管理，而 Core Plugin 插件与相应的代理是一一对应的。如果选择 Linux Bridge 插件，则只能选择 Linux Bridge 代理，必须在 OpenStack 的所有节点上使用 Linux Bridge 作为虚拟交换机。另一个问题是开发新的插件的工作量太大，而所有传统的 Core Plugin 之间存在大量反复代码。

为解决这两个问题，从 OpenStack 的 Havana 版本开始，Neutron 实现了 ML2（Module Layer 2）插件，旨在取代所有的 Core Plugin，允许在 OpenStack 网络中同时使用多种二层网络技术，不同的节点可以使用不同的网络实现机制。ML2 能够与现有的代理无缝集成，以前使用的代理无须变更，只需将传统的 Core Plugin 替换成 ML2。ML2 使得对新的网络技术的支持更为简单，无须从头开发 Core Plugin，只需要开发相应的机制驱动，大大减少了编写和维护的代码。

2）Linux Bridge 代理

Linux Bridge 是成熟可靠的 Neutron 二层网络虚拟化技术，支持 Local、Flat、VLAN 和 VXLAN 这 4 种网络类型。

Linux Bridge 可以将一台主机上的多个网卡桥接起来，充当一台交换机。它既可以桥接物理网卡，也可以桥接虚拟网卡。用于桥接虚拟机网卡（虚拟网卡）的是 Tap 接口，这是一个虚拟出来的网络设备，称为 Tap 设备，作为网桥的一个端口。Tap 接口在逻辑上与物理接口具有相同的功能，可以接收和发送数据包。

3）Open vSwitch 代理

与 Linux Bridge 相比，Open vSwitch（OVS）具有集中管控功能，而且性能更加优化，支持更多的功能，目前在 OpenStack 领域成为主流。它支持 Local、Flat、VLAN、VXLAN、GRE 和 GENEVE 等所有网络类型。

4）DHCP 代理

DHCP 代理的主要任务包括：

- 定期报告 DHCP 代理的网络状态，通过 RPC 报告给 Neutron - Server，然后通过 Core Plugin 报告给数据库并进行更新网络状态。
- 启动 DNSmasq 进程，检测 qdhcp - ××××名称空间（namespace）中的 ns - ×××× 端口接收到的 DHCP。

DHCP 代理工作机制：

- 创建实例时，Neutron 随机生成 MAC 并从配置数据中分配一个固定 IP 地址，一起保存到 DNSmasq 的 hosts 文件中，让 DNSmasq 进程做好准备。
- 与此同时，nova - compute 会设置虚拟机网卡的 MAC 地址。
- 实例启动，发出 DHCP DISCOVER 广播报文，该广播报文在整个网络中都可以被收到。

- 广播到达 DNSmasq 监听 Tap 接口。DNSmasq 收到后，检查其 host 文件，发现有对应项，它以 DHCP OFFER 消息将 IP 和网关 IP 发回到虚拟机实例。
- 虚拟机实例发回 DHCP REQUEST 消息，确认接收 DHCP OFFER。
- DNSmasq 发回确认消息 DHCP ACK，整个过程结束。

5）Neutron 路由器

Neutron 路由器是一个三层网络的抽象，模拟物理路由器，为用户提供路由、NAT 等服务。在 OpenStack 网络中，不同子网之间和项目网络与外部网络之间都需要通过路由器进行通信。Neutron 提供虚拟路由器，也支持物理路由器。Neutron 的虚拟路由器使用软件模拟物理路由器，路由实现机制相同。

6）L3 代理

在 Neutron 中，L3 代理不仅提供虚拟路由器，还通过 iptables 提供地址转换、浮动地址和安全组功能。L3 代理利用 Linux IP 栈、路由和 iptables 来实现内部网络中不同网络的虚拟机实例之间的通信，以及虚拟机实例和外部网络之间的网络流量的路由和转发。L3 代理可以部署在控制节点或者网络节点上。

7）虚拟防火墙

防火墙即服务（FireWall as a Service，FWaaS）是一种基于 L3 代理的虚拟防火墙，是 Neutron 的一个高级服务。通过它，OpenStack 可以将防火墙应用到项目、路由器、路由器端口和虚拟机端口，在子网边界上对三层和四层的流量进行过滤。

Neutron 路由器上应用防火墙规则，控制进出项目网络的数据。防火墙必须关联某个策略。FWaaS 的应用对象是虚拟路由器，可以在安全组之前控制从外部注入的流量，但是对同一个子网内的流量不做限制，安全组保护的是实例，而 FWaaS 保护的是子网，两者互为补充。

【任务实施】

OpenStack 的 Neutron 服务对虚拟网络基础设施（VNI）和物理网络基础设施的访问层进行管理，同时，Neutron 使项目能够创建高级虚拟网络拓扑，其中可能包括防火墙、负载平衡器和虚拟专用网络（VPN）等服务。Neutron 提供网络、子网和路由器的抽象，每个抽象对象都具备其对应的物理功能。

Neutron 需要进行 controller 节点和 compute 节点双节点安装。

1. 计算节点安装

1）Neutron 数据库创建及授权

在安装配置 Neutron 前，需要创建数据库，并对数据库进行授权。

（1）以 root 用户身份连接到数据库服务器，连接时，输入环境准备阶段安装 MariaDB 时的密码。

```
[root@ controller ~]# mysql -u root -p
Enter password:
Welcome to the MariaDB monitor. Commands end with;or \g.
```

```
Your MariaDB connection id is 38
Server version:10.1.20 -MariaDB MariaDB Server

Copyright(c)2000,2016,Oracle,MariaDB Corporation Ab and others.

Type 'help;' or '\h' for help. Type '\c' to clear the current input statement.

MariaDB[(none)]>
```

（2）创建 Neutron 数据库。

```
MariaDB[(none)]>CREATE DATABASE neutron;
Query OK,1 row affected(0.00 sec)
```

（3）为 Neutron 数据库授权。

```
MariaDB[(none)]>GRANT ALL PRIVILEGES ON neutron.*  TO 'neutron'@ 'localhost'\
    ->  IDENTIFIED BY 'neutron';
Query OK,0 rows affected(0.00 sec)
MariaDB[(none)]>GRANT ALL PRIVILEGES ON neutron.*  TO 'neutron'@ '%'\
    ->  IDENTIFIED BY 'neutron';
Query OK,0 rows affected(0.00 sec)
```

（4）退出数据库。

```
MariaDB[(none)]>exit
```

2）创建服务证书

（1）执行 admin-openstack.sh 脚本，获取管理员的命令行执行权限。

```
[root@ controller ~]# source./admin-openstack.sh
```

（2）创建 neutron 用户。

#需要输入 neutron 用户密码，这里设置的是 neutron。

```
[root@ controller ~]# openstack user create --domain default --password -
prompt neutron
User Password:
Repeat User Password:
+---------------------+-----------------------------------+
| Field               | Value                             |
+---------------------+-----------------------------------+
| domain_id           | default                           |
| enabled             | True                              |
| id                  | 5680d79de42f445c91b3e7aa2517c28d  |
| name                | neutron                           |
```

```
| options              | {}                                   |
| password_expires_at  | None                                 |
+----------------------+--------------------------------------+
```

（3）为 neutron 用户添加管理员（admin）角色。

```
[root@ controller ~]# openstack role add --project service --user neutron admin
```

（4）创建 neutron 服务实体。

```
[root@ controller ~]# openstack service create --name neutron \
>--description "OpenStack Networking" network
+-------------+----------------------------------+
| Field       | Value                            |
+-------------+----------------------------------+
| description | OpenStack Networking             |
| enabled     | True                             |
| id          | b97497ff5d0548a099c4cb3cb020e051 |
| name        | neutron                          |
| type        | network                          |
+-------------+----------------------------------+
```

（5）创建网络服务的 API 端点。

①创建公共端点。

```
[root@ controller ~]# openstack endpoint create --region RegionOne \
>network public http://controller:9696
+--------------+----------------------------------+
| Field        | Value                            |
+--------------+----------------------------------+
| enabled      | True                             |
| id           | ed17341614c74fb5af1b6a410031662c |
| interface    | public                           |
| region       | RegionOne                        |
| region_id    | RegionOne                        |
| service_id   | b97497ff5d0548a099c4cb3cb020e051 |
| service_name | neutron                          |
| service_type | network                          |
| url          | http://controller:9696           |
+--------------+----------------------------------+
```

②创建内部端点。

```
[root@ controller ~]# openstack endpoint create --region RegionOne \
>network internal http://controller:9696
```

```
+--------------+-----------------------------------+
| Field        | Value                             |
+--------------+-----------------------------------+
| enabled      | True                              |
| id           | a44d5e17312d4b24bf6a24f0866c7ca4  |
| interface    | internal                          |
| region       | RegionOne                         |
| region_id    | RegionOne                         |
| service_id   | b97497ff5d0548a099c4cb3cb020e051  |
| service_name | neutron                           |
| service_type | network                           |
| url          | http://controller:9696            |
+--------------+-----------------------------------+
```

③创建管理端点。

```
[root@ controller ~]# openstack endpoint create --region RegionOne \
>network admin http://controller:9696
+--------------+-----------------------------------+
| Field        | Value                             |
+--------------+-----------------------------------+
| enabled      | True                              |
| id           | 0d6cee20d50846bc877c4fe02d1fc362  |
| interface    | admin                             |
| region       | RegionOne                         |
| region_id    | RegionOne                         |
| service_id   | b97497ff5d0548a099c4cb3cb020e051  |
| service_name | neutron                           |
| service_type | network                           |
| url          | http://controller:9696            |
+--------------+-----------------------------------+
```

3）安装并配置服务组件

（1）安装 Neutron 相关服务组件。

```
[root@ controller ~]# yum install openstack-neutron openstack-neutron-ml2 \
>  openstack-neutron-linuxbridge ebtables
```

（2）编辑/etc/neutron/neutron.conf 文件，文件配置如下：

①配置数据库访问。

```
[database]
#…
connection=mysql+pymysql://neutron:neutron@ controller/neutron
```

②配置 ML2 插件、路由服务和重叠 IP 地址。

```
[DEFAULT]
```

```
#...
core_plugin = ml2
service_plugins = router
allow_overlapping_ips = true
```

③配置消息队列访问。

```
[DEFAULT]
#...
transport_url = rabbit://openstack:openstack@ controller
```

④配置身份认证访问。

```
[DEFAULT]
#...
auth_strategy = keystone
[keystone_authtoken]
#...
www_authenticate_uri = http://controller:5000
auth_url = http://controller:5000
memcached_servers = controller:11211
auth_type = password
project_domain_name = default
user_domain_name = default
project_name = service
username = neutron
password = neutron
```

⑤配置网络通知拓扑结构变化。

```
[DEFAULT]
#...
notify_nova_on_port_status_changes = true
notify_nova_on_port_data_changes = true
[nova]
#...
auth_url = http://controller:5000
auth_type = password
project_domain_name = default
user_domain_name = default
region_name = RegionOne
project_name = service
username = nova
password = nova
```

⑥配置锁定路径。

```
[oslo_concurrency]
```

```
#...
lock_path = /var/lib/neutron/tmp
```

（3）配置 ML2 插件。

ML2 插件利用 Linux 网桥机制在实例间构建二层虚拟网络架构。

编辑/etc/neutron/plugins/ml2/ml2_conf. ini 文件，文件内容如下：

①开启 Flat、VLAN 和 VXLAN。

```
[ml2]
#...
type_drivers = flat,vlan,vxlan
```

②开启 VXLAN 自服务网络。

```
[ml2]
#...
tenant_network_types = vxlan
```

③开启 Linux 网桥和二层同步机制。

```
[ml2]
#...
mechanism_drivers = linuxbridge,l2population
```

④开启端口安全扩展驱动。

```
[ml2]
#...
extension_drivers = port_security
```

⑤配置虚拟网络为 Flat 网络。

```
[ml2_type_flat]
#...
flat_networks = provider
```

⑥配置 VXLAN 网络表示范围。

```
[ml2_type_vxlan]
#...
vni_ranges = 1:1000
```

⑦开启 ipset 提高安全组规则效率。

```
[securitygroup]
#...
enable_ipset = true
```

（4）配置 Linux 网桥代理。

Linux 网桥代理在实例间构建二层虚拟网络并处理安全组。

编辑/etc/neutron/plugins/ml2/linuxbridge_agent. ini 文件，文件内容如下：

①虚拟网络映射到物理网络接口，这里的 ens33 为映射的网卡。

```
[linux_bridge]
physical_interface_mappings = provider:ens33
```

②启用 VXLAN 重叠网络，配置处理覆盖网络的物理网络接口的 IP 地址，并启用第 2 层填充。

```
[vxlan]
enable_vxlan = true
local_ip = 192.168.16.100
l2_population = true
```

③启用安全组并配置 Linux 桥接 iptables 防火墙驱动程序。

```
[securitygroup]
enable_security_group = true
firewall_driver = neutron.agent.linux.iptables_firewall.IptablesFirewallDriver
```

④将所有 sysctl 值设置为 1 并进行验证，以确保 Linux 操作系统内核支持网桥过滤器。

```
[root@ controller ~]# modprobe br_netfilter
[root@ controller ~]# ls/proc/sys/net/bridge
```

在/etc/sysctl. conf 中添加：

```
net.bridge.bridge-nf-call-ip6tables = 1
net.bridge.bridge-nf-call-iptables = 1
```

执行生效：

```
[root@ controller ~]# sysctl -p
```

（5）配置三层代理。

三层代理为自服务（self - service）网络提供路由和 NAT 服务。

编辑/etc/neutron/l3_agent. ini 文件，配置如下。

配置 Linux 桥接接口驱动程序和外部网桥：

```
[DEFAULT]
interface_driver = linuxbridge
```

（6）配置 DHCP 代理。

DHCP 代理为虚拟网络提供 DHCP 服务。

编辑/etc/neutron/dhcp_agent. ini 文件，配置如下：

配置 Linux 桥接接口驱动程序、DNSmasq DHCP 驱动程序，并启用隔离的元数据，以便提供商网络上的实例可以通过网络访问元数据。

```
[DEFAULT]
interface_driver = linuxbridge
dhcp_driver = neutron.agent.linux.dhcp.Dnsmasq
enable_isolated_metadata = true
```

（7）配置元数据代理。

元数据代理为虚拟机提供配置信息。

编辑/etc/neutron/metadata_agent.ini 文件，配置如下。

配置 metadata 主机和共享密钥：

```
[DEFAULT]
nova_metadata_host = controller
metadata_proxy_shared_secret = lycloud
```

#lycloud 为 Neutron 和 Nova 之间通信的密码（自行设置）。

（8）配置计算服务（Nova）使用网络服务。

编辑/etc/nova/nova.conf 文件，配置如下。

配置访问参数，启用 metadata 代理并配置密码：

```
[neutron]
url = http://controller:9696
auth_url = http://controller:5000
auth_type = password
project_domain_name = default
user_domain_name = default
region_name = RegionOne
project_name = service
username = neutron
password = neutron
service_metadata_proxy = true
metadata_proxy_shared_secret = lycloud
```

4）安装完成

（1）网络服务初始化脚本需要一个/etc/neutron/plugin.ini 指向 ML2 插件配置文件的符号链接/etc/neutron/plugins/ml2/ml2_conf.ini。如果此符号链接不存在，使用以下命令创建：

```
[root@ controller ~]# ln - s/etc/neutron/plugins/ml2/ml2_conf.ini/etc/neutron/plugin.ini
```

（2）填充数据库，这里需要用到 neutron.conf 和 ml2_conf.ini。

```
[root@ controller ~]# su - s/bin/sh - c "neutron - db - manage -- config - file/etc/neutron/neutron.conf -- config - file/etc/neutron/plugins/ml2/ml2_conf.ini upgrade head" neutron
```

（3）重启 Nova 计算服务，因为修改了它的配置文件。

```
[root@ controller ~]# systemctl restart openstack - nova - api.service
```

（4）启动网络服务并将其配置为在系统引导时启动。

```
[root@ controller ~]# systemctl start neutron - server.service \
>   neutron - linuxbridge - agent.service neutron - dhcp - agent.service \
>   neutron - metadata - agent.service
```

```
[root@ controller ~]# systemctl enable neutron-server.service \
> neutron-linuxbridge-agent.service neutron-dhcp-agent.service \
> neutron-metadata-agent.service
```

（5）如果网络配置选择的是配置自服务（self-service）网络，需要配置三层网络为系统引导时启动，并启动三层网络。

```
[root@ controller ~]# systemctl start neutron-l3-agent.service
[root@ controller ~]# systemctl enable neutron-l3-agent.service
```

2. compute 节点部署

1）安装并配置组件

（1）安装 Neutron 相关组件。

```
[root@ compute ~]# yum install openstack-neutron-linuxbridge ebtables ipset
```

（2）配置公共组件。

网络公共组件包括认证机制、消息队列和插件。

编辑/etc/neutron/neutron.conf 文件，配置如下。

①在［database］部分，注释掉所有的 connection 选项（compute 节点不直接访问数据库）。

②配置 RabbitMQ 消息队列访问。

```
[DEFAULT]
#...
transport_url = rabbit://openstack:openstack@ controller
```

③配置身份服务访问。

```
[DEFAULT]
#...
auth_strategy = keystone
[keystone_authtoken]
#...
www_authenticate_uri = http://controller:5000
auth_url = http://controller:5000
memcached_servers = controller:11211
auth_type = password
project_domain_name = default
user_domain_name = default
project_name = service
username = neutron
password = neutron
```

④配置锁定路径。

```
[oslo_concurrency]
```

```
#...
lock_path = /var/lib/neutron/tmp
```

（3）配置 Linux 网桥代理。

Linux 网桥代理为实例构建 2 层（桥接和交换）虚拟网络基础架构，并处理安全组编辑 /etc/neutron/plugins/ml2/linuxbridge_agent. ini 文件，配置如下。

①将虚拟网络映射到物理网络接口。

```
[linux_bridge]
physical_interface_mappings = provider:ens33
```

②启用 VXLAN 覆盖网络。

```
[vxlan]
#...
enable_vxlan = true
```

③配置处理覆盖的物理接口的 IP 地址。

```
[vxlan]
#...
local_ip =192.168.16.200
```

④启用二层同步。

```
[vxlan]
#...
l2_population = true
```

⑤启用安全组。

```
[securitygroup]
#...
enable_security_group = true
```

⑥配置 Linux 网桥 iptables 防火墙驱动程序。

```
[securitygroup]
#...
firewall_driver = neutron.agent.linux.iptables_firewall.IptablesFirewallDriver
```

⑦将所有 sysctl 值设置为 1 并进行验证，以确保 Linux 操作系统内核支持网桥过滤器。

```
[root@ controller ~]# modprobe br_netfilter
[root@ controller ~]# ls/proc/sys/net/bridge
```

在/etc/sysctl. conf 中添加：

```
net.bridge.bridge - nf - call - ip6tables =1
net.bridge.bridge - nf - call - iptables =1
```

执行生效：

```
[root@ controller ~]# sysctl - p
```

(4) 配置计算服务使用网络服务。

编辑/etc/nova/nova. conf 文件，配置如下。

```
[neutron]
#...
url = http://controller:9696
auth_url = http://controller:5000
auth_type = password
project_domain_name = default
user_domain_name = default
region_name = RegionOne
project_name = service
username = neutron
password = neutron
```

2) 完成安装

(1) 重启计算服务。

```
[root@ compute ~]# systemctl restart openstack - nova - compute. service
```

(2) 启用 Linux 网桥代理，并设置随系统启动。

```
[root@ compute ~]# systemctl start neutron - linuxbridge - agent. service
[root@ compute ~]# systemctl enable neutron - linuxbridge - agent. service
```

3. 验证

(1) 运行 admin - openstack. sh 脚本，获取管理员的命令行执行权限。

```
[root@ controller ~]# source. /admin - openstack. sh
```

(2) 列出已加载的网络扩展，验证 neutron - server 是否加载成功。

```
[root@ controller ~]# openstack extension list - network
```

(3) 列出代理，验证 Neutron 代理已加载成功，显示结果中包含 4 个 controller 节点的代理、1 个 compute 节点的代理。

```
[root@ controller ~]# openstack network agent list
+----------+----------------+---------+----------------+----+--
+-------------------------+
|ID         |Agent Type          |Host        |Availability Zone |Alive |State
|Binary                        |
+----------+----------------+---------+----------------+----+--
+-------------------------+
|00…3069    |Metadata agent      |controller  |None              |: - )  |UP
|neutron - metadata - agent    |
|13…49fa    |Linux bridge agent  |compute     |None              |: - )  |UP
|neutron - linuxbridge - agent |
```

```
|90…941c      |Linux bridge agent  |controller  |None              |: - )  |UP
|neutron - linuxbridge - agent     |
|b2…9734      |L3 agent            |controller  |nova              |: - )  |UP
|neutron - l3 - agent              |
|ce…cae2      |DHCP agent          |controller  |nova              |: - )  |UP
|neutron - dhcp - agent            |
+----------+---------------+---------+--------------+----+--
+-------------------------+
```

【任务工单】

工单号：3－5

<table>
<tr><td colspan="2">项目名称：OpenStack 云平台部署</td><td colspan="2">任务名称：网络服务 Neutron 部署</td></tr>
<tr><td colspan="2">班级：</td><td>学号：</td><td>姓名：</td></tr>
<tr><td>任务安排</td><td colspan="3">□梳理 Nova 架构及各组件的作用
□controller 节点部署 Neutron 服务
□Neutron 数据库的创建与授权
□创建网络服务的凭证，包括创建 Neutron 用户并授权、创建 Neutron 服务实体和 Endpoint
□安装 Neutron 组件并配置 Neutron 的配置文件完成部署
□配置/etc/neutron/neutron. conf 文件
□配置 ML2 插件
□配置 Linux 网桥代理
□配置 L3、DHCP、元数据代理
□配置计算服务使用 Neutron
□compute 节点部署 Neutron 服务
□安装 Neutron 组件并配置 Neutron 的公共组件
□配置 Linux 网桥代理
□配置计算服务使用 Neutron
□验证 neutron－server 是否加载成功，验证 controller 和 compute 节点的 Neutron 代理已加载成功</td></tr>
<tr><td>成果交付形式</td><td colspan="3">查看验证结果，controller 节点的 Metadata Agent、Linuxbridge Agent、L3 Agent、DHCP Agent 和 compute 节点的 Linuxbridge Agent 是否加载成功，并且状态是 UP</td></tr>
<tr><td>任务实施总结</td><td colspan="3">任务自评（0～10 分）：
任务收获：______
______
改进点：______
______</td></tr>
<tr><td>成果验收</td><td colspan="3">□完全满足任务要求
□基本满足任务要求
Neutron 部署完成，但 controller 和 compute 节点的个别代理加载失败或状态为 DOWN：
______
______
□不能满足需求
Neutron 部署完成，但 controller 和 compute 节点的代理加载失败：
______
______</td></tr>
</table>

【知识巩固】

1. 以下是 Neutron 组件功能的是（　　）。

A. 二层交换　　B. 三层路由　　C. 负载均衡　　D. 防火墙

2. 以下是 Neutron 网络类型的是（　　）。

A. Local　　B. Flat　　C. GRE　　D. GVRP

【小李的反思】

业精于勤，荒于嬉；行成于思，毁于随。

源自韩愈《进学解》，意思是说，学业由于勤奋而专精，由于玩乐而荒废；德行由于独立思考而有所成就，由于因循随俗而败坏。

苏洵是宋朝有名的文学家，唐宋八大家之一。他小时候很贪玩，直到二十七岁才理解到念书很重要，以后开始发奋读书，抓紧一切时间进修。有一年端午节，苏洵从晚上起来就扎在书房里念书。他的妻子端了一盘粽子和一碟白糖送进了书房。将近午时，夫人料理盘碟时，发现粽子已经吃完了，碟里的白糖却原封未动，而左右砚台上竟有不少糯米粒。原来，苏洵只顾埋头念书，误把砚台当成了糖碟。恰是凭着这种细心刻苦的精力，苏洵成为文学大家。

习近平总书记在党的二十大报告中指出："全党同志务必不忘初心、牢记使命，务必谦虚谨慎、艰苦奋斗，务必敢于斗争、善于斗争，坚定历史自信，增强历史主动，谱写新时代中国特色社会主义更加绚丽的华章。"

任务6 Web 界面 Horizon 部署

【任务描述】

小李已经完成了计算服务 Neutron 的部署并经过了验证，但现在小李苦恼的是，一切操作都要通过命令行界面查看，这样查看起来不直观，Windows 操作系统提供了人机交互的图形界面，如果 OpenStack 也能提供一个图形界面进行资源管理岂不是更好？带着这个疑问，小李通过网络查询资料得知，OpenStack 中的 Horizon 组件为其提供 dashboard 界面，现小李参考官方文档（https://docs.openstack.org/horizon/rocky/install/install - rdo.html）进行 Web 界面 Horizon 组件的部署工作，包括 controller 节点部署配置 Horizon 组件和主机端 Web 验证等。

【知识要点】

1. Horizon 的作用

Horizon 为 OpenStack 提供一个 Web 前端的管理界面（UI 服务），通过 Horizon 所提供的 dashboard 服务，管理员可以使用通过 Web UI 对 OpenStack 整体云环境进行管理，并可直观地看到各种操作结果与运行状态。

2. Horizon 中的概念

1）区域（Region）

地理上的概念，可以理解为一个独立的数据中心，每个所定义的区域有自己独立的 Endpoint；区域之间是完全隔离的，但多个区域之间共享同一个 Keystone 和 dashboard（目前 OpenStack 中的 dashboard 还不支持多个区域）；除了提供隔离的功能外，区域的设计更多地侧重于地理位置的概念，用户可以选择离自己更近的区域来部署自己的服务，选择不同的区域主要是考虑哪个区域更靠近自己，如用户在上海，可以选择离上海更近的区域。区域的概念是由 Amazon 在 AWS 中提出的，主要是解决容错能力和可靠性。

2）可用性区域（Availability Zone）

AZ 是在 Region 范围内的再次切分，例如，可以把一个机架上的服务器划分为一个 AZ，划分 AZ 是为了提高容灾能力和提供廉价的隔离服务；AZ 主要是通过冗余来解决可用性的问题，在 Amazon 的声明中，Instance 不可用是指用户的所有 AZ 中的同一个 Instance 都不可达才表明不可用；AZ 是用户可见的一个概念，并可选择，是物理隔离的，一个 AZ 不可用不会影响其他的 AZ，用户在创建 Instance 的时候可以选择创建到哪些 AZ 中。

【任务实施】

Dashboard 是 OpenStack 中提供的一个 Web 前端控制台，以此来展示 OpenStack 的功能。

Horizon 为 OpenStack 提供一个 Web 前端的管理界面（UI 服务），管理员可以通过 Web 界面对 OpenStack 整体云环境进行管理，并可直观地看到各种操作结果与运行状态。

1. 安装并配置组件

（1）安装 Horizon 组件。

```
[root@ controller ~]# yum install openstack-dashboard
```

（2）配置 dashboard，编辑/etc/openstack-dashboard/local_settings 文件，配置如下：

①配置仪表板，以在 controller 节点上使用 OpenStack 服务。

```
OPENSTACK_HOST = "controller"
```

②配置允许访问的主机列表，如果有其他 ALLOWED_HOSTS 选项，注释掉。

```
ALLOWED_HOSTS = [ '* ']
```

③配置 memcached 会话存储服务。

```
SESSION_ENGINE = 'django.contrib.sessions.backends.cache'
CACHES = {
    'default':{
        'BACKEND':'django.core.cache.backends.memcached.MemcachedCache',
        'LOCATION':'controller:11211',
    }
}
```

④启用 Identity API 版本 3。

```
OPENSTACK_KEYSTONE_URL = "http://%s:5000/v3" % OPENSTACK_HOST
```

⑤启用对多域的支持。

```
OPENSTACK_KEYSTONE_MULTIDOMAIN_SUPPORT = True
```

⑥配置 API 版本。

```
OPENSTACK_API_VERSIONS = {
    "identity":3,
    "image":2,
    "volume":2,
}
```

⑦配置 Default 为通过仪表板创建的用户的默认域。

```
OPENSTACK_KEYSTONE_DEFAULT_DOMAIN = "Default"
```

⑧配置 myrole 为通过仪表板创建的用户的默认角色。

```
OPENSTACK_KEYSTONE_DEFAULT_ROLE = "myrole"
```

⑨配置时区（可选）。

```
TIME_ZONE = "Asia/Shanghai "
```

（3）配置 WSGI 应用程序组属于全局组，在/etc/httpd/conf. d/openstack－dashboard. conf 配置文件中添加如下信息：

```
WSGIApplicationGroup %{GLOBAL}
```

（4）重启 httpd 和 memcached 服务。

```
[root@ controller ~]# systemctl restart httpd.service memcached.service
```

2. 验证

（1）在浏览器中输入 http://192. 168. 16. 100/dashboard 来访问 OpenStack 的 dashboard，如图 3－32 所示。

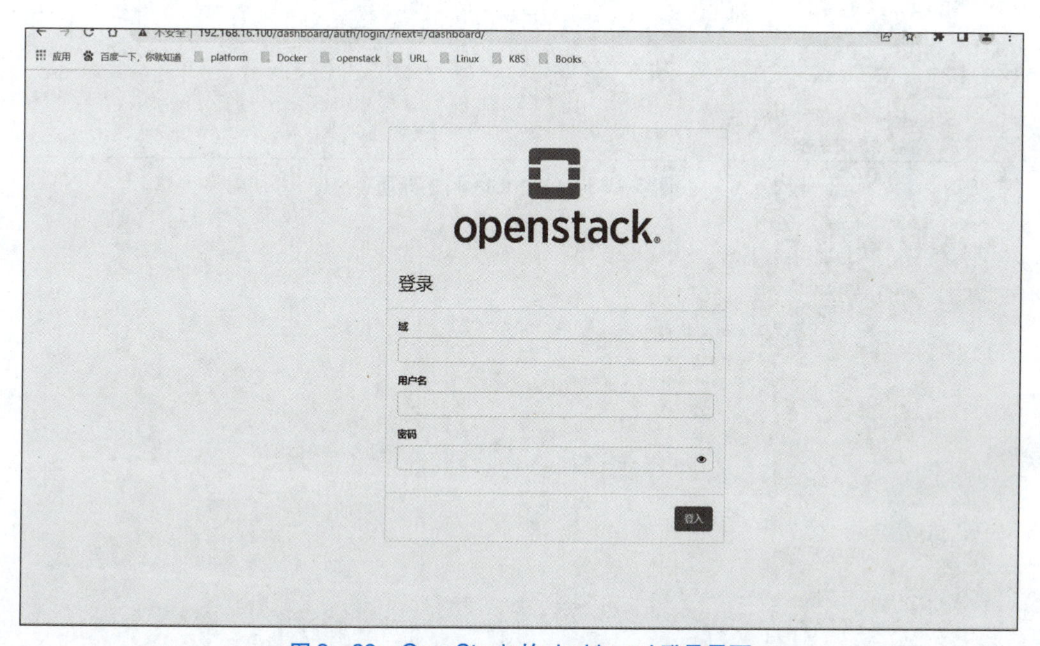

图 3－32　OpenStack 的 dashboard 登录界面

（2）输入域（default）、用户名（admin）和密码（admin）可以正常登录 OpenStack 平台，如图 3－33 所示。

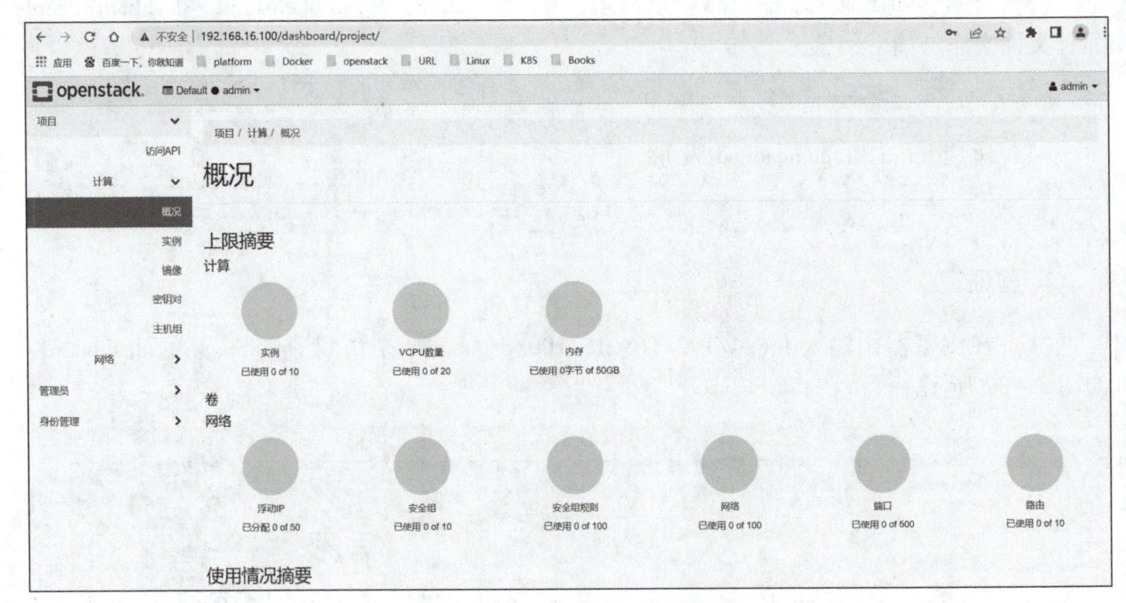

图 3－33　OpenStack 主界面

【任务工单】

工单号：3－6

<table>
<tr><td colspan="2">项目名称：OpenStack 云平台部署</td><td colspan="2">任务名称：Web 界面 Horizon 部署</td></tr>
<tr><td colspan="2">班级：</td><td>学号：</td><td>姓名：</td></tr>
<tr><td>任务安排</td><td colspan="3">□安装 Horizon 组件
□配置 Horizon 组件相关配置文件，使 Horizon 可以正常使用
□配置 WSGI 应用程序组
□重启 httpd 和 memcached 服务
□Web 端登录云平台验证 Horizon 部署正确性</td></tr>
<tr><td>成果交付形式</td><td colspan="3">通过 Web 端登录云平台，并查看概况信息及镜像界面，查验 Web 端能否查看到已上传的镜像</td></tr>
<tr><td>任务实施总结</td><td colspan="3">任务自评（0～10 分）：
任务收获：

改进点：</td></tr>
<tr><td>成果验收</td><td colspan="3">□完全满足任务要求
□基本满足任务要求
Horizon 部署完成，可以正常登录，但显示信息存在一些问题：

□不能满足需求
Web 界面无法正常登录：</td></tr>
</table>

【知识巩固】

1. Horizon 组件的作用是什么？

2. Horizon 组件既可以部署在控制节点，也可以部署在计算节点，但不能部署在存储节点。(　　)

A. 对　　　　B. 错

【小李的反思】

万人操弓，共射一招，招无不中。

源自吕不韦《吕氏春秋》，意思是说，众人拿着弓箭，共同射向一个目标，这个目标没有射不中的。正说明了做任何事情，团结的重要性，同学们在学习过程中也要注重培养自己的团队凝聚力，紧密团结在一起。

2020 年疫情爆发后，广大医护人员及无数志愿者团结一心，全力奋战在抗疫一线，与死神抢时间救病人。

习近平总书记在党的二十大报告中殷切寄语广大青年，“要坚定不移听党话、跟党走，怀抱梦想又脚踏实地，敢想敢为又善作善成，立志做有理想、敢担当、能吃苦、肯奋斗的新时代好青年”，我们作为青年一代学子，要将“小我”成长融入“大我”奋斗之中，融入社会主义现代化建设之中，让青春在全面建设社会主义现代化国家的火热实践中绽放绚丽之花。

任务 7 块存储 Cinder 部署

【任务描述】

小李已经完成了计算服务 Horizon 的部署并经过了验证，已经可以看到基本的私有云平台的 Web 界面了，小李想到自己的计算机都是有硬盘存储的，那么云平台上要存储很多的资源，通过什么存储呢？是否也有一个组件能和计算机的硬盘一样能提供云平台的存储呢？带着这个疑问，小李通过网络查询资料得知，OpenStack 中的 Cinder 组件为其提供块存储服务，现小李参考官方文档（https://docs.openstack.org/cinder/rocky/install/index - rdo.html）进行块存储服务 Cinder 组件的部署工作，包括 controller 节点部署配置 Cinder 组件、存储节点部署配置 Cinder 组件和进行配置结果验证等。

【知识要点】

1. Cinder 组件作用

Cinder 块存储是存储虚拟机镜像文件及虚拟机使用的数据的基础。前身是 nova - volume，OpenStack 中的实例是不能持久化的，一旦实例被关闭、重启或删除，该实例的数据就会全部丢失。实现持久化的方法就是使用 Cinder 块存储，挂载卷之后，在卷中实现持久化，为运行实例提供稳定的数据块存储服务，提供对卷从创建到删除整个生命周期的管理。

操作系统获得存储空间的方式一般有两种：

- 通过某种协议（SAS、SCSI、SAN、iSCSI 等）挂接裸硬盘，然后分区、格式化、创建文件系统；或者直接使用裸硬盘存储数据（数据库）。
- 通过 NFS、CIFS 等协议 mount 远程的文件系统。

第一种裸硬盘的方式叫作 Block Storage（块存储），每个裸硬盘通常也称作 volume（卷）；第二种叫作文件系统存储，NAS 服务器、NFS 服务器及各种分布式文件系统提供的都是这种存储。

块存储服务提供对卷从创建到删除整个生命周期的管理。从实例的角度看，挂载的每一个卷都是一块硬盘。OpenStack 提供的块存储服务是 Cinder，其具体功能是：

- 提供 REST API，使用户能够查询和管理卷、快照以及卷类型。
- 提供 scheduler，调度卷创建请求，合理优化存储资源的分配。
- 通过 driver 架构支持多种 back - end（后端）存储方式，包括 LVM、NFS、Ceph 和其他诸如 EMC、IBM 等商业存储产品与方案。

2. Cinder 核心架构

Cinder 的各核心组件提供不同的功能，组件间相互配合，完成 Cinder 的块存储管理功能，架构如图 3 - 34 所示。

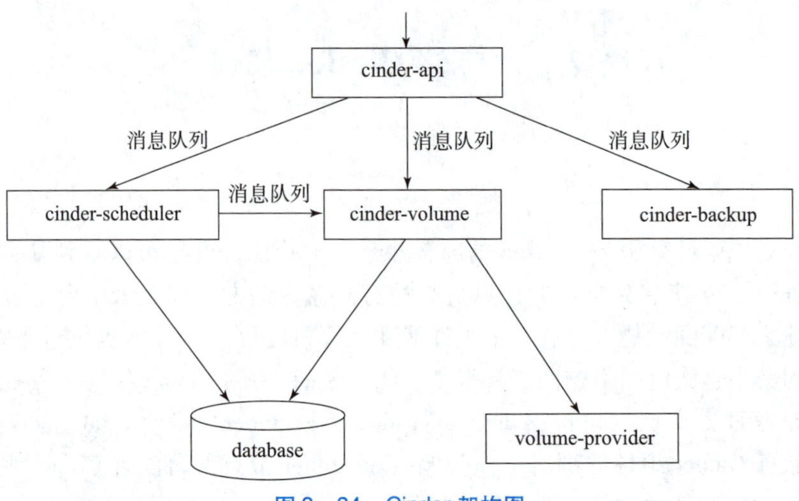

图 3－34　Cinder 架构图

Cinder 各组件的功能见表 3－5。

表 3－5　Cinder 各组件的功能

cinder－api	接收 API 请求，调用 cinder－volume 执行操作
cinder－volume	与 volume－provider 协调工作，管理 volume 生命周期
cinder－scheduler	通过调度算法选择最合适的存储节点创建 volume
volume－provider	数据存储设备，为 volume 提供物理存储空间
message－queue	Cinder 各子服务通过消息队列实现进程间通信和相互协作
database	存储数据文件的数据库

（1）cinder－api：接收 API 请求，调用 cinder－volume 执行操作。cinder－api 对接收到的 HTTP API 请求会做如下处理。

- 检查客户端传入的参数是否合法有效。
- 调用 Cinder 其他子服务处理的客户端请求。
- 将 Cinder 其他子服务返回的结果序列化并返回给客户端。

（2）cinder－scheduler：Cinder 可以有多个存储节点，当需要创建 volume 时，cinder－scheduler 会根据存储节点的属性和资源使用情况选择一个最合适的节点来创建 volume。

（3）cinder－volume：cinder－volume 在存储节点上运行，OpenStack 对 volume 的操作最后都是交给 cinder－volume 来完成的。cinder－volume 自身并不管理真正的存储设备，存储设备是由 volume－provider 管理的。cinder－volume 与 volume provider 一起实现 volume 生命周期的管理。cinder－volume 的主要功能如下：

- 通过 Driver 架构支持多种 volume－provider。
- 定期向 OpenStack 报告计算节点的状态。cinder－volume 会定期向 Cinder 报告存储节点的空闲容量来做筛选。
- 实现 volume 生命周期管理。Cinder 对 volume 的生命周期的管理最终都是通过 cinder－volume 完成的，包括卷的创建（create）、扩展（extend）、绑定（attach）、快照（snapshot）、

删除（delete）等。

（4）volume－provider：数据的存储设备，为 volume 提供物理存储空间。Cinder 提供了一个驱动架构，为这些存储设备定义了统一接口，第三方存储设备只需要实现这些接口，就可以驱动的形式加入 OpenStack。volume－provider 通过自己的 driver 与 cinder－volume 协调工作的关系图如图 3－35 所示。

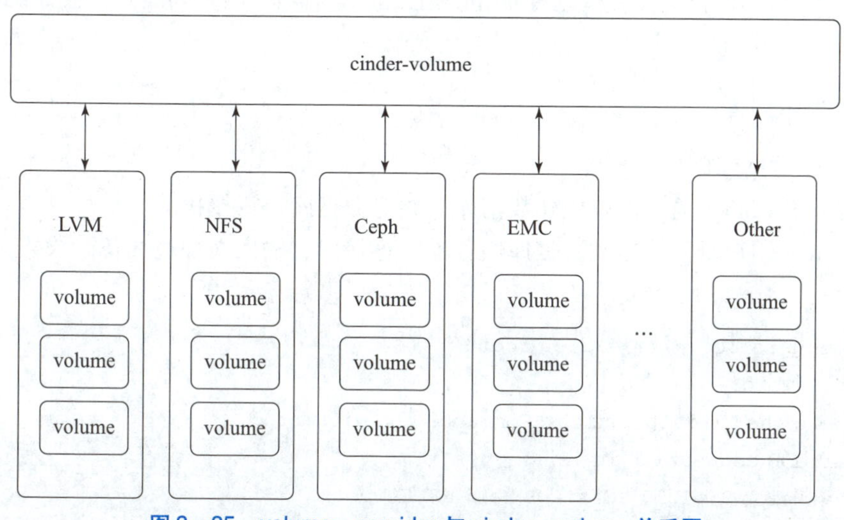

图 3－35　volume－provider 与 cinder－volume 关系图

（5）message－queue：Cinder 各个子服务通过消息队列实现进程间通信和相互协作。因为有了消息队列，子服务之间实现了解耦，这种松散的结构也是分布式系统的重要特征。

3. Cinder 工作流程

Cinder 工作流程如图 3－36 所示。

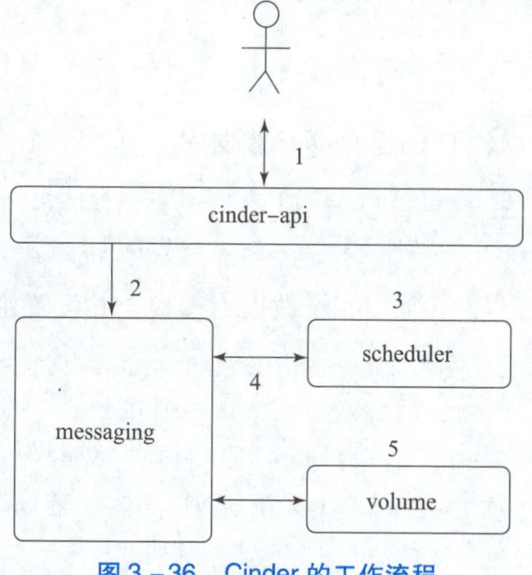

图 3－36　Cinder 的工作流程

- 用户向 cinder－api 发送创建 volume 的请求。
- cinder－api 对请求做一些必要处理后，通过 messaging 将创建消息发送给 cinder－scheduler。
- cinder－scheduler 从 messaging 获取到 cinder－api 发给它的消息，然后执行调度算法，从若干存储节点中选出节点 A。
- cinder－scheduler 通过 messaging 将创建消息发送给存储节点 A。
- 存储节点 A 的 cinder－volume 从 messaging 中获取到 cinder－scheduler 发给它的消息，然后通过 driver 在 volume－provider 上创建 volume。

【任务实施】

块存储服务（Cinder）为实例提供块存储设备。配置和使用存储的方法由块存储驱动程序或多后端配置驱动程序决定。有多种可用的驱动程序：NAS/SAN、NFS、iSCSI、Ceph 等。

块存储 API 和调度程序服务通常在控制器节点上运行。根据使用的驱动程序，卷服务可以在控制器节点、计算节点或独立存储节点上运行。本任务将 Cinder 卷服务配置在计算节点。

1. 控制节点部署

1）Cinder 数据库创建及授权

在安装配置 Cinder 前，需要创建数据库，并对数据库进行授权。

（1）以 root 用户身份连接到数据库服务器，连接时，输入在环境准备阶段安装 MariaDB 时的密码。

```
[root@ controller ~]# mysql -u root -p
Enter password:
Welcome to the MariaDB monitor. Commands end with;or \g.
Your MariaDB connection id is 38
Server version:10.1.20 -MariaDB MariaDB Server

Copyright(c)2000,2016,Oracle,MariaDB Corporation Ab and others.

Type 'help;' or '\h' for help. Type '\c' to clear the current input statement.

MariaDB[(none)]>
```

（2）创建 Cinder 数据库。

```
MariaDB[(none)]>CREATE DATABASE cinder;
Query OK,1 row affected(0.00 sec)
```

（3）为 Cinder 数据库授权，并设置密码为 cinder。

```
MariaDB[(none)]>GRANT ALL PRIVILEGES ON cinder.* TO 'cinder'@ 'localhost'\
    ->IDENTIFIED BY 'cinder';
Query OK,0 rows affected(0.00 sec)
MariaDB[(none)]>GRANT ALL PRIVILEGES ON cinder.* TO 'cinder'@ '%'\
    ->IDENTIFIED BY 'cinder';
Query OK,0 rows affected(0.00 sec)
```

（4）退出数据库。

```
MariaDB[(none)]>exit
```

2）创建服务证书

（1）运行 admin - openstack. sh 脚本，获取管理员的命令行执行权限。

```
[root@ controller ~]# source. /admin - openstack. sh
```

（2）在 default 域内创建 cinder 用户，设置密码为 cinder。

```
[root@ controller ~] # openstack user create - - domain default - - password -
prompt cinder
User Password:
Repeat User Password:
+---------------------+-----------------------------------+
| Field               | Value                             |
+---------------------+-----------------------------------+
| domain_id           | default                           |
| enabled             | True                              |
| id                  | 2d90fd1ba066426b8775e37c560aea47  |
| name                | cinder                            |
| options             | {}                                |
| password_expires_at | None                              |
+---------------------+-----------------------------------+
```

（3）为 cinder 用户添加 admin 权限。

```
[root@ controller ~]# openstack role add --project service --user cinder admin
```

（4）创建 cinderv2 服务实体。

```
[root@ controller ~]# openstack service create --name cinderv2 \
>   --description "OpenStack Block Storage" volumev2
+-------------+----------------------------------+
| Field       | Value                            |
+-------------+----------------------------------+
| description | OpenStack Block Storage          |
| enabled     | True                             |
| id          | 7dd8ab90835a406aa3d9b26a309a3bbe |
| name        | cinderv2                         |
| type        | volumev2                         |
+-------------+----------------------------------+
```

（5）创建 cinderv3 服务实体。

```
[root@ controller ~]# openstack service create --name cinderv3 \
>   --description "OpenStack Block Storage" volumev3
+-------------+----------------------------------+
| Field       | Value                            |
+-------------+----------------------------------+
```

```
| description   | OpenStack Block Storage          |
| enabled       | True                             |
| id            | 4d04c9881f304b7b86112ca64745b2cf |
| name          | cinderv3                         |
| type          | volumev3                         |
+---------------+----------------------------------+
```

（6）创建块存储服务的 API 服务端点。

①创建 volumev2 公共端点。

```
[root@ controller ~]# openstack endpoint create --region RegionOne \
>volumev2 public http://controller:8776/v2/% \(project_id\)s
+--------------+------------------------------------------+
| Field        | Value                                    |
+--------------+------------------------------------------+
| enabled      | True                                     |
| id           | ce3e09f103aa4c33b53126fd59af5e67         |
| interface    | public                                   |
| region       | RegionOne                                |
| region_id    | RegionOne                                |
| service_id   | 7dd8ab90835a406aa3d9b26a309a3bbe         |
| service_name | cinderv2                                 |
| service_type | volumev2                                 |
| url          | http://controller:8776/v2/%(project_id)s |
+--------------+------------------------------------------+
```

②创建 volumev2 内部端点。

```
[root@ controller ~]# openstack endpoint create --region RegionOne \
>  volumev2 internal http://controller:8776/v2/% \(project_id\)s
+--------------+------------------------------------------+
| Field        | Value                                    |
+--------------+------------------------------------------+
| enabled      | True                                     |
| id           | 7f9da2b476a34e14a38b50b6a0484c11         |
| interface    | internal                                 |
| region       | RegionOne                                |
| region_id    | RegionOne                                |
| service_id   | 7dd8ab90835a406aa3d9b26a309a3bbe         |
| service_name | cinderv2                                 |
| service_type | volumev2                                 |
| url          | http://controller:8776/v2/%(project_id)s |
+--------------+------------------------------------------+
```

③创建 volumev2 管理端点。

```
[root@ controller ~]# openstack endpoint create --region RegionOne \
>  volumev2 admin http://controller:8776/v2/% \(project_id\)s
+--------------+------------------------------------------+
| Field        | Value                                    |
+--------------+------------------------------------------+
| enabled      | True                                     |
| id           | 8824150931c14bc6a0ba9b5e8b14f907         |
| interface    | admin                                    |
| region       | RegionOne                                |
| region_id    | RegionOne                                |
| service_id   | 7dd8ab90835a406aa3d9b26a309a3bbe         |
| service_name | cinderv2                                 |
| service_type | volumev2                                 |
| url          | http://controller:8776/v2/%(project_id)s |
+--------------+------------------------------------------+
```

④创建 volumev3 公共端点。

```
[root@ controller ~]# openstack endpoint create --region RegionOne \
>volumev3 public http://controller:8776/v3/% \(project_id\)s
+--------------+------------------------------------------+
| Field        | Value                                    |
+--------------+------------------------------------------+
| enabled      | True                                     |
| id           | 8511da1e0e434e68814fe11eb0822869         |
| interface    | public                                   |
| region       | RegionOne                                |
| region_id    | RegionOne                                |
| service_id   | 4d04c9881f304b7b86112ca64745b2cf         |
| service_name | cinderv3                                 |
| service_type | volumev3                                 |
| url          | http://controller:8776/v3/%(project_id)s |
+--------------+------------------------------------------+
```

⑤创建 volumev3 内部端点。

```
[root@ controller ~]# openstack endpoint create --region RegionOne \
>  volumev3 internal http://controller:8776/v3/% \(project_id\)s
+--------------+------------------------------------------+
| Field        | Value                                    |
+--------------+------------------------------------------+
| enabled      | True                                     |
| id           | 232a15aa441e44b991d7534ebe7dd780         |
```

```
| interface    | internal                                     |
| region       | RegionOne                                    |
| region_id    | RegionOne                                    |
| service_id   | 4d04c9881f304b7b86112ca64745b2cf             |
| service_name | cinderv3                                     |
| service_type | volumev3                                     |
| url          | http://controller:8776/v3/%(project_id)s     |
+--------------+----------------------------------------------+
```

⑥创建 volumev3 管理端点。

```
[root@ controller ~]# openstack endpoint create --region RegionOne \
>volumev3 admin http://controller:8776/v3/% \(project_id\)s
+--------------+----------------------------------------------+
| Field        | Value                                        |
+--------------+----------------------------------------------+
| enabled      | True                                         |
| id           | 2983fb5945a647d2adf25245b4aaa126             |
| interface    | admin                                        |
| region       | RegionOne                                    |
| region_id    | RegionOne                                    |
| service_id   | 4d04c9881f304b7b86112ca64745b2cf             |
| service_name | cinderv3                                     |
| service_type | volumev3                                     |
| url          | http://controller:8776/v3/%(project_id)s     |
+--------------+----------------------------------------------+
```

3）安装并配置 Cinder 组件

（1）安装 Cinder 服务相关组件。

```
[root@ controller ~]# yum install openstack-cinder
```

（2）编辑/etc/cinder/cinder. conf 文件，文件配置内容如下：

①配置数据库访问。

```
[database]
#...
connection=mysql+pymysql://cinder:cinder@ controller/cinder
```

②配置 RabbitMQ 消息队列访问。

```
[DEFAULT]
#...
transport_url=rabbit://openstack:openstack@  controller
```

③配置身份服务访问。

```
[DEFAULT]
```

```
#...
auth_strategy = keystone
[keystone_authtoken]
#...
www_authenticate_uri = http://controller:5000
auth_url = http://controller:5000
memcached_servers = controller:11211
auth_type = password
project_domain_id = default
user_domain_id = default
project_name = service
username = cinder
password = cinder
```

④配置 IP 地址，此处 IP 地址为控制节点的管理 IP 地址。

```
[DEFAULT]
#...
my_ip = 192.168.16.100
```

⑤配置锁定路径。

```
[oslo_concurrency]
#...
lock_path = /var/lib/cinder/tmp
```

（3）同步块存储数据库。

```
[root@ controller ~]# su - s/bin/sh - c "cinder - manage db sync" cinder
Deprecated:Option "logdir" from group "DEFAULT" is deprecated. Use option "log - dir" from group "DEFAULT".
```

（4）验证数据库同步是否成功。

```
[root@ controller ~]# mysql - ucinder - pcinder - e "use cinder;show tables;"
+-----------------------------+
| Tables_in_cinder            |
+-----------------------------+
| attachment_specs            |
| backup_metadata             |
| backups                     |
| ……                          |
| workers                     |
+-----------------------------+
```

4）配置计算服务使用块存储

编辑/ect/nova/nova. conf 文件，文件配置内容如下：

```
[cinder]
```

```
os_region_name = RegionOne
```

5）完成安装

（1）重启计算 API 服务。

```
[root@ controller ~]# systemctl restart openstack - nova - api. service
```

（2）启动块存储服务并配置为随系统启动。

```
[root@ controller ~]# systemctl start openstack - cinder - api. service openstack - cinder - scheduler. service
[root@ controller ~]# systemctl enable openstack - cinder - api. service openstack - cinder - scheduler. service
Created symlink from/etc/systemd/system/multi - user. target. wants/openstack - cinder - api. service to/usr/lib/systemd/system/openstack - cinder - api. service.
Created symlink from/etc/systemd/system/multi - user. target. wants/openstack - cinder - scheduler. service to/usr/lib/systemd/system/openstack - cinder - scheduler. service.
```

2. 存储节点部署

受设备限制，此处将计算节点作为存储节点共用，为 compute 节点添加一个/dev/sdb 的硬盘，将 Cinder 的 volume 服务安装在计算节点上。

1）LVM 配置

（1）安装 LVM 软件包。

```
[root@ compute ~]# yum install lvm2 device - mapper - persistent - data
```

（2）开启 LVM 元数据服务，并设置为随系统启动。

```
[root@ compute ~]# systemctl start lvm2 - lvmetad. service
[root@ compute ~]# systemctl enable lvm2 - lvmetad. service
```

（3）创建物理卷/dev/sdb。

```
[root@ compute ~]# pvcreate/dev/sdb
  Physical volume "/dev/sdb" successfully created.
```

（4）创建 LVM 卷组 cinder - volumes，块存储服务在此卷组中创建逻辑卷。

```
[root@ compute ~]# vgcreate cinder - volumes/dev/sdb
  Volume group "cinder - volumes" successfully created
```

（5）配置只有实例可以访问块存储卷。

默认情况下，LVM 卷扫描工具会在/dev 目录中扫描包含卷的块存储设备。如果项目在其卷上使用 LVM，则扫描工具会检测到这些卷并尝试对其进行缓存，这可能导致基础操作系统卷和项目卷出现各种问题，需要将 LVM 重新配置为仅扫描包含 cinder - volumes 卷组的设备。如果存储节点在操作系统磁盘上使用了 LVM，还必须添加相关的设备到过滤器中。类似地，如果计算节点在操作系统磁盘上使用了 LVM，也必须修改这些节点上/etc/lvm/lvm. conf 文件中的过滤器，将操作系统磁盘包含到过滤器中。

每个过滤器组中的元素都以 a 开头，即为 accept，或以 r 开头，即为 reject，并且包括一个设备名称的正则表达式规则。过滤器组必须以 r/. * /结束，过滤所有保留设备。

编辑/etc/lvm/lvm. conf 文件，在 devices 部分添加一个过滤器，由于计算节点的系统盘使用了 LVM，配置过滤器只接受/dev/sda 和/dev/sdb 设备，拒绝其他所有设备。

```
devices {
#...
filter=["a/sda/","a/sdb/","r/. * /"]
```

2）安装并配置 Cinder 组件

（1）安装软件包。

```
[root@ compute ~]# yum install openstack-cinder targetcli python-keystone
```

（2）编辑/etc/cinder/cinder. conf 文件，文件配置内容如下：

①配置数据库访问。

```
[database]
#...
connection=mysql+pymysql://cinder:cinder@ controller/cinder
```

②配置 RabbitMQ 消息队列访问。

```
[DEFAULT]
#...
transport_url=rabbit://openstack:openstack@ controller
```

③配置身份认证服务访问。

```
[DEFAULT]
#...
auth_strategy=keystone
[keystone_authtoken]
#...
www_authenticate_uri=http://controller:5000
auth_url=http://controller:5000
memcached_servers=controller:11211
auth_type=password
project_domain_id=default
user_domain_id=default
project_name=service
username=cinder
password=cinder
```

④配置 IP 地址，此处配置为存储节点（compute）管理 IP 地址。

```
[DEFAULT]
#...
my_ip=192.168.16.200
```

⑤为 LVM 后端配置 LVM 驱动程序、cinder - volumes 卷组、iSCSI 协议和适当的 iSCSI 服务。如果该［lvm］部分不存在，自行创建。

```
[lvm]
volume_driver = cinder.volume.drivers.lvm.LVMVolumeDriver
volume_group = cinder - volumes
iscsi_protocol = iscsi
iscsi_helper = lioadm
```

⑥启用 LVM 后端配置。

```
[DEFAULT]
#...
enabled_backends = lvm
```

⑦配置镜像服务 API 访问地址。

```
[DEFAULT]
#...
glance_api_servers = http://controller:9292
```

⑧配置锁定路径。

```
[oslo_concurrency]
#...
lock_path = /var/lib/cinder/tmp
```

3）完成安装

启用块存储卷服务及其依赖组件，并设置为随开机自启动。

```
[root@ compute ~]# systemctl start openstack - cinder - volume.service target.service
[root@ compute ~]# systemctl enable openstack - cinder - volume.service target.service
Created symlink from/etc/systemd/system/multi - user.target.wants/openstack - cinder - volume.service to/usr/lib/systemd/system/openstack - cinder - volume.service.
Created symlink from/etc/systemd/system/multi - user.target.wants/target.service to/usr/lib/systemd/system/target.service.
```

3. 验证 Cinder 操作

在 controller 节点查询卷服务，确定 Cinder 是否部署成功。

（1）执行 admin - openstack.sh 脚本，获取 admin 权限。

```
[root@ controller ~]# source./admin - openstack.sh
```

（2）查看卷服务列表，确认每个进程是否加载成功，列表中会显示 cinder - scheduler 和 cinder - volume 两个服务。

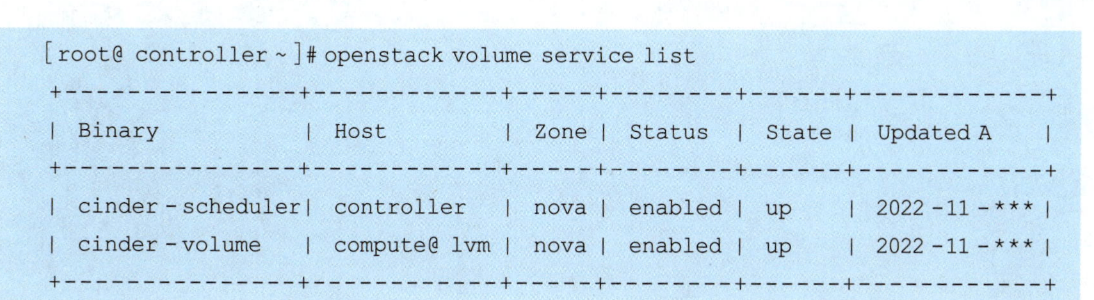

```
[root@ controller ~]# openstack volume service list
+----------------+------------+-----+--------+------+------------+
| Binary         | Host       | Zone | Status  | State | Updated A   |
+----------------+------------+-----+--------+------+------------+
| cinder-scheduler| controller  | nova | enabled | up    | 2022-11-*** |
| cinder-volume   | compute@ lvm | nova | enabled | up    | 2022-11-*** |
+----------------+------------+-----+--------+------+------------+
```

【任务工单】

工单号：3 – 7

<table>
<tr><td colspan="2">项目名称：OpenStack 云平台部署</td><td colspan="2">任务名称：块存储 Cinder 部署</td></tr>
<tr><td colspan="2">班级：</td><td>学号：</td><td>姓名：</td></tr>
<tr><td>任务安排</td><td colspan="3">□控制节点部署 Cinder
□创建 Cinder 数据库并授权
□创建相关服务证书，包括 Cinder 用户、Cinder 服务实体和 endpoint 等
□安装并配置 Cinder 配置文件
□配置计算服务使用 Cinder
□部署存储节点
□配置 LVM
□安装并配置 Cinder 组件
□验证 Cinder 操作</td></tr>
<tr><td>成果交付</td><td colspan="3">查看卷服务列表，确认每个进程是否加载成功，列表中会显示 cinder – scheduler 和 cinder – volume 两个服务</td></tr>
<tr><td>任务实施总结</td><td colspan="3">任务自评（0 ~ 10 分）：
任务收获：

改进点：</td></tr>
<tr><td>成果验收</td><td colspan="3">□完全满足任务要求
□基本满足任务要求
Cinder 部署完成，cinder – scheduler 和 cinder – volume 两个服务可以显示，但部分状态为 DOWN：

□不能满足需求
Cinder 部署未完成，或 cinder – scheduler 和 cinder – volume 两个服务未完全显示出来：</td></tr>
</table>

【知识巩固】

Cinder 组件的作用是（　　）。

A. 提供对象存储服务　　　　B. 提供镜像管理服务

C. 提供块存储服务　　　　D. 提供计算服务

【小李的反思】

满招损，谦受益。

源自《书·大禹谟》，意思是说，骄傲自满必然招致损失，谦虚谨慎就会获得好处。历史上的许多事实证明了这句名言的正确性。唐太宗能虚心听取臣下的意见，国家治理得很好，出现了“贞观之治”的太平盛世；但到了晚年，骄傲自满起来，在攻打高丽（朝鲜）时惨遭失败。唐玄宗刚登基做皇帝的日子里也算英明，在政治、经济方面取得成绩后，就骄傲起来，导致了“安史之乱”，唐朝从此衰落。

1948 年，24 岁的张富清离开陕西汉中洋县的家，光荣入伍，成为西北野战军第二纵队三五九旅七一八团二营六连的一名战士。先后在人民解放战争和“三大战役”之一的淮海战役子战场冒着枪林弹雨，不顾自身安危保家卫国，先后荣立一等功三次、二等功一次，被西北野战军记“特等功”，两次获得“战斗英雄”荣誉称号。1955 年，张富清退役转业，主动选择到湖北省最偏远的来凤县工作，为贫困山区奉献一生。

习近平总书记在党的二十大报告中指出，“中国共产党已走过百年奋斗历程。我们党立志于中华民族千秋伟业，致力于人类和平与发展崇高事业，责任无比重大，使命无上光荣。全党同志务必不忘初心、牢记使命，务必谦虚谨慎、艰苦奋斗，务必敢于斗争、善于斗争，坚定历史自信，增强历史主动，谱写新时代中国特色社会主义更加绚丽的华章。”作为青年学生，同学们更应该保持谦虚谨慎的学习和工作态度，善于汲取他人经验，丰富自我修养，必能在未来闯出一片天地。

项目评价

项目名称：OpenStack 云平台部署					
班级：		学号：	姓名：		
评价指标		评价等级及分值	学生自评	组内互评	教师评分
素质目标达成情况（30%）	精益求精的工匠精神（10%）	A（10 分）：云平台部署过程中能不断修正问题，精益求精，追求完美 B（7 分）：能够完成云平台部署的任务，但对过程中存在的问题，修改积极性不高 C（3 分）：对云平台部署过程中存在的问题放任不管			

续表

评价指标		评价等级及分值	学生自评	组内互评	教师评分
素质目标达成情况（30%）	乐于探索的学习精神（10%）	A（10 分）：自我学习热情高涨，积极和同学探讨学习问题 B（7 分）：自我学习热情较好，能够自主完成学习 C（3 分）：自我学习热情一般，学习积极性不高			
	云计算工程师职业素养（10%）	A（10 分）：云平台部署过程中，细致认真，积极主动，具备团队协作精神和很强的责任心，具有良好的语言表达能力、沟通能力、分析能力 B（7 分）：云平台部署过程中，不够细致，有一定的团队协作精神和很强的责任心，语言表达能力、沟通能力和分析能力稍有欠缺 C（3 分）：云平台部署过程中，由于粗心大意，出现了一些本不该出现的错误，缺少团队合作，很少与同组同学沟通交流			
知识目标达成情况（40%）	任务实施完成情况（20%）	A（20 分）：云平台部署完成，且满足使用要求 B（16 分）：云平台部署任务基本完成，但存在一些小的问题 C（10 分）：云平台部署任务部分完成，存在一些影响任务实施结果的问题			
	测验作业完成情况（10%）	A（10 分）：测验作业全部完成，知识理解透彻 B（7 分）：测验作业大部分完成，能基本完成知识的理解 C（3 分）：测验作业部分完成，对知识的理解较为片面			
	课上活动（10%）	A（10 分）：积极参与课上抢答、提问、主题讨论等 B（7 分）：能够参与课上抢答、提问、主题讨论等，但积极性不够高 C（3 分）：很少参与抢答、提问、主题讨论等			

续表

评价指标		评价等级及分值	学生自评	组内互评	教师评分
能力目标达成情况（30%）	任务实施完成质量（20%）	A（20 分）：任务实施完成质量优秀 B（16 分）：任务实施完成质量良好 C（10 分）：任务实施完成质量一般			
	超凡脱俗（10%）	A（10 分）：能够帮助同组同学解决云平台部署过程中存在的问题，并能整理问题解决手册 B（7 分）：能够帮助同组同学解决云平台部署过程中存在的问题 C（3 分）：能够规定时间内完成学习任务			

项目总结

本项目主要讲述基于 OpenStack 的云平台的搭建，包括 OpenStack 部署环境的准备和 Keystone、Glance、Nova、Neutron、Horizon、Cinder 基本组件的部署。在 OpenStack 部署环境准备任务中，主要讲述云平台部署的规划、网络的部署、数据库的安装、防火墙等安全配置、时间服务器 NTP 配置、消息队列配置、memcached 配置和 etcd 配置；在认证服务 Keystone 部署任务中，主要讲述 Keystone 组件的部署和域、项目、用户、角色等的创建与使用；在镜像服务 Glance 部署任务中，主要讲述 Glance 组件的部署和配置，并对配置结果通过上传镜像进行验证；在计算服务 Nova 部署任务中，主要讲述 controller 节点和 compute 节点部署 Nova 组件的过程，并通过查看服务列表和检查与其他服务的连接情况验证 Nova 组件部署的正确性，并给出部署过程中的一个排错案例；在网络服务 Neutron 组件部署任务中，主要讲述 controller 和 compute 节点部署并配置 Neutron 组件的过程，以及在各节点上配置 ML2、L3、DHCP 代理、元数据代理等；在 Web 界面 Horizon 部署任务中，主要讲述 Horizon 组件的部署和配置，已经通过 Web 端登录云平台并进行基本的操作；在块存储 Cinder 部署任务中，主要讲述 Cinder 组件的部署、配置和验证；通过部署 Keystone、Glance、Nova、Neutron、Horizon 和 Cinder 组件，完成基于 OpenStack 的云平台的基本功能的构建。

项目四

云主机运行

项目导入

小李已经搭建了 OpenStack 私有云平台，并学会了基本组件的应用，小李决定尝试从云平台创建云主机实例，在服务器的 CLI 界面可以创建，项目实施过程中也安装了 Horizon 图形界面化组件，同样，也可以在 Web 端创建云主机实例。本项目将随着小李的项目实施脚步，通过命令行创建云主机实例和 Web 端创建云主机实例。

项目目标

【素质目标】

- 培养学生精益求精的工匠精神
- 激发学生知学 - 好学 - 乐学的学习热情
- 培养学生面对挫折与失败，仍能积极乐观的态度

【知识目标】

- 掌握创建云主机实例的流程
- 掌握在命令行端和 Web 端创建云主机实例的方法

【能力目标】

- 能够在命令行端创建云主机实例
- 能够在 Web 端创建云主机实例
- 能够远程连接到云主机实例并正常使用

任务 1 命令行创建云主机

【任务描述】

小李已经搭建了 OpenStack 私有云平台，并学会了基本组件的应用，现在小李非常激动地想要开始创建第一个云主机实例。小李决定在 controller 节点上以命令行形式创建一个云主机实例，包含创建虚拟网络、创建密钥对、添加安全组规则、加载实例、挂载块存储等。

【任务实施】

1. 创建网络

1）创建 provider 网络及其子网

（1）执行 admin - openstack. sh 脚本，获取管理员的命令行执行权限。

```
[root@ controller ~]# source. /admin - openstack. sh
```

（2）创建 provider 网络，如图 4 - 1 所示。

```
[root@ controller ~]# openstack network create --share --external \
>   --provider - physical - network provider \
>   --provider - network - type flat provider
```

-- share 允许所有项目使用虚拟网络。

-- external 将虚拟网络定义为外部网络，如果要创建内部网络，则可以使用 -- internal。

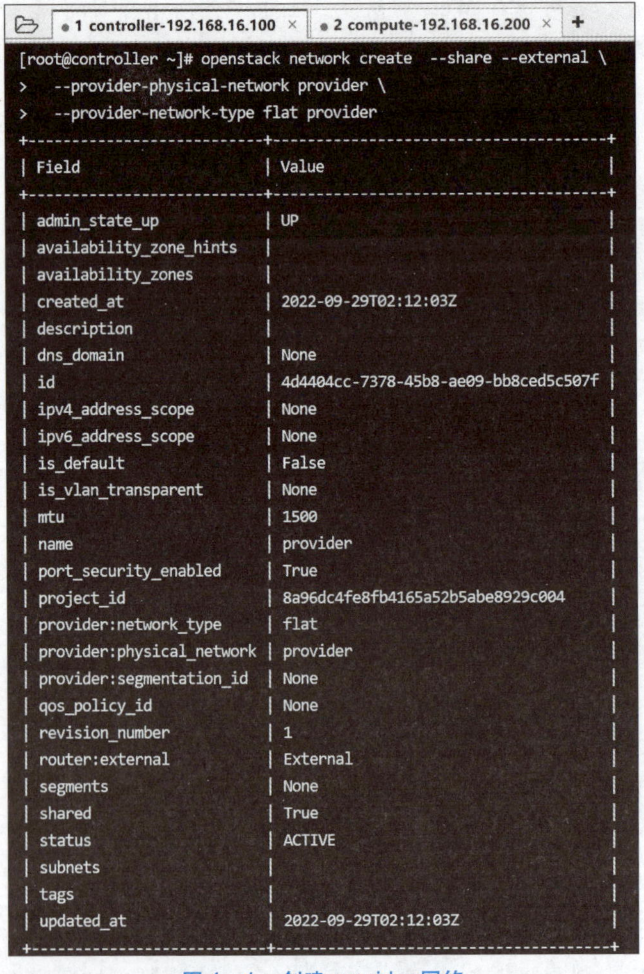

图 4 - 1　创建 provider 网络

（3）创建子网，如图 4-2 所示。

```
[root@ controller ~]# openstack subnet create --network provider \
>   --allocation-pool start=192.168.16.20,end=192.168.16.60 \
>   --dns-nameserver 8.8.4.4 --gateway 192.168.16.2 \
>   --subnet-range 192.168.16.0/24 provider
```

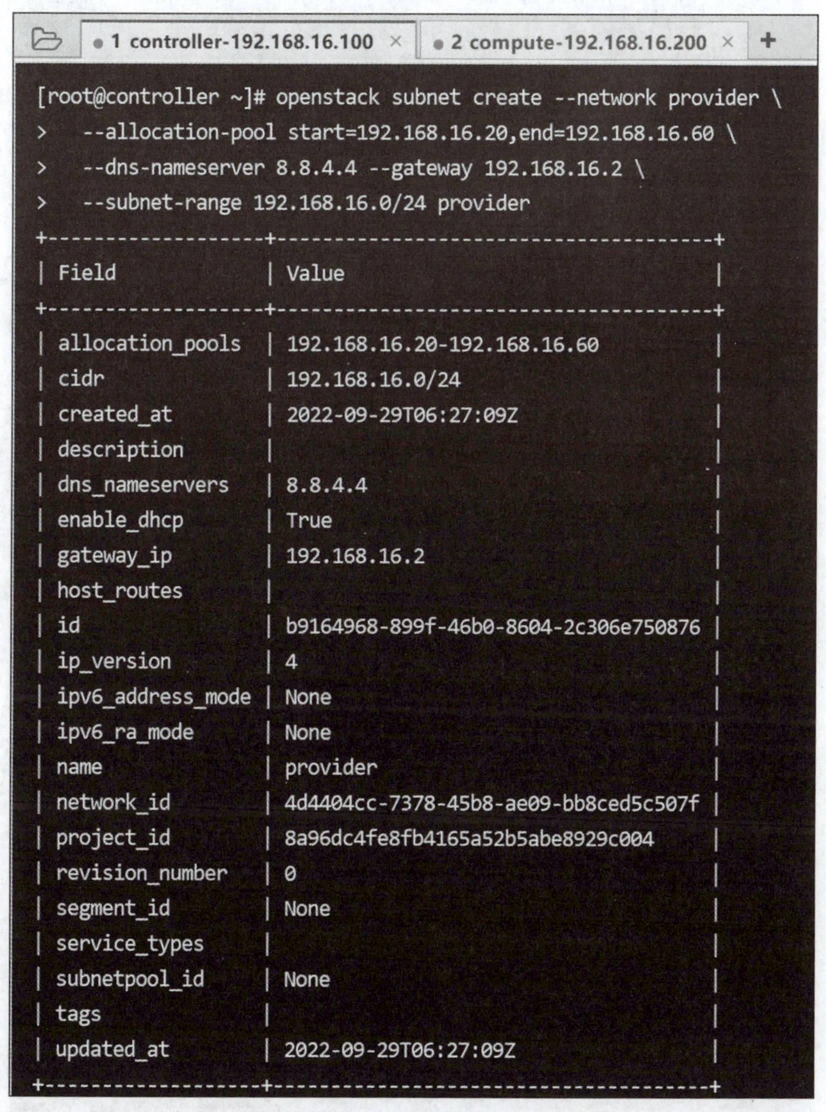

```
1 controller-192.168.16.100    2 compute-192.168.16.200

[root@controller ~]# openstack subnet create --network provider \
>   --allocation-pool start=192.168.16.20,end=192.168.16.60 \
>   --dns-nameserver 8.8.4.4 --gateway 192.168.16.2 \
>   --subnet-range 192.168.16.0/24 provider
+-------------------+--------------------------------------+
| Field             | Value                                |
+-------------------+--------------------------------------+
| allocation_pools  | 192.168.16.20-192.168.16.60          |
| cidr              | 192.168.16.0/24                      |
| created_at        | 2022-09-29T06:27:09Z                 |
| description       |                                      |
| dns_nameservers   | 8.8.4.4                              |
| enable_dhcp       | True                                 |
| gateway_ip        | 192.168.16.2                         |
| host_routes       |                                      |
| id                | b9164968-899f-46b0-8604-2c306e750876 |
| ip_version        | 4                                    |
| ipv6_address_mode | None                                 |
| ipv6_ra_mode      | None                                 |
| name              | provider                             |
| network_id        | 4d4404cc-7378-45b8-ae09-bb8ced5c507f |
| project_id        | 8a96dc4fe8fb4165a52b5abe8929c004     |
| revision_number   | 0                                    |
| segment_id        | None                                 |
| service_types     |                                      |
| subnetpool_id     | None                                 |
| tags              |                                      |
| updated_at        | 2022-09-29T06:27:09Z                 |
+-------------------+--------------------------------------+
```

图 4-2　创建子网

--allocation-pool：设置 IP 地址池。

--dns-nameserver：设置 DNS 服务器。

--gateway：设置网关。

--subnet-range：设置网段。

2）创建 self-service 网络

（1）执行 demo-openstack.sh 脚本，获取 myuser 用户的命令行执行权限。

```
[root@ controller ~]# source. /demo - openstack. sh
```

（2）创建 self - service 网络，如图 4 - 3 所示。

```
[root@ controller ~]# openstack network create selfservice
```

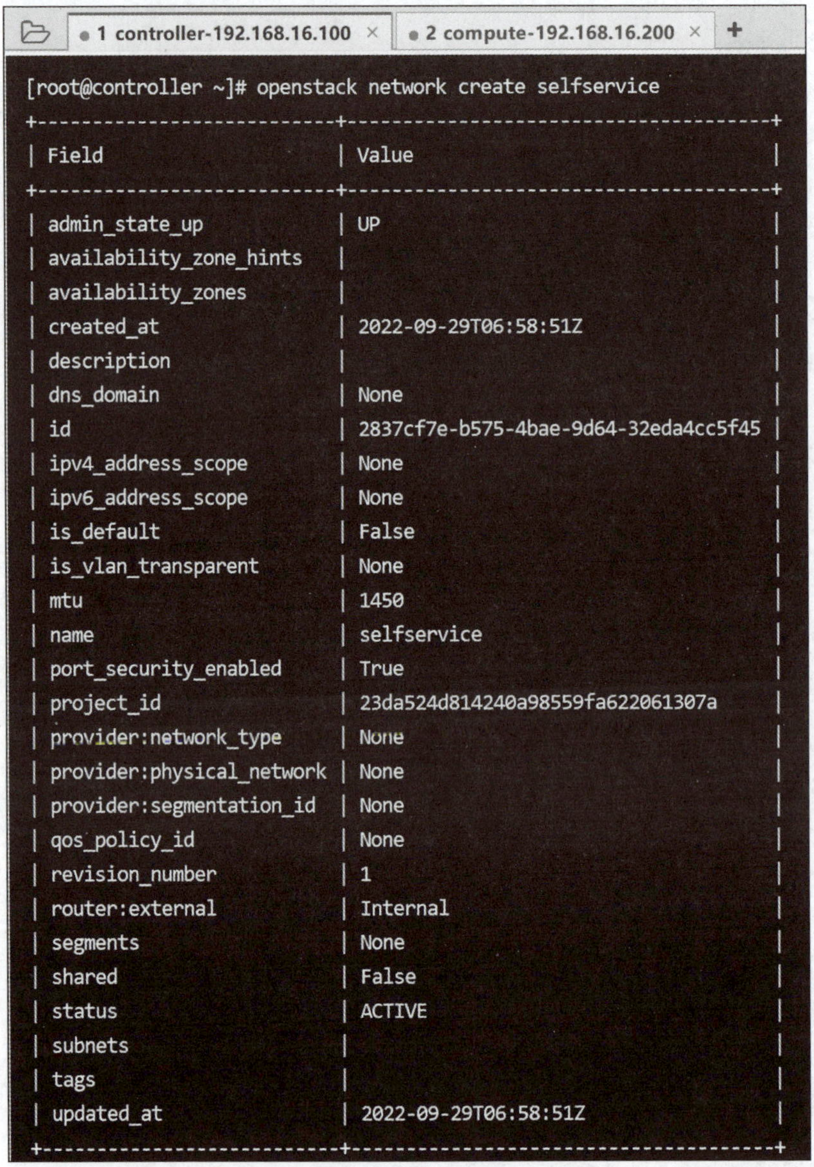

```
1 controller-192.168.16.100    2 compute-192.168.16.200

[root@controller ~]# openstack network create selfservice
+---------------------------+--------------------------------------+
| Field                     | Value                                |
+---------------------------+--------------------------------------+
| admin_state_up            | UP                                   |
| availability_zone_hints   |                                      |
| availability_zones        |                                      |
| created_at                | 2022-09-29T06:58:51Z                 |
| description               |                                      |
| dns_domain                | None                                 |
| id                        | 2837cf7e-b575-4bae-9d64-32eda4cc5f45 |
| ipv4_address_scope        | None                                 |
| ipv6_address_scope        | None                                 |
| is_default                | False                                |
| is_vlan_transparent       | None                                 |
| mtu                       | 1450                                 |
| name                      | selfservice                          |
| port_security_enabled     | True                                 |
| project_id                | 23da524d814240a98559fa622061307a     |
| provider:network_type     | None                                 |
| provider:physical_network | None                                 |
| provider:segmentation_id  | None                                 |
| qos_policy_id             | None                                 |
| revision_number           | 1                                    |
| router:external           | Internal                             |
| segments                  | None                                 |
| shared                    | False                                |
| status                    | ACTIVE                               |
| subnets                   |                                      |
| tags                      |                                      |
| updated_at                | 2022-09-29T06:58:51Z                 |
+---------------------------+--------------------------------------+
```

图 4 - 3　创建 self - service 网络

（3）创建子网，如图 4 - 4 所示。

```
[root@ controller ~]# openstack subnet create --network selfservice \
>   --dns - nameserver 8.8.4.4 --gateway 172.16.1.1 \
>   --subnet - range 172.16.1.0/24 selfservice
```

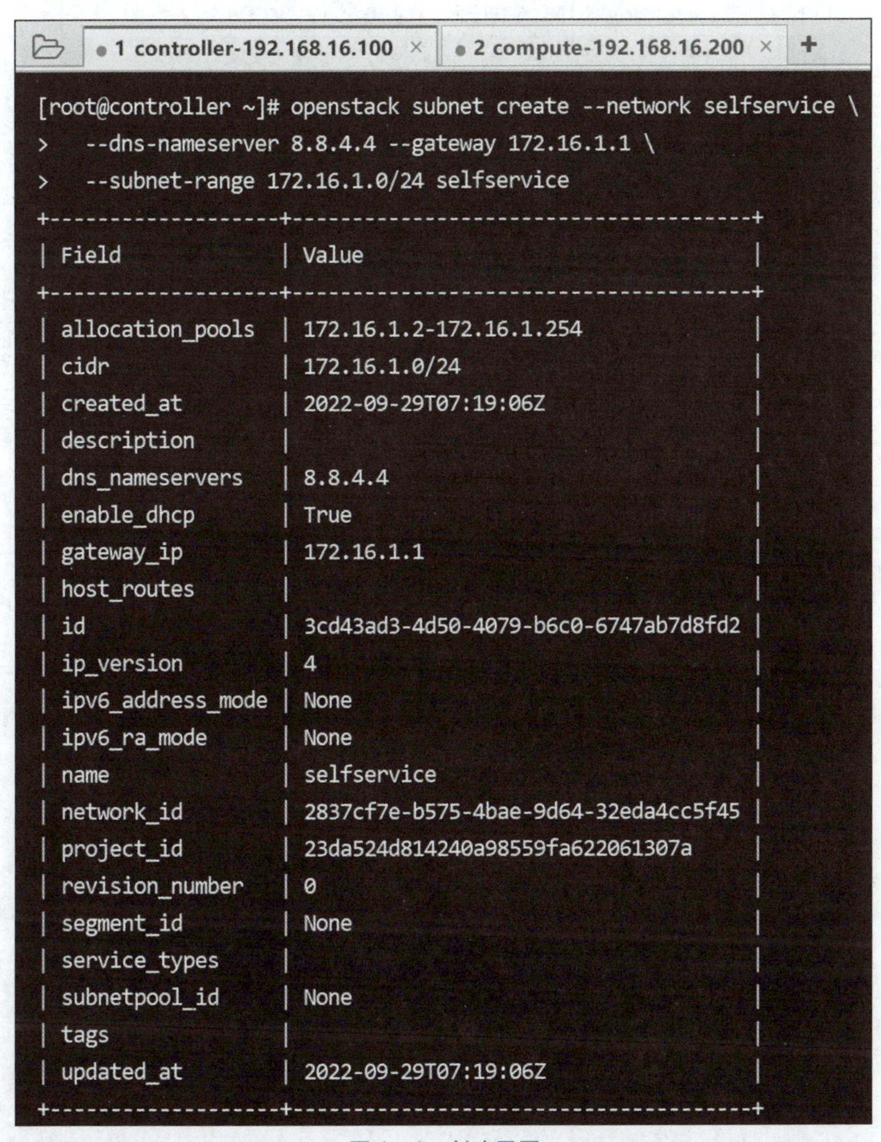

```
1 controller-192.168.16.100    2 compute-192.168.16.200

[root@controller ~]# openstack subnet create --network selfservice \
>   --dns-nameserver 8.8.4.4 --gateway 172.16.1.1 \
>   --subnet-range 172.16.1.0/24 selfservice
+-------------------+--------------------------------------+
| Field             | Value                                |
+-------------------+--------------------------------------+
| allocation_pools  | 172.16.1.2-172.16.1.254              |
| cidr              | 172.16.1.0/24                        |
| created_at        | 2022-09-29T07:19:06Z                 |
| description       |                                      |
| dns_nameservers   | 8.8.4.4                              |
| enable_dhcp       | True                                 |
| gateway_ip        | 172.16.1.1                           |
| host_routes       |                                      |
| id                | 3cd43ad3-4d50-4079-b6c0-6747ab7d8fd2 |
| ip_version        | 4                                    |
| ipv6_address_mode | None                                 |
| ipv6_ra_mode      | None                                 |
| name              | selfservice                          |
| network_id        | 2837cf7e-b575-4bae-9d64-32eda4cc5f45 |
| project_id        | 23da524d814240a98559fa622061307a     |
| revision_number   | 0                                    |
| segment_id        | None                                 |
| service_types     |                                      |
| subnetpool_id     | None                                 |
| tags              |                                      |
| updated_at        | 2022-09-29T07:19:06Z                 |
+-------------------+--------------------------------------+
```

图 4-4　创建子网

（4）创建路由器，如图 4-5 所示。

```
[root@controller ~]# openstack router create router
```

（5）将 self-service 网络的子网添加到路由器端口。

```
[root@controller ~]# openstack router add subnet router selfservice
```

（6）设置路由器的 provider 网络的网关。

```
[root@controller ~]# openstack router set router --external-gateway provider
```

3）验证网络

（1）执行 admin-openstack. sh 脚本，获取管理员的命令行执行权限。

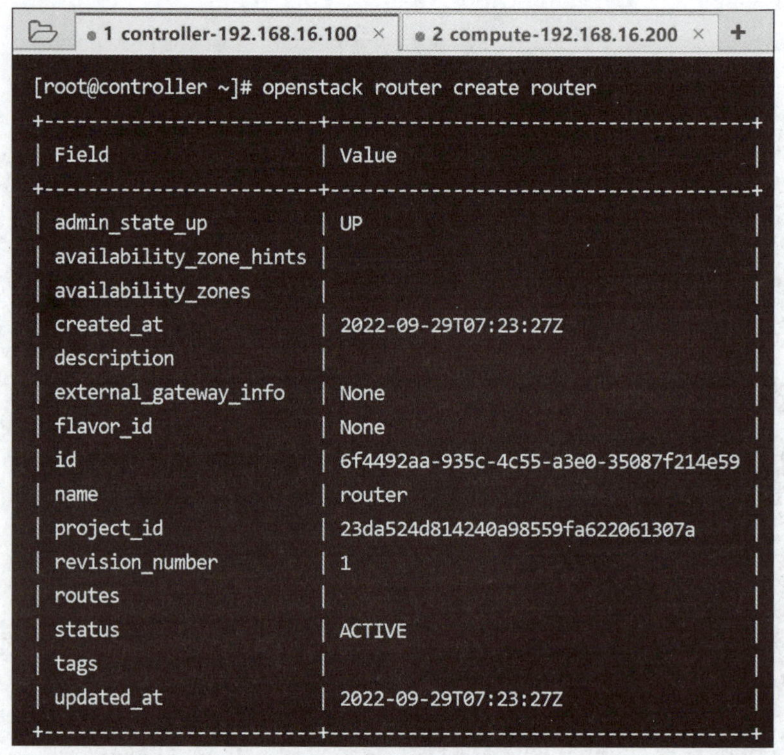

图 4-5　创建路由器

```
[root@ controller ~]# source. /admin-openstack. sh
```

（2）查看网络命名空间，查询结果应包含 1 个 qrouter 命名空间和 2 个 qdhcp 命名空间。

```
[root@ controller ~]# ip netns
qrouter-6f4492aa-935c-4c55-a3e0-35087f214e59(id:2)
qdhcp-2837cf7e-b575-4bae-9d64-32eda4cc5f45(id:1)
qdhcp-4d4404cc-7378-45b8-ae09-bb8ced5c507f(id:0)
```

（3）列出路由器上的端口，以确定提供商网络上的网关 IP 地址，如图 4-6 所示。

```
[root@ controller ~]# openstack port list --router router
```

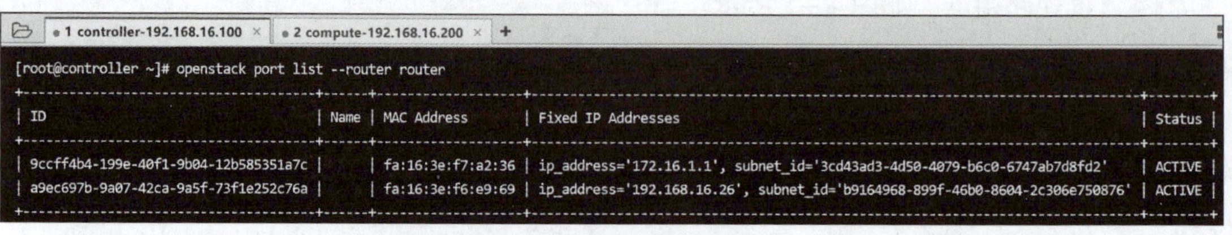

图 4-6　列出路由器上的端口

（4）从 controller 节点或物理网络上的任何主机 ping 此 IP 地址。

①controller 节点 ping 192. 168. 16. 26。

```
[root@ controller ~]# ping -c 4 192. 168. 16. 26
```

```
PING 192.168.16.26(192.168.16.26)56(84)bytes of data.
64bytes from 192.168.16.26:icmp_seq=1 ttl=64 time=0.273ms
64bytes from 192.168.16.26:icmp_seq=2 ttl=64 time=0.064ms
64bytes from 192.168.16.26:icmp_seq=3 ttl=64 time=0.070ms
64bytes from 192.168.16.26:icmp_seq=4 ttl=64 time=0.030ms
---192.168.16.26 ping statistics---
4 packets transmitted,4 received,0% packet loss,time 3003ms
rtt min/avg/max/mdev=0.030/0.109/0.273/0.096 ms
```

②物理机 ping 192.168.16.26。

```
C:\Users\ly>ping 192.168.16.26
```

正在 ping 192.168.16.26 具有 32 字节的数据：
来自 192.168.16.26 的回复：字节=32 时间<1 ms TTL=64
来自 192.168.16.26 的回复：字节=32 时间<1 ms TTL=64
来自 192.168.16.26 的回复：字节=32 时间=1 ms TTL=64
来自 192.168.16.26 的回复：字节=32 时间<1 ms TTL=64
192.168.16.26 的 ping 统计信息：
数据包：已发送=4，已接收=4，丢失=0（0% 丢失）
往返行程的估计时间（以毫秒为单位）：
最短=0 ms，最长=1 ms，平均=0 ms

2. 创建云主机实例

（1）创建一台 1 核 CPU、64 MB 内存、1 GB 硬盘的实例类型，如图 4-7 所示。

```
[root@ controller ~]# openstack flavor create --id 0 --vcpus 1 --ram 64 --disk 1 m1.nano
```

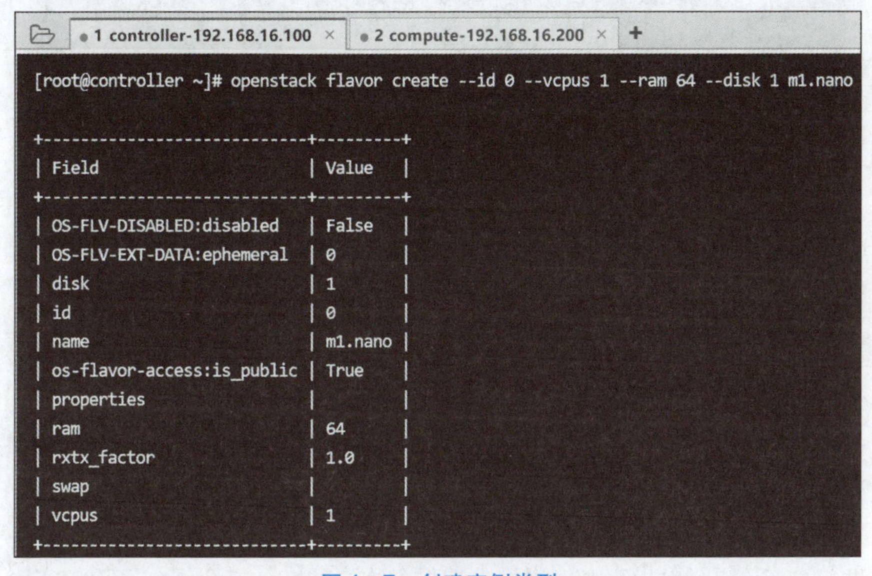
```
[root@controller ~]# openstack flavor create --id 0 --vcpus 1 --ram 64 --disk 1 m1.nano

+----------------------------+---------+
| Field                      | Value   |
+----------------------------+---------+
| OS-FLV-DISABLED:disabled   | False   |
| OS-FLV-EXT-DATA:ephemeral  | 0       |
| disk                       | 1       |
| id                         | 0       |
| name                       | m1.nano |
| os-flavor-access:is_public | True    |
| properties                 |         |
| ram                        | 64      |
| rxtx_factor                | 1.0     |
| swap                       |         |
| vcpus                      | 1       |
+----------------------------+---------+
```

图 4-7　创建实例类型

(2) 创建密钥对。

①执行 demo-openstack.sh 脚本，获取 myuser 用户的命令行执行权限。

```
[root@ controller ~]# source./demo-openstack.sh
```

②创建密钥对，此处直接按 Enter 键默认密钥存放位置。

```
[root@ controller ~]# ssh-keygen -q -N ""
Enter file in which to save the key(/root/.ssh/id_rsa):
```

③创建一个公钥，如图 4-8 所示。

```
[root@ controller ~]# openstack keypair create --public-key ~/.ssh/id_rsa.pub mykey
```

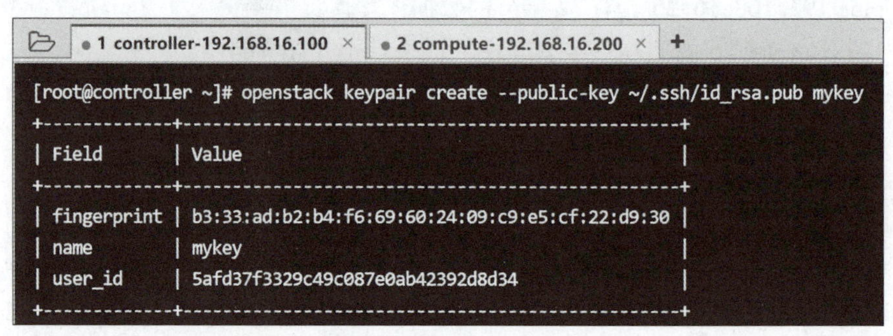

图 4-8 创建一个公钥

④查询密钥对，如图 4-9 所示。

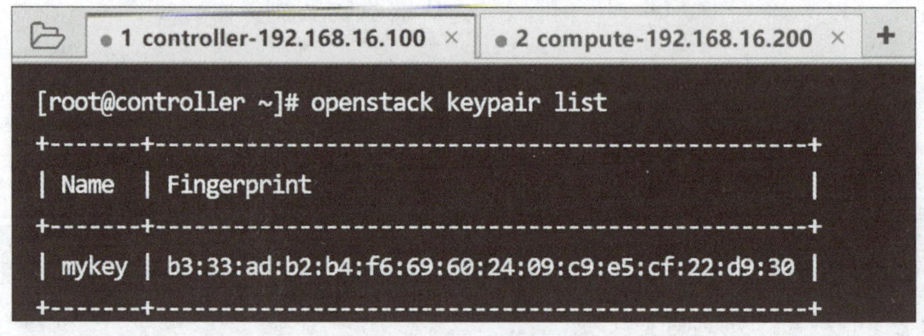

图 4-9 查询密钥对

(3) 添加安全组规则。

①向 default 安全组添加允许 ICMP 协议通行的规则，如图 4-10 所示。

```
[root@ controller ~]# openstack security group rule create --proto icmp default
```

②向 default 安全组添加允许 SSH 协议通行的规则，如图 4-11 所示。

```
[root@ controller ~]# openstack security group rule create --proto tcp --dst-port 22 default
```

(4) 启动实例。

要启动实例，必须至少指定实例类型、映像名称、网络、安全组、密钥和实例名称。

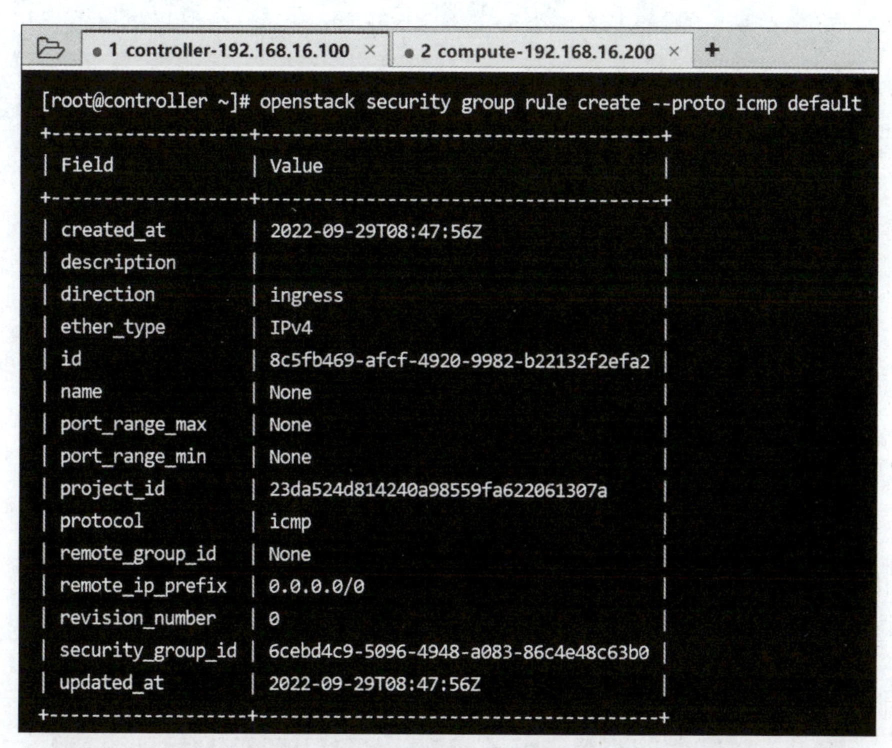

图 4-10　default 安全组添加允许 ICMP 协议通行的规则

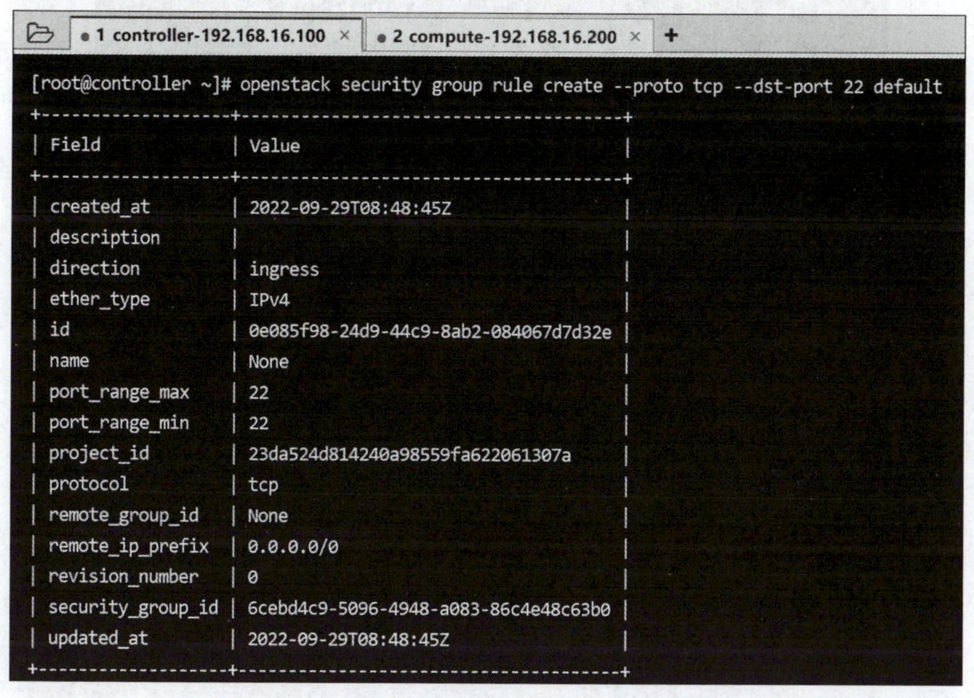

图 4-11　default 安全组添加允许 SSH 协议通行的规则

①执行 demo-openstack. sh 脚本，获取 myuser 用户的命令行执行权限。

```
[root@ controller ~]# source. /demo - openstack. sh
```

②列出可用实例类型，如图 4 – 12 所示。

```
[root@ controller ~]# openstack flavor list
```

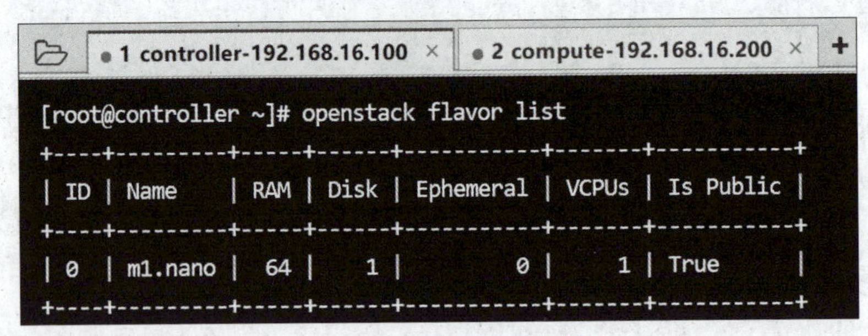

```
[root@controller ~]# openstack flavor list
+----+---------+-----+------+-----------+-------+-----------+
| ID | Name    | RAM | Disk | Ephemeral | VCPUs | Is Public |
+----+---------+-----+------+-----------+-------+-----------+
| 0  | m1.nano |  64 |    1 |         0 |     1 | True      |
+----+---------+-----+------+-----------+-------+-----------+
```

图 4 – 12　列出可用实例类型

③列出可用镜像列表，如图 4 – 13 所示。

```
[root@ controller ~]# openstack image list
```

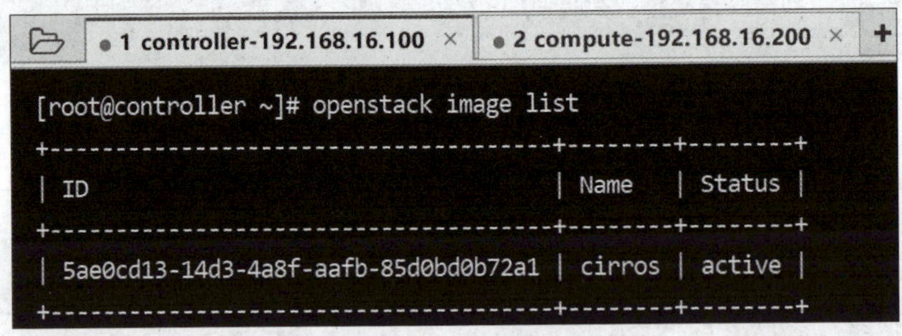

```
[root@controller ~]# openstack image list
+--------------------------------------+--------+--------+
| ID                                   | Name   | Status |
+--------------------------------------+--------+--------+
| 5ae0cd13-14d3-4a8f-aafb-85d0bd0b72a1 | cirros | active |
+--------------------------------------+--------+--------+
```

图 4 – 13　列出可用镜像列表

④列出可用网络列表，如图 4 – 14 所示。

```
[root@ controller ~]# openstack network list
```

```
[root@controller ~]# openstack network list
+--------------------------------------+-------------+--------------------------------------+
| ID                                   | Name        | Subnets                              |
+--------------------------------------+-------------+--------------------------------------+
| 2837cf7e-b575-4bae-9d64-32eda4cc5f45 | selfservice | 3cd43ad3-4d50-4079-b6c0-6747ab7d8fd2 |
| 4d4404cc-7378-45b8-ae09-bb8ced5c507f | provider    | b9164968-899f-46b0-8604-2c306e750876 |
+--------------------------------------+-------------+--------------------------------------+
```

图 4 – 14　列出可用网络列表

⑤列出可用安全组，如图 4 – 15 所示。

```
[root@ controller ~]# openstack security group list
```

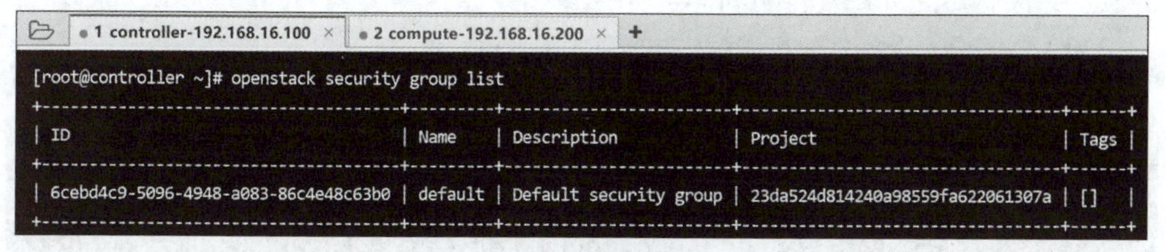

```
[root@controller ~]# openstack security group list
+--------------------------------------+---------+------------------------+----------------------------------+------+
| ID                                   | Name    | Description            | Project                          | Tags |
+--------------------------------------+---------+------------------------+----------------------------------+------+
| 6cebd4c9-5096-4948-a083-86c4e48c63b0 | default | Default security group | 23da524d814240a98559fa622061307a | []   |
+--------------------------------------+---------+------------------------+----------------------------------+------+
```

图 4－15　列出可用安全组

⑥启动实例，如图 4－16 所示。

```
[root@ controller ~ ]# openstack server create --flavor m1. nano --image cirros \
> --nic net-id=2837cf7e-b575-4bae-9d64-32eda4cc5f45 --security-group default \
> --key-name mykey selfservice-instance
```

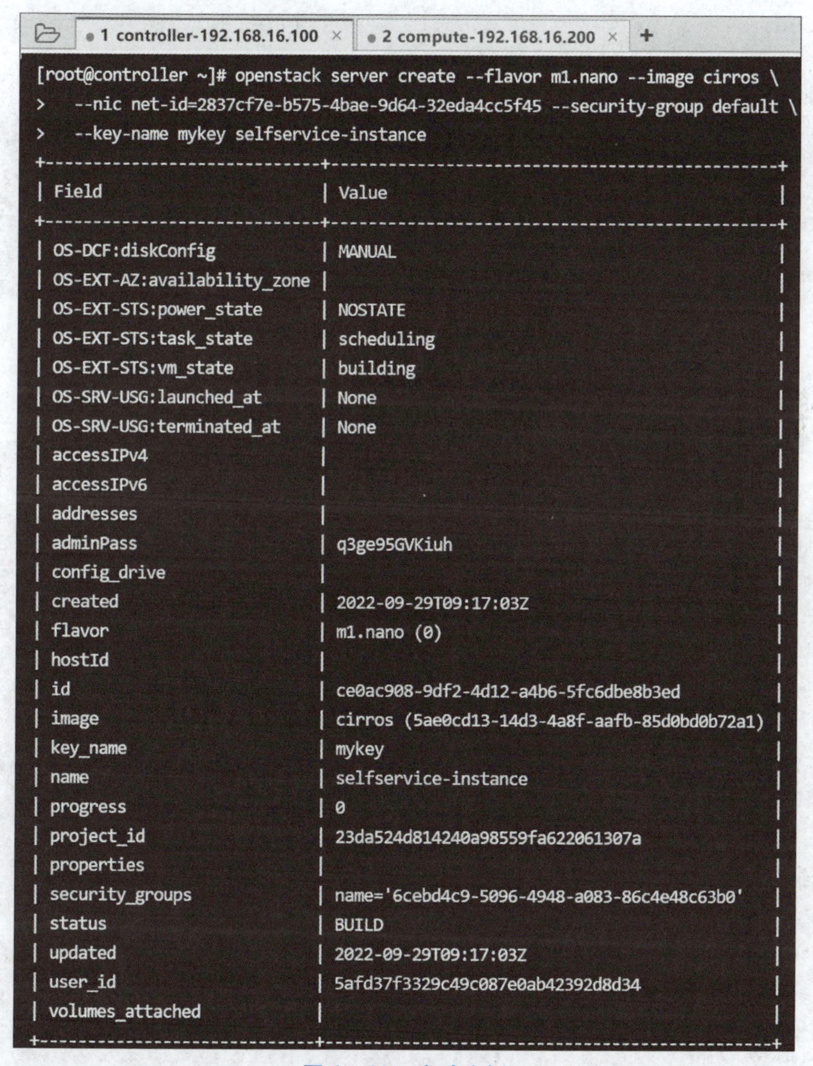

```
[root@controller ~]# openstack server create --flavor m1.nano --image cirros \
>   --nic net-id=2837cf7e-b575-4bae-9d64-32eda4cc5f45 --security-group default \
>   --key-name mykey selfservice-instance
+-----------------------------+-----------------------------------------------+
| Field                       | Value                                         |
+-----------------------------+-----------------------------------------------+
| OS-DCF:diskConfig           | MANUAL                                        |
| OS-EXT-AZ:availability_zone |                                               |
| OS-EXT-STS:power_state      | NOSTATE                                       |
| OS-EXT-STS:task_state       | scheduling                                    |
| OS-EXT-STS:vm_state         | building                                      |
| OS-SRV-USG:launched_at      | None                                          |
| OS-SRV-USG:terminated_at    | None                                          |
| accessIPv4                  |                                               |
| accessIPv6                  |                                               |
| addresses                   |                                               |
| adminPass                   | q3ge95GVKiuh                                  |
| config_drive                |                                               |
| created                     | 2022-09-29T09:17:03Z                          |
| flavor                      | m1.nano (0)                                   |
| hostId                      |                                               |
| id                          | ce0ac908-9df2-4d12-a4b6-5fc6dbe8b3ed          |
| image                       | cirros (5ae0cd13-14d3-4a8f-aafb-85d0bd0b72a1) |
| key_name                    | mykey                                         |
| name                        | selfservice-instance                          |
| progress                    | 0                                             |
| project_id                  | 23da524d814240a98559fa622061307a              |
| properties                  |                                               |
| security_groups             | name='6cebd4c9-5096-4948-a083-86c4e48c63b0'   |
| status                      | BUILD                                         |
| updated                     | 2022-09-29T09:17:03Z                          |
| user_id                     | 5afd37f3329c49c087e0ab42392d8d34              |
| volumes_attached            |                                               |
+-----------------------------+-----------------------------------------------+
```

图 4－16　启动实例

⑦检查实例状态，如图 4－17 所示。

```
[root@ controller ~]# openstack server list
```

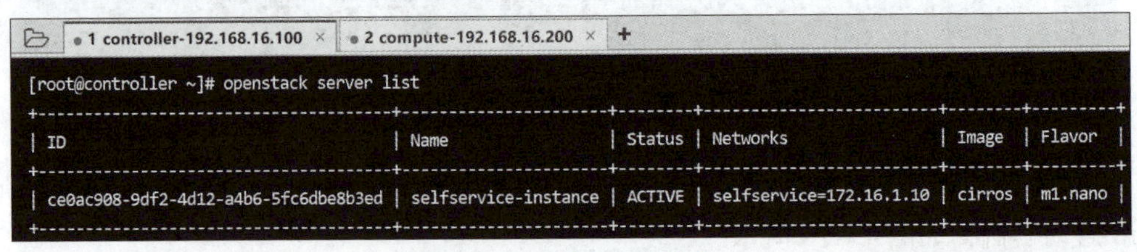

图 4－17　检查实例状态

3. 应用实例

1）Web 端访问实例

（1）查看从 Web 端访问实例的 URL，如图 4－18 所示。

```
[root@ controller ~]# openstack console url show selfservice-instance
```

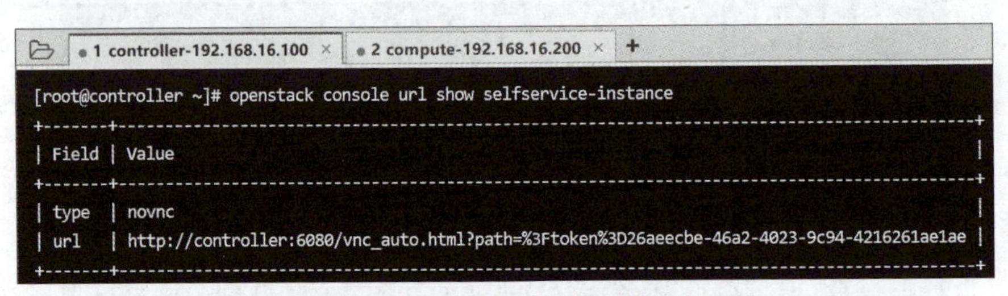

图 4－18　查看从 Web 端访问实例的 URL

（2）通过该 URL 在网页端进入该实例，如图 4－19 所示。

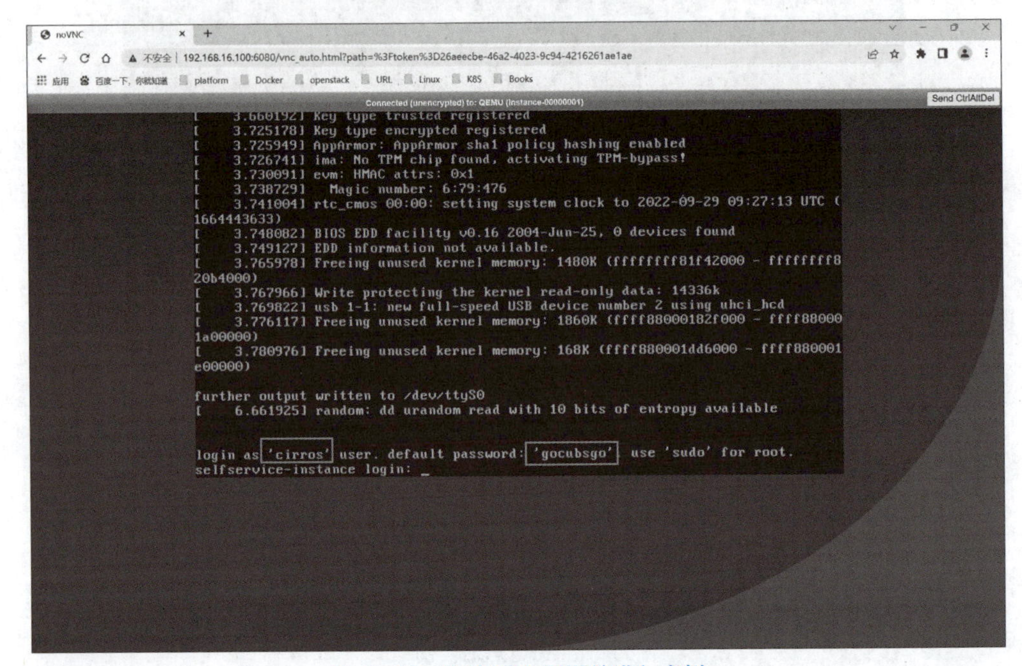

图 4－19　通过 URL 网页端进入实例

如果网页端无法解析 controller，可以用 IP 地址替换，或者将 compute 节点/etc/nova/nova. conf 文件中的［vnc］下的 novncproxy_base_url = http://controller:6080/vnc_auto. html 中的 controller 换成 controller 节点管理网络的 IP 地址，重启 openstack - nova - compute 服务。

（3）以 cirrors 镜像提供的用户名和密码登录虚拟机，测试是否可以访问 Self - network 网络的网关，如图 4 - 20 所示。

```
Connected (unencrypted) to: QEMU (instance-00000001)
login as 'cirros' user. default password: 'gocubsgo'. use 'sudo' for root.
selfservice-instance login: cirrors
Password: [  228.874169] random: nonblocking pool is initialized

Login incorrect
selfservice-instance login: cirrors
Password:
Login incorrect
selfservice-instance login: cirros
Password:
$
$
$
$ ping 172.16.1.1
PING 172.16.1.1 (172.16.1.1): 56 data bytes
64 bytes from 172.16.1.1: seq=0 ttl=64 time=11.723 ms
64 bytes from 172.16.1.1: seq=1 ttl=64 time=2.323 ms
64 bytes from 172.16.1.1: seq=2 ttl=64 time=0.874 ms
64 bytes from 172.16.1.1: seq=3 ttl=64 time=0.872 ms
^C
--- 172.16.1.1 ping statistics ---
4 packets transmitted, 4 packets received, 0% packet loss
round-trip min/avg/max = 0.872/3.948/11.723 ms
$
```

图 4 - 20　测试是否可以访问网关

（4）测试是否可以访问外网，如图 4 - 21 所示。

```
Connected (unencrypted) to: QEMU (instance-00000001)
$ ping www.baidu.com
PING www.baidu.com (180.101.49.11): 56 data bytes
64 bytes from 180.101.49.11: seq=0 ttl=127 time=27.523 ms
64 bytes from 180.101.49.11: seq=1 ttl=127 time=27.803 ms
64 bytes from 180.101.49.11: seq=2 ttl=127 time=28.042 ms
64 bytes from 180.101.49.11: seq=3 ttl=127 time=27.253 ms
64 bytes from 180.101.49.11: seq=4 ttl=127 time=27.296 ms
64 bytes from 180.101.49.11: seq=5 ttl=127 time=27.873 ms
64 bytes from 180.101.49.11: seq=6 ttl=127 time=28.147 ms
64 bytes from 180.101.49.11: seq=7 ttl=127 time=27.415 ms
64 bytes from 180.101.49.11: seq=8 ttl=127 time=27.629 ms
64 bytes from 180.101.49.11: seq=9 ttl=127 time=27.557 ms
64 bytes from 180.101.49.11: seq=10 ttl=127 time=28.250 ms
64 bytes from 180.101.49.11: seq=11 ttl=127 time=27.752 ms
64 bytes from 180.101.49.11: seq=12 ttl=127 time=28.307 ms
64 bytes from 180.101.49.11: seq=13 ttl=127 time=28.420 ms
64 bytes from 180.101.49.11: seq=14 ttl=127 time=28.219 ms
64 bytes from 180.101.49.11: seq=15 ttl=127 time=29.733 ms
64 bytes from 180.101.49.11: seq=16 ttl=127 time=27.878 ms
64 bytes from 180.101.49.11: seq=17 ttl=127 time=27.359 ms
^C
--- www.baidu.com ping statistics ---
18 packets transmitted, 18 packets received, 0% packet loss
round-trip min/avg/max = 27.253/27.914/29.733 ms
$ _
```

图 4 - 21　测试是否可以访问外网

2）远程连接实例

（1）创建浮动 IP 地址，如图 4 – 22 所示。

```
[root@ controller ~]# openstack floating ip create provider
```

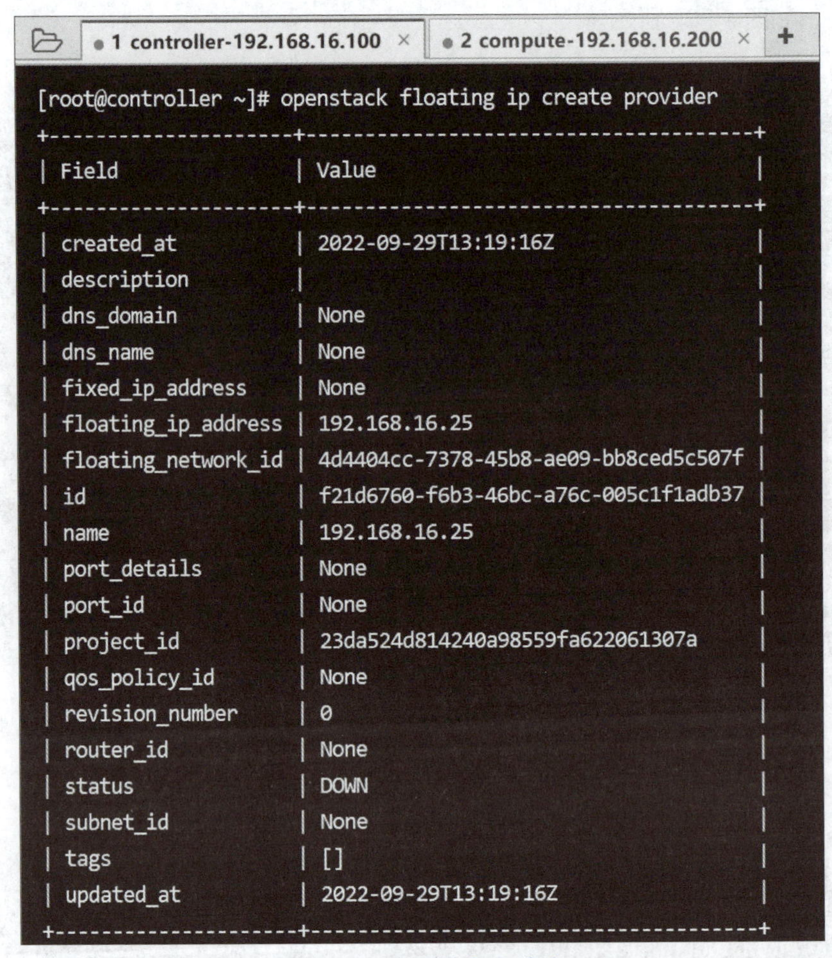

```
[root@controller ~]# openstack floating ip create provider
+---------------------+--------------------------------------+
| Field               | Value                                |
+---------------------+--------------------------------------+
| created_at          | 2022-09-29T13:19:16Z                 |
| description         |                                      |
| dns_domain          | None                                 |
| dns_name            | None                                 |
| fixed_ip_address    | None                                 |
| floating_ip_address | 192.168.16.25                        |
| floating_network_id | 4d4404cc-7378-45b8-ae09-bb8ced5c507f |
| id                  | f21d6760-f6b3-46bc-a76c-005c1f1adb37 |
| name                | 192.168.16.25                        |
| port_details        | None                                 |
| port_id             | None                                 |
| project_id          | 23da524d814240a98559fa622061307a     |
| qos_policy_id       | None                                 |
| revision_number     | 0                                    |
| router_id           | None                                 |
| status              | DOWN                                 |
| subnet_id           | None                                 |
| tags                | []                                   |
| updated_at          | 2022-09-29T13:19:16Z                 |
+---------------------+--------------------------------------+
```

图 4 – 22　创建浮动 IP 地址

（2）将浮动 IP 地址绑定到 selfservice – instance 实例，如图 4 – 23 所示。

```
[root@ controller ~]# openstack server add floating ip selfservice-instance 192.168.16.25
```

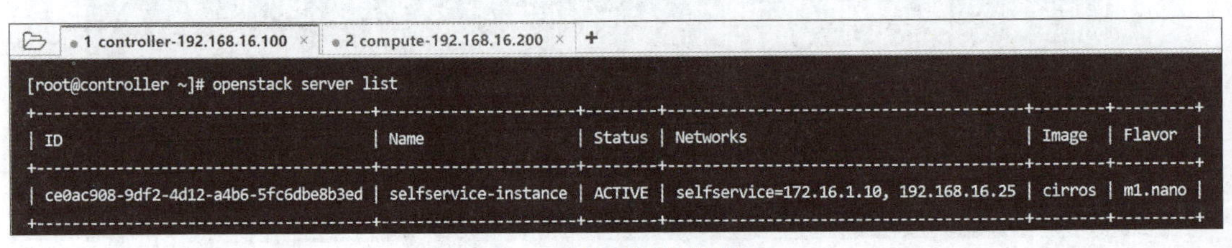

```
[root@controller ~]# openstack server list
+--------------------------------------+----------------------+--------+----------------------------------------+--------+---------+
| ID                                   | Name                 | Status | Networks                               | Image  | Flavor  |
+--------------------------------------+----------------------+--------+----------------------------------------+--------+---------+
| ce0ac908-9df2-4d12-a4b6-5fc6dbe8b3ed | selfservice-instance | ACTIVE | selfservice=172.16.1.10, 192.168.16.25 | cirros | m1.nano |
+--------------------------------------+----------------------+--------+----------------------------------------+--------+---------+
```

图 4 – 23　绑定浮动 IP 地址

(3) 测试 controller 及其他物理主机上的节点和浮动 IP 的连通性。

```
[root@ controller ~]# ping -c 4 192.168.16.25
PING 192.168.16.25(192.168.16.25)56(84)bytes of data.
64 bytes from 192.168.16.25:icmp_seq=1 ttl=63 time=8.05 ms
64 bytes from 192.168.16.25:icmp_seq=2 ttl=63 time=0.699 ms
64 bytes from 192.168.16.25:icmp_seq=3 ttl=63 time=0.919 ms
64 bytes from 192.168.16.25:icmp_seq=4 ttl=63 time=0.988 ms

---192.168.16.25 ping statistics ---
4 packets transmitted,4 received,0% packet loss,time 3004ms
rtt min/avg/max/mdev=0.699/2.666/8.059/3.115 ms
```

(4) 在 controller 节点通过 ssh 连接到实例，至此，已完成通过命令行创建实例并可以正常联网使用和远程登录。

```
[root@ controller ~]# ssh cirros@ 192.168.16.25
The authenticity of host '192.168.16.25(192.168.16.25)' can't be established.
ECDSA key fingerprint is SHA256:EbZqxLu9q4xj7BiZx0Ozq2f6wFkLgWMRVv54dWbXBCQ.
ECDSA key fingerprint is MD5:fd:8b:95:2c:87:60:16:a7:56:5d:3f:f1:74:ae:cd:ae.
Are you sure you want to continue connecting(yes/no)? yes
Warning:Permanently added '192.168.16.25'(ECDSA)to the list of known hosts.
$
```

【任务工单】

工单号：4－1

<table>
<tr><td colspan="2">项目名称：云主机运行</td><td colspan="2">任务名称：命令行创建云主机</td></tr>
<tr><td colspan="2">班级：</td><td>学号：</td><td>姓名：</td></tr>
<tr><td>任务安排</td><td colspan="3">□创建云主机使用的网络，包括 provider 网络和 self－service 网络
□创建云主机实例，包括实例类型、密钥对、安全组规则和实例等的创建
□通过 Web 端和远程访问两种方式连接到云主机实例并测试和外网的连通情况</td></tr>
<tr><td>成果交付</td><td colspan="3">同组同学互相远程连接云主机，并测试是否可以连通外网</td></tr>
<tr><td>任务实施总结</td><td colspan="3">任务自评（0～10 分）：
任务收获：

改进点：

</td></tr>
<tr><td>成果验收</td><td colspan="3">□完全满足任务要求
□基本满足任务要求
要求全部完成，但云主机创建的操作不熟练，所用时间较长：

□不能满足需求
要求不能独立完成，部分任务需要在老师指导下才能完成：

</td></tr>
</table>

【知识巩固】

创建浮动 IP 地址的作用是什么？

【小李的反思】

知者行之始，行者知之成。

源自王阳明《传习录》，意思是说，实践是获取知识的开始，获取知识是实践的成果。实践是获取认知的必需途径，向社会学习、向实践学习，知行合一，只有实践才能出真知。

战国时期，赵国大将赵奢的儿子赵括，从小熟读兵书，张口爱谈军事，别人往往说不过他，因此他很骄傲，自以为天下无敌。公元前 259 年，秦军又来犯，秦国施行了反间计，派人到赵国散布“秦军最害怕赵奢将军的儿子赵括”的话。赵王上当受骗，派赵括替代了廉颇。赵括自认为很会打仗，死搬兵书上的条文，到长平后完全改变了廉颇的作战方案，结果四十多万赵军尽被歼灭，他自己也被秦军箭射身亡。这个事例告诉我们，知行合一才是做事最基本的原则。

党的二十大报告明确提出了“六个坚持”，坚持守正创新正是其中的一个重要方面。我们作为青年学子，在前进道路上，要坚持守正创新、不断超越自己，博采众长、不断完善自己，就一定能不断开辟新境界，谱写属于自己的华章。

任务2　Web端创建云主机

【任务描述】

小李已经可以通过命令行从云平台创建云主机实例了，但对于非云计算，专业人员并不会使用命令行操作，通过图形界面操作将会更加便捷。在部署阶段已经部署了Horizon图形化界面，接下来就跟随小李开始学习如何通过图形界面化方式创建云主机实例。

【任务实施】

（1）在Web端输入dashboard的URL:http://192.168.16.100/dashboard（IP地址为controller节点的管理IP），以myuser用户登录到dashboard界面，如图4-24所示。

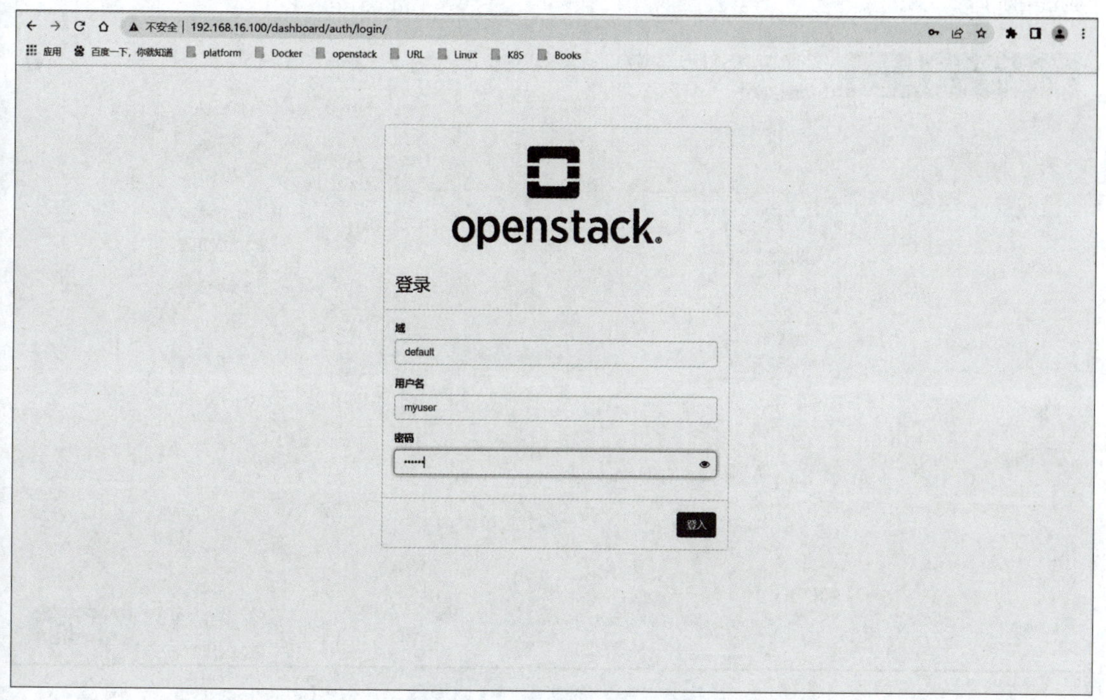

图4-24　dashboard登录界面

（2）在dashboard的实例页面单击“创建实例”按钮，进入实例创建引导界面，如图4-25所示。

（3）在“详情”选项下输入实例名称、实例描述、可用域及实例的数量后，单击“下一步”按钮，如图4-26所示。

（4）在“源”选项下选择镜像源、设置卷大小并添加可用镜像，如图4-27所示。

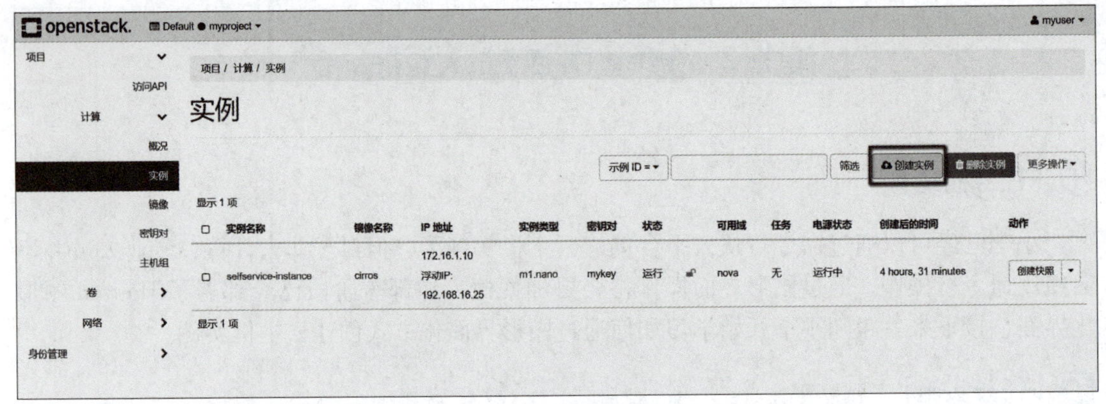

图 4-25　实例创建引导界面

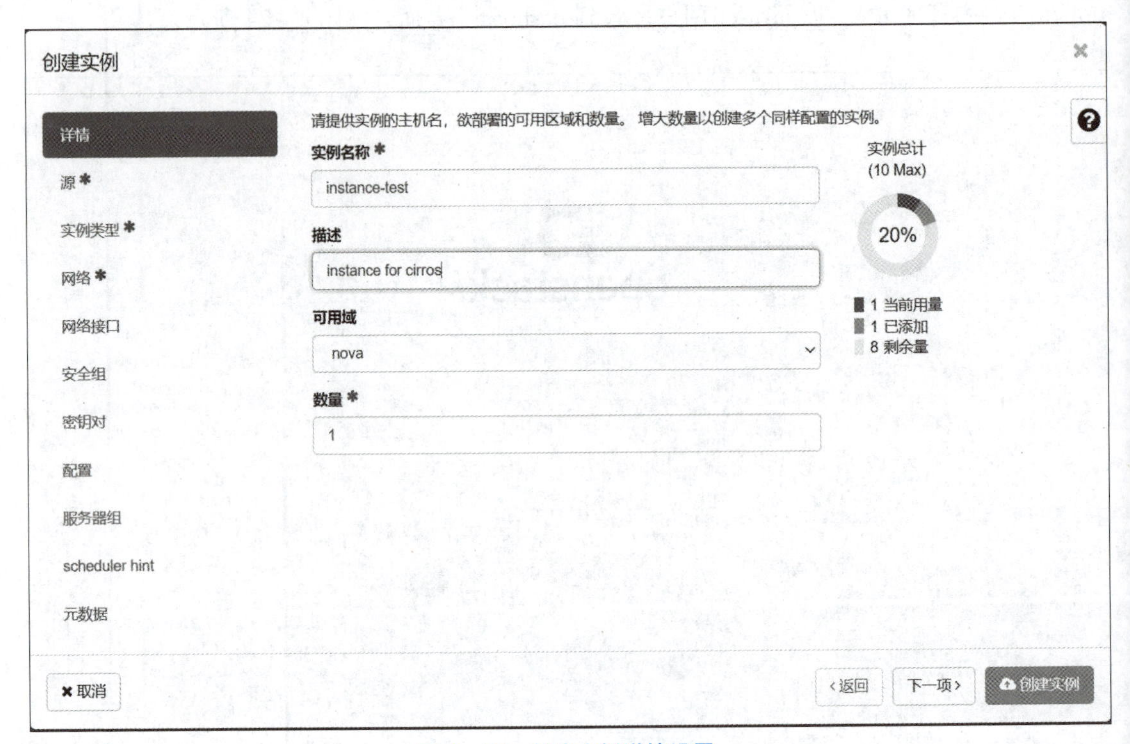

图 4-26　创建实例详情设置

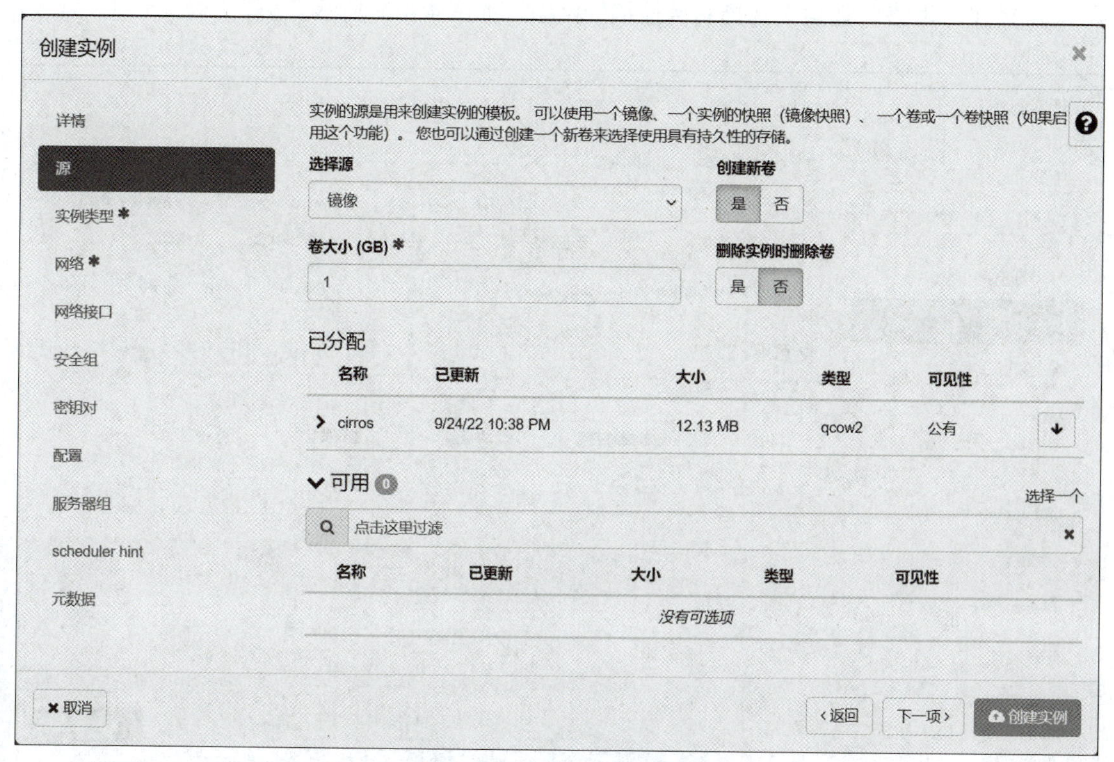

图 4－27　设置实例镜像源

（5）在“实例类型”选项下分配可用实例，如图 4－28 所示。

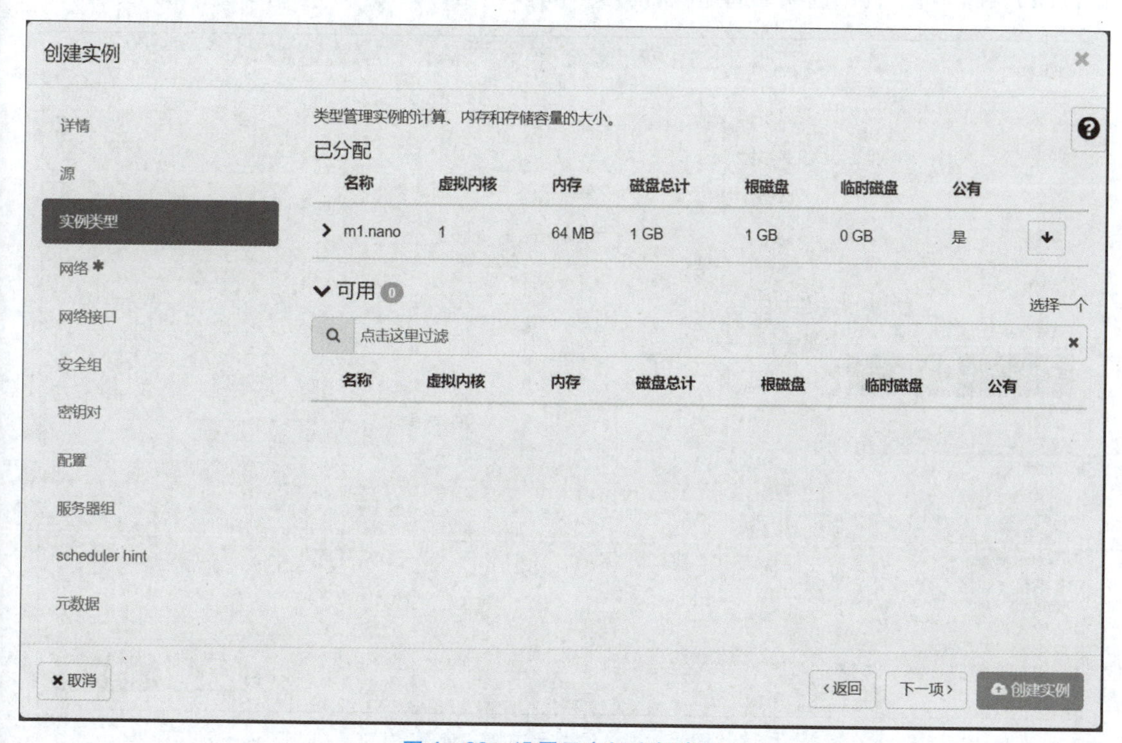

图 4－28　设置云主机实例类型

（6）在“网络”选项下选择实例连接的网络，如图 4－29 所示。

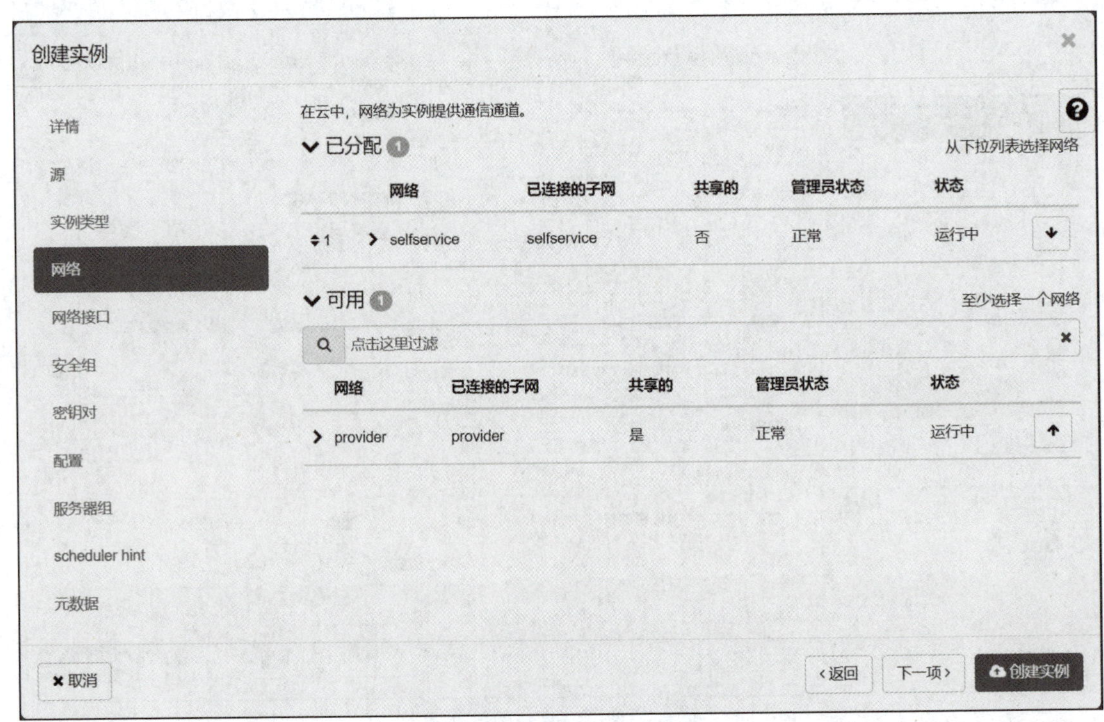

图 4－29　设置云主机网络

（7）在“安全组”选项下选择该启动实例的安全组，如图 4－30 所示。

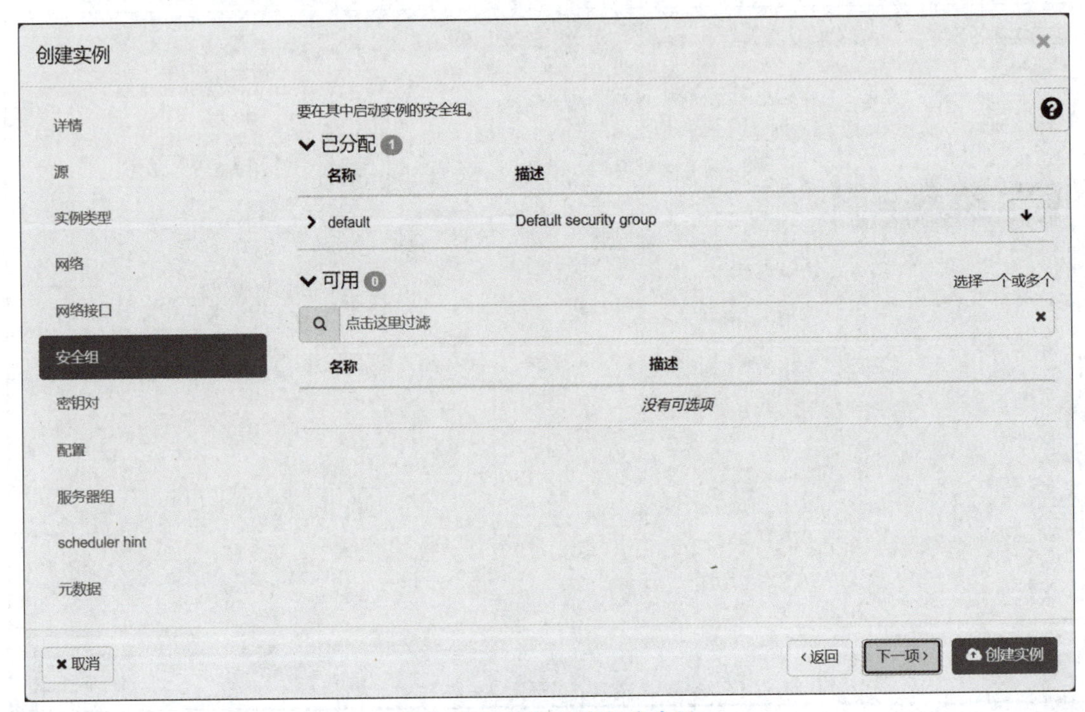

图 4－30　设置云主机安全组

（8）选择该实例的密钥对，单击“创建实例”按钮，如图 4－31 所示。

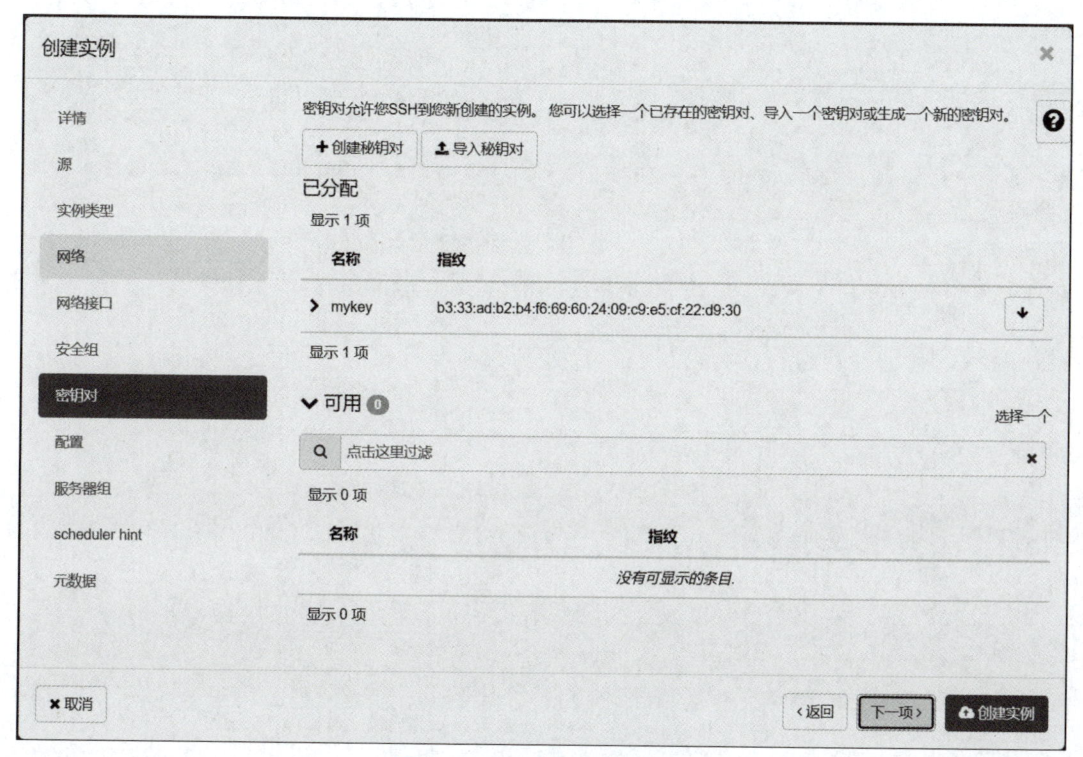

图 4－31　设置云主机密钥对

（9）此时，进入实例创建阶段，待创建完成后，状态会变成运行，便可以正常使用实例了，如图 4－32 所示。

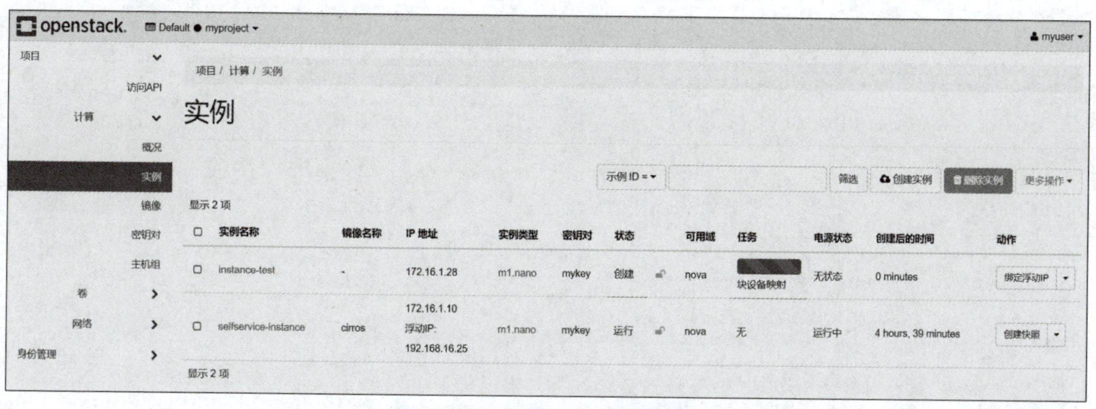

图 4－32　完成云主机实例创建

【任务工单】

工单号：4－2

<table>
<tr><td colspan="2">项目名称：云主机运行</td><td colspan="2">任务名称：Web 端创建云主机</td></tr>
<tr><td colspan="2">班级：</td><td>学号：</td><td>姓名：</td></tr>
<tr><td>任务安排</td><td colspan="3">□Web 端创建云主机实例，包括镜像源的选择、实例类型的设置、网络设置、安全组设置、密钥对设置等。</td></tr>
<tr><td>成果交付</td><td colspan="3">Web 端创建的云主机处于运行状态</td></tr>
<tr><td>任务实施总结</td><td colspan="3">任务自评（0～10 分）：
任务收获：________________
____
____
改进点：________________
____
________________</td></tr>
<tr><td>成果验收</td><td colspan="3">□完全满足任务要求
□基本满足任务要求
要求全部完成，但云主机创建的操作不熟练，所用时间较长：
____
____
____
□不能满足需求
要求不能独立完成，部分任务需要在老师指导下才能完成：
____
____
________________</td></tr>
</table>

【知识巩固】

描述 Web 端创建云主机实例的流程。

【小李的反思】

路漫漫其修远兮，吾将上下而求索。

源自屈原《离骚》，意思是说，在追寻真理方面，前方的道路还很漫长，但我将百折不挠，不遗余力地去追求和探索。在学习的道路上，知识是无尽的，充满艰难险阻的，但再困难的事我们也要不遗余力上下求索。植根于中华优秀传统文化的愚公移山精神，是一种伟大的民族精神，更是党和人民的“传家宝”。千百年来，中华儿女发扬愚公移山的精神，一次又一次不畏艰苦、勇于探索，创造出了无数的世界闻名。

党的二十大报告明确提出了“六个坚持”，也提出了“广大青年要坚定不移听党话、跟党走，怀抱梦想又脚踏实地，敢想敢为又善作善成，立志做有理想、敢担当、能吃苦、肯奋斗的新时代好青年”，这正是党和国家在激励我们青年一代学子，不断奋发勇为，不断追求发展，为实现人生真理而努力奋斗。

项目评价

项目名称：云主机运行					
班级：		学号：	姓名：		
评价指标		评价等级及分值	学生自评	组内互评	教师评分
素质目标达成情况（30%）	精益求精的工匠精神（10%）	A（10 分）：云主机运行过程中能不断修正问题，精益求精，追求完美 B（7 分）：能够完成云主机运行的任务，但对过程中存在的问题修改积极性不高 C（3 分）：对云主机运行过程中存在的问题放任不管			
	知学－好学－乐学的学习热情（10%）	A（10 分）：自我学习热情高涨，积极和同学探讨学习问题 B（7 分）：自我学习热情较好，能够自主完成学习 C（3 分）：自我学习热情一般，学习积极性不高			

续表

<table>
<tr><th colspan="2">评价指标</th><th>评价等级及分值</th><th>学生自评</th><th>组内互评</th><th>教师评分</th></tr>
<tr><td>素质目标达成情况（30%）</td><td>面对挫折与失败，仍能积极乐观的态度（10%）</td><td>A（10 分）：云主机运行出现问题时，能积极乐观面对，分析问题出现的原因，并解决问题
B（7 分）：云主机运行出现问题时，能够主动分析并解决问题，但面对问题态度不够坚决
C（3 分）：云主机运行出现问题时放任不管</td><td></td><td></td><td></td></tr>
<tr><td rowspan="3">知识目标达成情况（40%）</td><td>任务实施完成情况（20%）</td><td>A（20 分）：云主机运行任务全部完成，且满足任务要求
B（16 分）：云主机运行任务基本完成，但存在一些小的问题
C（10 分）：云主机运行任务部分完成，存在一些影响云主机使用的问题</td><td></td><td></td><td></td></tr>
<tr><td>测验作业完成情况（10%）</td><td>A（10 分）：测验作业全部完成，知识理解透彻
B（7 分）：测验作业大部分完成，能基本完成知识的理解
C（3 分）：测验作业部分完成，对知识的理解较为片面</td><td></td><td></td><td></td></tr>
<tr><td>课上活动（10%）</td><td>A（10 分）：积极参与课上抢答、提问、主题讨论等
B（7 分）：能够参与课上抢答、提问、主题讨论等，但积极性不够高
C（3 分）：很少参与课上抢答、提问、主题讨论等</td><td></td><td></td><td></td></tr>
<tr><td rowspan="2">能力目标达成情况（30%）</td><td>任务实施完成质量（20%）</td><td>A（20 分）：任务实施完成质量优秀
B（16 分）：任务实施完成质量良好
C（10 分）：任务实施完成质量一般</td><td></td><td></td><td></td></tr>
<tr><td>超凡脱俗（10%）</td><td>A（10 分）：能够帮助同组同学解决云主机运行过程中存在的问题，并能整理问题解决手册
B（7 分）：能够帮助同组同学解决云主机运行过程中存在的问题
C（3 分）：能够在规定时间内完成学习任务</td><td></td><td></td><td></td></tr>
</table>

项目总结

本项目主要讲述云主机创建的过程，在前面的项目中已经完成了云平台的搭建，搭建云平台的目标就是为了能够在平台上根据用户的需求运行云主机，本项目通过命令行和 Web 端两种方式创建云主机。在命令行创建云主机任务中，主要包括云主机实例的创建、网络的创建、安全组创建及流量方行等，对创建好的云主机通过 Web 端登录，为云主机绑定浮动 IP 并远程连接；在 Web 端创建云主机任务中，主要通过 Web 端的创建方式了解云主机创建的流程及云主机运行需要进行的配置。

项目五

OpenStack 云平台运维

项目导入

小李已经通过参考官方文档自学完成了私有云平台的初步部署工作，在 OpenStack 框架内部署了 Keystone 认证服务、Glance 镜像服务、Nova 计算服务、Neutron 网络服务等，并通过 dashboard 可以访问到云平台的主界面，但对云平台各组件的进一步管理还有所欠缺，小李觉得下一步的计划是学习 OpenStack 各组件的管理。

本项目将随着小李的学习脚步，学习 Keystone 身份认证服务组件的管理，包括用户管理、项目管理、域管理、角色管理等；CentOS 7 镜像的制作和 Glance 镜像的基本管理操作；自服务网络实例的部署。

项目目标

【素质目标】

- 培养学生精益求精的工匠精神
- 激发学生知学 - 好学 - 乐学的学习热情
- 培养学生云计算工程师的职业素养

【知识目标】

- 掌握 Keystone 的用户、项目、域、服务等的管理
- 掌握 CentOS 7 镜像的制作方法
- 掌握 Glance 镜像的基本管理
- 掌握自服务网络实例的部署

【能力目标】

- 能够完成 Keystone 的用户、项目、域、服务等的管理
- 能够自行创建 CentOS 7 镜像
- 能够创建自服务网络实例

任务 1　身份认证服务 Keystone 管理

【任务描述】

小李已经完成 OpenStack 私有云平台的搭建，并能够通过云平台运行云主机，但是对云平台的运维工作还不太熟练，现在他想要学习一下各个组件的运维，在部署时，他部署的第一个服务是 Keystone 服务，在学习组件使用时，他也决定从 Keystone 组件开始。在云平台部署时，涉及了角色、项目、域等概念，所以他决定这次的任务是学习云平台中角色、项目、域、服务、服务目录等的管理。

【知识要点】

1. 用户管理

1）创建用户

```
openstack user create
[ --domain <domain> ]
[ --project <project> ]
[ --project-domain <project-domain> ]
[ --password <password> ]
[ --password-prompt]
[ --email <email-address> ]
[ --description <description> ]
<name>
```

--domain <domain>：用户所属的域。

--project <project>：用户所属项目。

--password <password>：设置用户密码。

--password-prompt：交互式设置用户密码。

--email <email-address>：设置用户的邮箱。

--description <description>：用户描述。

2）查看用户列表

```
openstack user list
[ --sort-column SORT_COLUMN]
[ --domain <domain> ]
[ --group <group> |--project <project> ]
[ --long]
```

--sort-column SORT_COLUMN：指定列表按照某个列排序。

--domain <domain>：按照域过滤用户。

--project <project>：按照项目过滤用户。

--long：列出其他字段。

3）修改当前用户密码

```
openstack user password set
[--password <new-password>]
[--original-password <original-password>]
```

--password <new-password>：指定新密码。

--original-password <original-password>：指定原始用户密码。

4）修改用户属性信息

```
openstack user set
[--name <name>]
[--domain <domain>]
[--project <project>]
[--password <password>]
[--password-prompt]
[--email <email-address>]
[--description <description>]
[--enable|--disable]
<user>
```

--name <name>：设置用户名。

--domain <domain>：设置用户所属的域。

--project <project>：设置用户所属的项目。

--password <password>：设置用户密码。

--password-prompt：交互式设置用户密码。

--email <email-address>：设置用户邮箱。

--description <description>：设置用户描述。

--enable：启用用户。

--disable：不启用用户。

5）查看用户详细信息

```
openstack user show[--domain <domain>] <user>
```

--domain <domain>：指定具体某个域下的用户。

6）删除用户

```
openstack user delete[--domain <domain>] <user>[<user>...]
```

--domain <domain>：指定删除某个特定域下的用户。

2. 项目管理

1）项目创建

```
openstack project create
```

```
[ --domain <domain> ]
[ --description <description> ]
[ --enable | --disable]
[ --or-show]
<name>
```

--domain <domain>：指定项目所属的域。

--description <description>：添加项目的描述信息。

--enable：启用项目（默认）。

--disable：不启用项目。

--or-show：如果创新的项目已经存在，不报失败，而是显示已存在项目的详细信息。

2）查看项目列表

```
openstack project list
[ --domain <domain> ]
[ --user <user> ]
[ --my-projects]
[ --long]
[ --sort <key>[:<direction>,<key>:<direction>,…]]
```

--domain <domain>：根据域筛选项目。

--user <user>：根据用户筛选项目（列出用户所属项目）。

--my-projects：列出当前认证用户的项目，取代其他过滤器。

--long：列出其他字段。

--sort <key>[:<direction>,<key>:<direction>,…]：输出结果按照指定字段排序，asc 表示升序（默认），desc 表示降序，也可以指定多个字段排序。

3）项目设置

```
openstack project set
[ --name <name> ]
[ --domain <domain> ]
[ --description <description> ]
[ --enable | --disable]
<project>
```

--name <name>：设置项目名字。

--domain <domain>：根据域过滤项目。

--description <description>：设置项目描述。

--enable：启用项目。

--disable：不启用项目。

4）显示项目详细信息

```
openstack project show
[ --domain <domain> ]
<project>
```

--domain <domain>：指定项目所属的域。

5）项目删除

```
openstack project delete
[ --domain <domain> ]
<project>[ <project>... ]
```

--domain <domain>：指定要删除的项目所属的域。

3. 域管理

1）域创建

```
openstack domain create
[ --description <description> ]
[ --enable | --disable]
[ --or-show]
<domain-name>
```

--description <description>：域的描述信息。

--enable：启用域（默认）。

--disable：不启用域。

--or-show：如果新创建的域已经存在，则显示域的详细信息。

2）查看域列表

```
openstack domain list[ --sort-column SORT_COLUMN]
```

--sort-column SORT_COLUMN：根据某个列排序。

3）设置域信息

```
openstack domain set
[ --name <name> ]
[ --description <description> ]
[ --enable | --disable]
<domain>
```

--name <name>：设置域名。

--description <description>：设置域的描述。

--enable：设置域为启用状态。

--disable：设置域为不启用状态。

4）显示域详细信息

```
openstack domain show <domain>
```

5）删除域

```
openstack domain delete <domain>[ <domain>…]
```

4. 角色管理

1）角色创建

```
openstack role create
[--or-show]
[--domain <domain>]
<name>
```

--domain <domain>：指定角色所属的域。

--or-show：如果角色已经存在，不新创建，而是返回角色的详细信息。

2）查看角色列表

```
openstack role list
--domain <domain>
```

--domain <domain>：指定查看指定域下的角色列表。

```
openstack role assignment list
[--role <role>]
[--user <user>]
[--domain <domain>]
[--project <project>]
[--names]
```

--role <role>：根据角色过滤。

--user <user>：根据用户过滤。

--domain <domain>：根据域过滤。

--project <project>：根据项目过滤。

--names：角色情况返回值以名字显示（默认返回的是 ID）。

3）为用户添加角色

```
openstack role add
--domain <domain> |--project <project>
--user <user>
<role>
```

--domain <domain>：为域指定角色。

--project <project>：为项目指定角色。

--user <user>：为用户指定角色。

4）设置角色属性

```
openstack role set
[--name <name>]
[--domain <domain>]
<role>
```

--name <name>：设置角色名称。

--domain <domain>：根据域过滤角色。

5）显示角色信息

```
openstack role show
```

```
[ --domain <domain> ]
<role>
```

--domain <domain>：根据域过滤角色。

6）移除用户角色

```
openstack role remove
--domain <domain> | --project <project>
--user <user>
<role>
```

--domain <domain>：角色所属的域。

--project <project>：角色所属项目。

--user <user>：指定用户。

7）删除角色

```
openstack role delete
<role>[ <role>... ]
[ --domain <domain> ]
```

--domain <domain>：指定用户所属的域。

5. 服务管理

1）服务创建

```
openstack service create
[ --name <name> ]
[ --description <description> ]
[ --enable | --disable]
<type>
```

--name <name>：设置服务名称。

--description <description>：设置服务描述。

--enable：启用服务。

--disable：不启用服务。

<type>：指定服务类型。

2）查看服务列表

```
Openstack service list
[ --long]
```

--long：输出结果中列出所有字段。

3）修改服务信息

```
openstack service set
[ --type <type> ]
[ --name <name> ]
```

```
[ --description <description> ]
[ --enable | --disable]
<service>
```

--type <type>：修改服务类型。

--name <name>：修改服务名称。

--description <description>：修改服务描述信息。

--enable | --disable：修改服务状态为启用或不启用。

4）显示服务信息

```
openstack service show
[ --catalog]
<service>
```

5）服务删除

```
openstack service delete
<service>[ <service>... ]
```

6. 目录管理

1）查看目录列表

```
openstack catalog list[ --sort-column SORT_COLUMN]
```

--sort-column SORT_COLUMN：指定根据某个字段排序（默认根据第一列排序）。

2）查看某个服务的目录详细信息

```
openstack catalog show <service>
```

【任务实施】

1. 认证用户管理

1）创建用户

（1）创建一个名称为“allen”的用户，以交互式方式设置密码为“user123456”，邮箱为 allen@ example. com。

```
[root@ controller ~]# openstack user create --password-prompt --email "allen@
example. com" --description "user allen" allen
User Password:
Repeat User Password:
+--------------------+------------------------------------+
| Field              | Value                              |
+--------------------+------------------------------------+
| description        | user allen                         |
| domain_id          | default                            |
| email              | allen@ example. com                |
```

```
| enabled              | True                             |
| id                   | 8a3c42b66776480089c5c2b93fb3f342 |
| name                 | allen                            |
| options              | {}                               |
| password_expires_at  | None                             |
+----------------------+----------------------------------+
```

（2）创建一个名称为“allen1”的用户，密码设置为“user123456”，归属 example 域。

```
[root@ controller ~]# openstack user create --domain example --password user123456 allen1
+----------------------+----------------------------------+
| Field                | Value                            |
+----------------------+----------------------------------+
| domain_id            | 7e4a2925e3ee47c1af8aa85f76db8c64 |
| enabled              | True                             |
| id                   | 248be8240d0a49ea9a71180c2e0c6ec5 |
| name                 | allen                            |
| options              | {}                               |
| password_expires_at  | None                             |
+----------------------+----------------------------------+
```

2）多维度查看用户列表

（1）显示用户列表。

```
[root@ controller ~]# openstack user list
+----------------------------------+-----------+
| ID                               | Name      |
+----------------------------------+-----------+
| 1ecd571c56b34ff78dbae47cd7b20c89 | myuser    |
| 248be8240d0a49ea9a71180c2e0c6ec5 | allen1    |
| 2d90fd1ba066426b8775e37c560aea47 | cinder    |
| 5680d79de42f445c91b3e7aa2517c28d | neutron   |
| 8a3c42b66776480089c5c2b93fb3f342 | allen     |
| 974ca43e23234f0b91847c162982961d | admin     |
| bfdd250781b949b88a5e7f0b77837552 | nova      |
| f2516aa1ead44ee999c900578739a96b | placement |
| fbf37de9362643ddb8f9f401b92fedd0 | glance    |
+----------------------------------+-----------+
```

（2）显示用户列表的所有字段。

```
[root@ controller ~]# openstack user list --long
+------+---------+-------+-------+-----------+--------+-------+
| ID   | Name    | Project | Domain | Description | Email  | Enabled |
```

```
+------+---------+-------+-------+-----------+---------+-------+
| 1e…89 | myuser   |      | default |            |          | True  |
| 24…c5 | allen1   |      | 7e…64  |            |          | True  |
| 2d…47 | cinder   |      | default |            |          | True  |
| 56…8d | neutron  |      | defaul  |            |          | True  |
| 8a…42 | allen    |      | default | user allen | al…. com | True  |
| 97…1d | admin    |      | default |            |          | True  |
| bf…52 | nova     |      | default |            |          | True  |
| f2…6b | placement|      | default |            |          | True  |
| fb…d0 | glance   |      | default |            |          | True  |
+------+---------+-------+-------+-----------+---------+-------+
```

（3）查看用户列表，按照用户名排序。

```
[root@ controller ~]# openstack user list --long --sort-column Name
+-------+---------+-------+-------+-----------+--------+--------+
| ID     | Name      | Project | Domain | Description | Email    | Enabled |
+-------+---------+-------+-------+-----------+--------+--------+
| 97…1d  | admin     |         | default|             |          | True    |
| 24…c5  | allen1    |         | 7e…64  |             |          | True    |
| 8a…42  | allen     |         | default| user allen  | al…. com| True    |
| 2d…47  | cinder    |         | default|             |          | True    |
| fb…d0  | glance    |         | default|             |          | True    |
| 1e…89  | myuser    |         | default|             |          | True    |
| 56…8d  | neutron   |         | default|             |          | True    |
| bf…52  | nova      |         | default|             |          | True    |
| f2…6b  | placement |         | default|             |          | True    |
+-------+---------+-------+-------+-----------+--------+--------+
```

（4）显示 default 域下的所有用户列表。

```
[root@ controller ~]# openstack user list --domain default
+----------------------------------+-----------+
| ID                               | Name      |
+----------------------------------+-----------+
| 1ecd571c56b34ff78dbae47cd7b20c89 | myuser    |
| 2d90fd1ba066426b8775e37c560aea47 | cinder    |
| 5680d79de42f445c91b3e7aa2517c28d | neutron   |
| 8a3c42b66776480089c5c2b93fb3f342 | allen     |
| 974ca43e23234f0b91847c162982961d | admin     |
| bfdd250781b949b88a5e7f0b77837552 | nova      |
| f2516aa1ead44ee999c900578739a96b | placement |
| fbf37de9362643ddb8f9f401b92fedd0 | glance    |
+----------------------------------+-----------+
```

3）修改当前用户密码

执行 demo - openstack. sh 脚本，使 OpenStack 的当前用户环境为 myuser，将当前的 myuser 用户密码修改为 user123456。

```
[root@ controller ~]# source. /demo - openstack. sh
[root@ controller ~]# openstack user password set --password user123456 --original-password myuser
```

4）修改用户信息

将 default 域内的 allen 用户的用户名改为 new_allen，密码改为 123456。

```
[root@ controller ~]# openstack user set --name new_allen --password 123456 --domain default allen
```

5）显示用户的详细信息

（1）显示 allen 用户的详细信息。

```
[root@ controller ~]# openstack user show allen
+---------------------+----------------------------------+
| Field               | Value                            |
+---------------------+----------------------------------+
| domain_id           | 7e4a2925e3ee47c1af8aa85f76db8c64 |
| enabled             | True                             |
| id                  | 957b9013639b4792891d58e8104bf202 |
| name                | allen                            |
| options             | {}                               |
| password_expires_at | None                             |
+---------------------+----------------------------------+
```

（2）当 default 域和 example 域中各存在一个 allen 用户时，想查询 default 域中的 allen 用户详细信息。

```
[root@ controller ~]# openstack user show --domain default allen
+---------------------+----------------------------------+
| Field               | Value                            |
+---------------------+----------------------------------+
| description         | user allen                       |
| domain_id           | default                          |
| email               | allen@ example. com              |
| enabled             | True                             |
| id                  | 8a3c42b66776480089c5c2b93fb3f342 |
| name                | allen                            |
| options             | {}                               |
| password_expires_at | None                             |
+---------------------+----------------------------------+
```

6）删除用户

（1）删除 allen1 用户。

```
[root@ controller ~]# openstack user delete allen1
```

用户名 allen1 也可以替换成用户的 id：

```
[root@ controller ~]# openstack user delete 248be8240d0a49ea9a71180c2e0c6ec5
```

（2）创建一个名称为“allen”的用户，密码设置为“user123456”，归属 example 域。

```
[root@ controller ~]# openstack user create --password user123456 --domain example allen
+---------------------+----------------------------------+
| Field               | Value                            |
+---------------------+----------------------------------+
| domain_id           | 7e4a2925e3ee47c1af8aa85f76db8c64 |
| enabled             | True                             |
| id                  | 957b9013639b4792891d58e8104bf202 |
| name                | allen                            |
| options             | {}                               |
| password_expires_at | None                             |
+---------------------+----------------------------------+
```

（3）删除 example 域中的 allen 用户。

```
[root@ controller ~]# openstack user delete --domain example allen
```

2. 项目管理

1）创建项目

（1）在 default 域内创建一个名为“acm”的项目，描述信息为“acm project”。

```
[root@ controller ~]# openstack project create --domain default --description "acm project" acm
+-------------+----------------------------------+
| Field       | Value                            |
+-------------+----------------------------------+
| description | acm project                      |
| domain_id   | default                          |
| enabled     | True                             |
| id          | eb5de405ec554385b97e8b6e6e45ccfd |
| is_domain   | False                            |
| name        | acm                              |
| parent_id   | default                          |
| tags        | []                               |
+-------------+----------------------------------+
```

（2）创建一个 acm 项目，当已存在项目时，新创建不报错。

```
[root@ controller ~]# openstack project create acm --or-show
```

```
+-------------+----------------------------------+
| Field       | Value                            |
+-------------+----------------------------------+
| description | acm project                      |
| domain_id   | default                          |
| enabled     | True                             |
| id          | eb5de405ec554385b97e8b6e6e45ccfd |
| is_domain   | False                            |
| name        | acm                              |
| parent_id   | default                          |
| tags        | []                               |
+-------------+----------------------------------+
```

2）显示项目列表

（1）显示所有项目。

```
[root@ controller ~]# openstack project list
+----------------------------------+-----------+
| ID                               | Name      |
+----------------------------------+-----------+
| 62b92970ed2a42cb9ea0e8f013004062 | admin     |
| 836801414b99461b91e31be9f1c94b97 | service   |
| e598a75ab1434d4ea2cfa4e0da5ae9ee | myproject |
| eb5de405ec554385b97e8b6e6e45ccfd | acm       |
+----------------------------------+-----------+
```

（2）显示 default 域下的项目。

```
[root@ controller ~]# openstack project list --domain default --long
+-------+-----------+-----------+---------------------+---------+
| ID    | Name      | Domain ID | Description         | Enabled |
+-------+-----------+-----------+---------------------+---------+
| eb…fd | acm       | default   | acm project         | True    |
| 62…62 | admin     | default   | Bootstrap… the cloud. | True  |
| e5…ee | myproject | default   | demo project        | True    |
| 83…97 | service   | default   | service project     | True    |
+-------+-----------+-----------+---------------------+---------+
```

（3）显示 nova 用户所在的项目。

```
[root@ controller ~]# openstack project list --user nova
+----------------------------------+---------+
| ID                               | Name    |
+----------------------------------+---------+
| 836801414b99461b91e31be9f1c94b97 | service |
+----------------------------------+---------+
```

3）修改项目信息

把 default 域下的 acm 项目改名为 new_acm，并且把项目描述改为“acm to new_acm”。

```
[root@ controller ~]# openstack project set --domain default --name new_acm --description "acm to new_acm" acm
[root@ controller ~]# openstack project list --long
+--------+-----------+-----------+--------------------+---------+
| ID     | Name      | Domain ID | Description        | Enabled |
+--------+-----------+-----------+--------------------+---------+
| 62…62  | admin     | default   | Bootstrap… cloud.  | True    |
| 83…97  | service   | default   | service project    | True    |
| e5…ee  | myproject | default   | demo project       | True    |
| eb…fd  | new_acm   | default   | acm to new_acm     | True    |
+--------+-----------+-----------+--------------------+---------+
```

4）显示项目的详细信息

（1）显示 acm 项目的详细信息。

```
[root@ controller ~]# openstack project show acm
+-------------+----------------------------------+
| Field       | Value                            |
+-------------+----------------------------------+
| description | acm project                      |
| domain_id   | default                          |
| enabled     | True                             |
| id          | eb5de405ec554385b97e8b6e6e45ccfd |
| is_domain   | False                            |
| name        | acm                              |
| parent_id   | default                          |
| tags        | []                               |
+-------------+----------------------------------+
```

（2）当 default 域和 example 域中各存在一个 acm 项目时，查询 example 域中的 acm 项目详细信息。

```
[root@ controller ~]# openstack project show --domain example acm
+-------------+----------------------------------+
| Field       | Value                            |
+-------------+----------------------------------+
| description |                                  |
| domain_id   | 7e4a2925e3ee47c1af8aa85f76db8c64 |
| enabled     | True                             |
| id          | b5438b42e024408a96ba9227f99288c0 |
| is_domain   | False                            |
| name        | acm                              |
| parent_id   | 7e4a2925e3ee47c1af8aa85f76db8c64 |
```

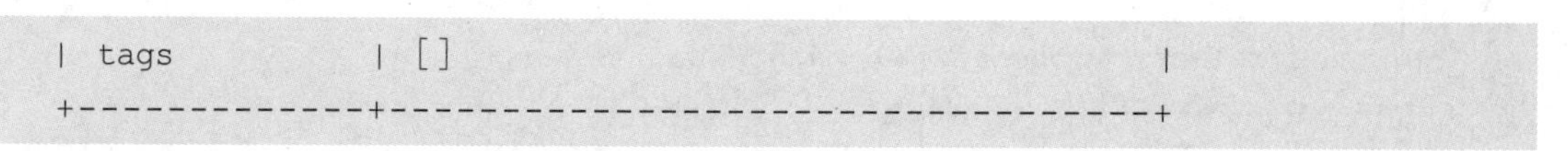

```
| tags          | []                                       |
+---------------+------------------------------------------+
```

5）删除项目

（1）当 default 域和 example 域都存在 acm 项目时，删除 example 域内的 acm 项目。

```
[root@ controller ~]# openstack project delete --domain example acm
```

（2）当只存在一个 acm 项目时，删除 acm 项目。

```
[root@ controller ~]# openstack project delete acm
```

也可以将项目名称更换成项目 ID：

```
[root@ controller ~]# openstack project delete eb5de405ec554385b97e8b6e6e45ccfd
```

3. 域管理

1）创建域

创建一个域，叫作 asia，描述信息为“domain asia”，如果存在 asia 域，则不新创建。

```
    [root@ controller ~]# openstack domain create --description "domain asia" --or-
show asia
    +-------------+----------------------------------+
    | Field       | Value                            |
    +-------------+----------------------------------+
    | description | domain asia                      |
    | enabled     | True                             |
    | id          | a77ea7a5bcc5491ca2553a9fade6de72 |
    | name        | asia                             |
    | tags        | []                               |
    +-------------+----------------------------------+
```

2）显示域列表

```
[root@ controller ~]# openstack domain list
+----------------------------------+---------+---------+--------------------+
| ID                               | Name    | Enabled | Description        |
+----------------------------------+---------+---------+--------------------+
| 7e4a2925e3ee47c1af8aa85f76db8c64 | example | True    | An example domain  |
| a77ea7a5bcc5491ca2553a9fade6de72 | asia    | True    | domain asia        |
| default                          | Default | True    | The default domain|
+----------------------------------+---------+---------+--------------------+
```

3）修改域的信息

将 asia 域名字改为 domain_asia，描述改为“asia to domain_asia”。

```
    [root@ controller ~]# openstack domain set --name domain_asia --description "a-
sia to domain_asia" asia
```

```
[root@ controller ~]# openstack domain list
+------------+------------+---------+---------------------+
| ID         | Name       | Enabled | Description         |
+------------+------------+---------+---------------------+
| 7e4a…8c64  | example    | True    | An example domain   |
| a77e…de72  | domain_asia | True   | asia to domain_asia |
| default    | Default    | True    | The default domain  |
+------------+------------+---------+---------------------+
```

4）显示 domain_asia 域的详细信息

```
[root@ controller ~]# openstack domain show domain_asia
+-------------+----------------------------------+
| Field       | Value                            |
+-------------+----------------------------------+
| description | asia to domain_asia              |
| enabled     | True                             |
| id          | a77ea7a5bcc5491ca2553a9fade6de72 |
| name        | domain_asia                      |
| tags        | []                               |
+-------------+----------------------------------+
```

5）删除 domain_asia 域

```
[root@ controller ~]# openstack domain delete domain_asia
Failed to delete domain with name or ID 'domain_asia':Cannot delete a domain that is enabled,please disable it first.(HTTP 403)(Request - ID:req - 3df2b33a - 31f9 - 468a - a943 - ac00af022dc5)
1 of 1 domains failed to delete.
```

提示报错：不能删除 enabled 的域，所以，删除一个域之前，需要先将其 disable 掉，再次执行删除命令即可删除 domain_asia 域。

```
[root@ controller ~]# openstack domain set --disable domain_asia
[root@ controller ~]# openstack domain delete domain_asia
```

4. 角色管理

1）创建角色 compute - user

如果已经存在该角色，显示角色详细信息不重新创建。

```
[root@ controller ~]# openstack role create compute - role -- or - show
+-----------+----------------------------------+
| Field     | Value                            |
+-----------+----------------------------------+
| domain_id | None                             |
| id        | 347e70d5540941d3b9c8250a1cb61c72 |
| name      | compute - role                   |
+-----------+----------------------------------+
```

2）查看角色列表

(1) 查看所有角色列表。

```
[root@ controller ~]# openstack role list
+------------------------------------+--------------+
| ID                                 | Name         |
+------------------------------------+--------------+
| 0506186fe40e41189f6a7a76dfaa3780   | admin        |
| 0694847e73ff48a4b06a819494c0008f   | myrole       |
| 347e70d5540941d3b9c8250a1cb61c72   | compute-role |
| 92da6921b83647e0a559d002d1503ffa   | reader       |
| d22065a1193a43d7bd6b5690752c3cd0   | member       |
+------------------------------------+--------------+
```

(2) 查看角色分配列表（角色、项目、用户等默认使用 ID 表示）。

```
[root@ controller ~]# openstack role assignment list
+--------+--------+-------+--------+-------+-------+----------+
| Role   | User   | Group | Project | Domain | System | Inherited |
+--------+--------+-------+--------+-------+-------+----------+
| 06…8f  | 1e…89 |       | e5…ee  |       |       | False    |
| 05…80  | 2d…47 |       | 83…97  |       |       | False    |
| 05…80  | 56…8d |       | 83…97  |       |       | False    |
| 05…80  | 97…1d |       | 62…62  |       |       | False    |
| 05…80  | bf…52 |       | 83…97  |       |       | False    |
| 05…80  | f2…6b |       | 83…97  |       |       | False    |
| 05…80  | fb…d0 |       | 83…97  |       |       | False    |
| 05…80  | 97…1d |       |        |       | all   | False    |
+--------+--------+-------+--------+-------+-------+----------+
```

(3) 查看 service 项目的角色分配情况，并且角色、用户、项目等用名字表示。

```
[root@ controller ~]# openstack role assignment list --project service -names
+-----+-----------------+-----+---------+------+------+---------+
| Role | User               | Group | Project | Domain | System | Inherited |
+-----+-----------------+-----+---------+------+------+---------+
| admin| cinder@ Default    |       | service@| Default |       | False     |
| admin| neutron@ Default   |       | service@| Default |       | False     |
| admin| nova@ Default      |       | service@| Default |       | False     |
| admin| placement@ Default |       | service@| Default |       | False     |
| admin| glance@ Default    |       | service@| Default |       | False     |
+-----+-----------------+-----+---------+------+------+---------+
```

(4) 查看 nova 的角色分配情况，并且角色、用户、项目等用名字表示。

```
[root@ controller ~]# openstack role assignment list --user nova --names
```

```
+-----+----------+------+-------------+------+-------+---------+
|Role |User      |Group |Project      |Domain |System |Inherited |
+-----+----------+------+-------------+------+-------+---------+
|admin |nova@ Default |      |service@ Default |      |       |False     |
+-----+----------+------+-------------+------+-------+---------+
```

3）为用户添加角色

为 allen 用户添加 myproject 项目的 myrole 角色：

```
[root@ controller ~]# openstack role add --project myproject --user allen myrole
[root@ controller ~]# openstack role assignment list --project myproject --user
allen --names
+------+------------+-----+-------------+------+------+--------+
|Role   |User          |Group |Project          |Domain |System |Inherited |
+------+------------+-----+-------------+------+------+--------+
|myrole |allen@ Default |      |myproject@ Default|       |       |False    |
+------+------------+-----+-------------+------+------+--------+
```

4）设置角色属性信息

将 compute - role 角色改名为 testrole：

```
[root@ controller ~]# openstack role set --name roletest compute-role
[root@ controller ~]# openstack role show roletest
+-----------+-----------------------------------+
| Field      | Value                              |
+-----------+-----------------------------------+
| domain_id  | None                               |
| id         | 347e70d5540941d3b9c8250a1cb61c72   |
| name       | roletest                           |
+-----------+-----------------------------------+
```

5）显示角色属性信息

显示 roletest 角色的详细信息：

```
[root@ controller ~]# openstack role show roletest
+-----------+-----------------------------------+
| Field      | Value                              |
+-----------+-----------------------------------+
| domain_id  | None                               |
| id         | 347e70d5540941d3b9c8250a1cb61c72   |
| name       | roletest                           |
+-----------+-----------------------------------+
```

6）移除用户的角色

移除 allen 用户在 myproject 项目的 myrole 角色：

```
[root@ controller ~]# openstack role assignment list --user allen --names
```

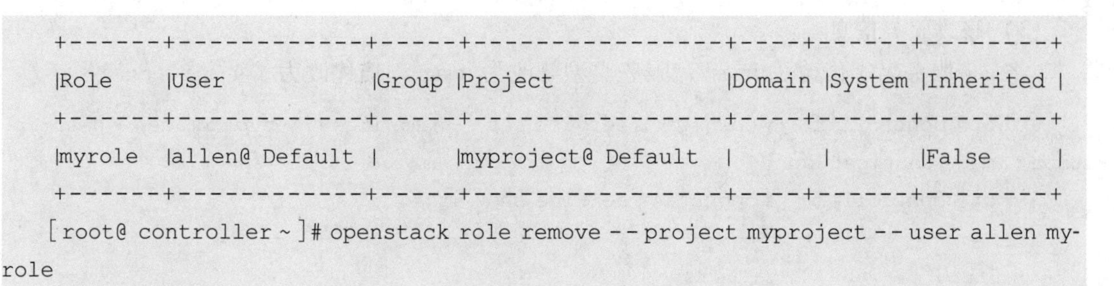

```
+------+------------+-----+---------------+----+-----+--------+
|Role   |User            |Group |Project            |Domain |System |Inherited |
+------+------------+-----+---------------+----+-----+--------+
|myrole  |allen@ Default |       |myproject@ Default  |       |        |False      |
+------+------------+-----+---------------+----+-----+--------+
[root@ controller ~]# openstack role remove --project myproject --user allen my-
role
```

7）删除角色

删除 roletest 角色：

```
[root@ controller ~]# openstack role delete roletest
```

5. 服务管理

1）创建服务

创建一个计算类型服务，服务名为 csv，描述为“compute service”。

```
[root@ controller ~]# openstack service create --name csv --description "compute
service" compute
+-------------+----------------------------------+
| Field       | Value                            |
+-------------+----------------------------------+
| description | compute service                  |
| enabled     | True                             |
| id          | 6f32b5941d074f668056e655f0a24c93 |
| name        | csv                              |
| type        | compute                          |
+-------------+----------------------------------+
```

2）查看服务列表

查看服务列表，并显示所有字段。

```
[root@ controller ~]# openstack service list --long
+--------+----------+-----------+-----------------------+-------+
| ID     | Name      | Type      | Description            | Enabled |
+--------+----------+-----------+-----------------------+-------+
| 19…53  | glance    | image     | OpenStack Image        | True    |
| 3f…0d  | nova      | compute   | OpenStack Compute      | True    |
| 4d…cf  | cinderv3  | volumev3  | OpenStack Block Storage | True   |
| 6f…93  | csv       | compute   | compute service        | True    |
| 7d…be  | cinderv2  | volumev2  | OpenStack Block Storage | True   |
| 85…54  | placement | placement | Placement API          | True    |
| b9…51  | neutron   | network   | OpenStack Networking   | True    |
| ef…e3  | keystone  | identity  |                        | True    |
+--------+----------+-----------+-----------------------+-------+
```

3）修改服务信息

将 csv 服务名字修改为 imgsv，服务类型修改为 image，描述改为“image service”。

```
[root@ controller ~]# openstack service set --name imgsv --type image --description "image service" csv
[root@ controller ~]# openstack service show imgsv
+-------------+----------------------------------+
| Field       | Value                            |
+-------------+----------------------------------+
| description | image service                    |
| enabled     | True                             |
| id          | 6f32b5941d074f668056e655f0a24c93 |
| name        | imgsv                            |
| type        | image                            |
+-------------+----------------------------------+
```

4）显示服务信息

显示 imgsv 服务详细信息：

```
[root@ controller ~]# openstack service show imgsv
+-------------+----------------------------------+
| Field       | Value                            |
+-------------+----------------------------------+
| description | image service                    |
| enabled     | True                             |
| id          | 6f32b5941d074f668056e655f0a24c93 |
| name        | imgsv                            |
| type        | image                            |
+-------------+----------------------------------+
```

5）删除服务

删除 imgsv 服务：

```
[root@ controller ~]# openstack service delete imgsv
```

6. 目录管理

1）查看目录列表

查看目录列表，并按照 Type 字段排序：

```
[root@ controller ~]# openstack catalog list --sort-column Type
+----------+----------+----------------------------------------+
| Name     | Type     | Endpoints                              |
+----------+----------+----------------------------------------+
| nova     | compute  | RegionOne                              |
|          |          |   public:http://controller:8774/v2.1   |
|          |          | RegionOne                              |
```

```
|           |           |   internal:http://controller:8774/v2.1       |
|           |           | RegionOne                                    |
|           |           |   admin:http://controller:8774/v2.1          |
|           |           |                                              |
| keystone  | identity  | RegionOne                                    |
|           |           |   internal:http://controller:5000/v3/        |
|           |           | RegionOne                                    |
|           |           |   admin:http://controller:5000/v3/           |
|           |           | RegionOne                                    |
|           |           |   public:http://controller:5000/v3/          |
|           |           |                                              |
| glance    | image     | RegionOne                                    |
|           |           |   admin:http://controller:9292               |
|           |           | RegionOne                                    |
|           |           |   public:http://controller:9292              |
|           |           | RegionOne                                    |
|           |           |   internal:http://controller:9292            |
|           |           |                                              |
| neutron   | network   | RegionOne                                    |
|           |           |   admin:http://controller:9696               |
|           |           | RegionOne                                    |
|           |           |   internal:http://controller:9696            |
|           |           | RegionOne                                    |
|           |           |   public:http://controller:9696              |
|           |           |                                              |
| placement | placement | RegionOne                                    |
|           |           |   public:http://controller:8778              |
|           |           | RegionOne                                    |
|           |           |   internal:http://controller:8778            |
|           |           | RegionOne                                    |
|           |           |   admin:http://controller:8778               |
|           |           |                                              |
| cinderv2  | volumev2  | RegionOne                                    |
|           |           |   internal:http://controller:8776/v2/62...62 |
|           |           | RegionOne                                    |
|           |           |   admin:http://controller:8776/v2/62...62    |
|           |           | RegionOne                                    |
|           |           |   public:http://controller:8776/v2/62...62   |
|           |           |                                              |
| cinderv3  | volumev3  | RegionOne                                    |
|           |           |   internal:http://controller:8776/v3/62...62 |
|           |           | RegionOne                                    |
|           |           |   admin:http://controller:8776/v3/62...62    |
```

```
|           |           | RegionOne                                  |
|           |           |   public:http://controller:8776/v3/62...62 |
|           |           |                                            |
+-----------+-----------+--------------------------------------------+
```

2）查看服务的目录信息

查看 Nova 服务的目录信息：

```
[root@ controller ~]# openstack catalog show nova
+-----------+------------------------------------------+
| Field     | Value                                    |
+-----------+------------------------------------------+
| endpoints | RegionOne                                |
|           | public:http://controller:8774/v2.1       |
|           | RegionOne                                |
|           | internal:http://controller:8774/v2.1     |
|           | RegionOne                                |
|           | admin:http://controller:8774/v2.1        |
|           |                                          |
| id        | 3ff9f76cc2c64bba9e4e6ae7dc2cdd0d         |
| name      | nova                                     |
| type      | compute                                  |
+-----------+------------------------------------------+
```

【任务工单】

工单号：5－1

<table>
<tr><td colspan="2">项目名称：OpenStack 云平台运维</td><td colspan="2">任务名称：身份认证服务 Keystone 管理</td></tr>
<tr><td colspan="2">班级：</td><td>学号：</td><td>姓名：</td></tr>
<tr><td>任务安排</td><td colspan="3">□用户管理，包括创建用户、修改信息、查看用户列表及查看用户详细信息等
□项目管理，包括创建项目、修改项目信息、查看项目列表及查看项目详细信息等
□域管理，包括创建域、修改域信息、查看域列表及查看域详细信息等
□角色管理，包括创建角色、修改信息、为用户添加角色等
□服务管理，包括创建服务、修改服务信息、查看服务列表及查看服务详细信息等</td></tr>
<tr><td>成果交付</td><td colspan="3">实验案例整理成操作指导交付文档</td></tr>
<tr><td>任务实施总结</td><td colspan="3">任务自评（0～10 分）：
任务收获：

改进点：

</td></tr>
<tr><td>成果验收</td><td colspan="3">□完全满足任务要求
□基本满足任务要求
要求全部完成，能够对用户、项目、域、服务、角色等进行管理，但操作不够熟练：

□不能满足需求
要求不能独立完成，对用户、项目、域、服务、角色等的操作管理比较混乱：

</td></tr>
</table>

【知识巩固】

执行 openstack user create -- password - prompt -- email alice@ example. com alice 命令创建用户时，-- password - prompt 的含义是（　　）。

A. 显示输入密码　　B. 交互式输入密码

C. 不需要设置密码　　D. 使用默认密码

【小李的反思】

患生于所忽，祸起于细微。

语出汉代刘向《说苑·说丛》，意谓忧患滋生于不重视，灾害起因于微小之处。在本任务中，身份认证服务 Keystone 包括 Keystone 的用户、项目、域、服务等的管理，虽然仅仅是一个用户或者密码，却关乎整个 OpenStack 云平台的安全，务必要引起足够重视，防微杜渐。

2022 年 6 月，西北工业大学发布《公开声明》称，西北工业大学电子邮件系统遭受网络攻击，有来自境外的黑客组织和不法分子向该校师生发送包含木马程序的钓鱼邮件，企图窃取相关师生邮件数据和公民个人信息。国家计算机病毒应急处理中心发布《西北工业大学遭美国 NSA 网络攻击事件调查报告（之一）》，初步判断对西北工业大学实施网络攻击行动的是美国国家安全局（NSA）信息情报部（代号 S）数据侦察局（代号 S3）下属 TAO（代号 S32）部门。

党的二十大报告指出，国家安全是民族复兴的根基，社会稳定是国家强盛的前提。《中华人民共和国密码法》于 2020 年 1 月 1 日起施行，密码直接关系国家经济安全、政治安全、国防安全和信息安全。密码可谓无处不在，维护着社会的正常运转，如身份认证、消费支付、网络交易、个人信息、财产保护等，不断增强公民、法人和其他组织的密码安全意识，不断提升密码工作科学化、规范化、法治化，为提高人民幸福感、获得感、安全感，实现中华民族伟大复兴的中国梦贡献力量。

任务 2 镜像服务 Glance 管理

【任务描述】

小李已经学会了 Keystone 组件的运维管理，云平台需要对学生分发虚拟机，因而涉及镜像的定制，从而能够根据上课需求给学生分发不同镜像，小李决定通过 Glance 组件制作自定义的 CentOS 7 镜像。

【知识要点】

1. 查看镜像列表

```
openstack image list
[ --public | --private | --community | --shared]
[ --name <name> ]
[ --status <status> ]
[ --long]
```

--public | --private | --community | --shared：根据镜像的可见性进行过滤。

--name <name>：根据镜像名字过滤。

--status <status>：根据镜像状态过滤。

--long：显示镜像全部信息。

2. 查看镜像详细信息

```
openstack image show
[ --human-readable]
<image>
```

--human-readable：镜像大小的信息以可读性更好的方式显示。

3. 镜像创建

```
openstack image create
[ --id <id> ]
[ --container-format <container-format> ]
[ --disk-format <disk-format> ]
[ --min-disk <disk-gb> ]
[ --min-ram <ram-mb> ]
[ --file <file> | --volume <volume> ]
[ --protected | --unprotected]
[ --public | --private | --community | --shared]
```

```
[ --project <project> ]
<image-name>
```

--id <id>：指定镜像的 ID。

--container-format <container-format>：指定镜像的容器格式，支持的格式包括 ami、ari、aki、bare、docker、ova、ovf。默认是 bare。

--disk-format <disk-format>：指定镜像硬盘格式，支持的格式包括 ami、ari、aki、vhd、vmdk、raw、qcow2、vhdx、vdi、iso 和 ploop。默认是 raw。

--min-disk <disk-gb>：启动镜像需要的最小硬盘，单位为 GB。

--min-ram <ram-mb>：启动镜像需要的最小内存，单位为 MB。

--file <file>：从本地文件上传镜像，<file>为具体的镜像路径和文件名。

--volume <volume>：从卷创建镜像。

--protected：镜像处于保护模式，不允许被删除。

--unprotected：镜像处于非保护模式，可以删除（默认为 unprotected）。

--public：镜像为公共镜像。

--private：镜像为私有镜像。

--community：镜像为社区镜像。

--shared：镜像为共享镜像。

--project <project>：指定镜像所属项目。

4. 共享镜像到项目

```
openstack image add project
<image>
<project>
```

5. 显示镜像关联的项目

```
openstack image member list
<image>
```

6. 解除镜像到项目的共享

```
openstack image remove project
[ --project-domain <project-domain> ]
<image>
<project>
```

7. 设置镜像属性

```
openstack image set
[ --name <name> ]
[ --container-format <container-format> ]
```

```
[ --disk-format <disk-format> ]
[ --protected | --unprotected ]
[ --public | --private | --community | --shared ]
[ --project <project> ]
[ --accept | --reject | --pending ]
<image>
```

8. 保存镜像

```
openstack image save
--file <filename>
<image>
```

【任务实施】

1. 镜像管理

1）下载镜像

（1）下载 cirros 镜像文件，cirros 是一个用来进行云上镜像测试的最小化 Linux 发行版，镜像下载地址为 https://docs.openstack.org/image-guide/obtain-images.html，如果没有安装 wget，先使用 yum install wget 命令安装 wget 工具。

使用 wget 下载，出现 Unable to establish SSL connection 问题，可能是部分网站不允许非浏览器方式下载文件，可以在 wget 语句后添加 --no-check-certificate 参数解决。

```
[root@ controller ~]# wget http://download.cirros-cloud.net/0.4.0/cirros-0.4.0-x86_64-disk.img --no-check-certificate
```

（2）查看文件信息。

```
[root@ controller ~]# file cirros-0.4.0-x86_64-disk.img
cirros-0.4.0-x86_64-disk.img:QEMU QCOW Image(v3),46137344 bytes
```

2）查看镜像列表

```
[root@ controller ~]# openstack image list
+--------------------------------------+--------+--------+
| ID                                   | Name   | Status |
+--------------------------------------+--------+--------+
| a58a2e2f-fd7e-4ef3-b8c1-bd4446f3d33d | cirros | active |
+--------------------------------------+--------+--------+
```

3）查看镜像详细信息

查看 cirros 镜像的详细信息：

```
[root@ controller ~]# openstack image show cirros --human-readable
```

```
+------------------+----------------------------------------------------+
| Field            | Value                                              |
+------------------+----------------------------------------------------+
| checksum         | 443b7623e27ecf03dc9e01ee93f67afe                   |
| container_format | bare                                               |
| created_at       | 2022-10-21T13:53:33Z                               |
| disk_format      | qcow2                                              |
| file             | /v2/images/a58…33d/file                            |
| id               | a58a2e2f-fd7e-4ef3-b8c1-bd4446f3d33d               |
| min_disk         | 0                                                  |
| min_ram          | 0                                                  |
| name             | cirros                                             |
| owner            | 62b92970ed2a42cb9ea0e8f013004062                   |
| properties       | os_hash_algo='sha512',…,os_hidden='False'          |
| protected        | False                                              |
| schema           | /v2/schemas/image                                  |
| size             | 12.7M                                              |
| status           | active                                             |
| tags             |                                                    |
| updated_at       | 2022-10-21T13:53:33Z                               |
| virtual_size     | None                                               |
| visibility       | public                                             |
+------------------+----------------------------------------------------+
```

container_format：表示镜像的容器格式，此处为 bare。OpenStack 还支持其他格式，如 ami、ari、aki、docker、ova、ovf。

created_at：表示镜像上传时间。

disk_format：表示镜像硬盘格式，此处为 qcow2。OpenStack 还支持其他格式，如 ami、ari、aki、vhd、vmdk、raw、vhdx、vdi、iso 和 ploop。

file：表示文件路径，实际存储在/var/lib/glance/images 下。

owner：表示该镜像所属的项目。

protected：表示该镜像是否是保护状态，保护状态下镜像不可以被删除，非保护状态下可以删除。

visibility：表示镜像的可见性，可以是公有镜像 public、私有镜像 private、社区镜像 community、共享镜像 shared。

4）创建镜像

（1）以 cirros-0.4.0-x86_64-disk.img 文件创建镜像 cirros-0.4.0，容器格式为 bare，硬盘格式为 qcow2。

```
[root@controller ~]# openstack image create --container-format bare --disk-format qcow2 --file cirros-0.4.0-x86_64-disk.img cirros-0.4.0
```

```
+--------------------+--------------------------------------------------+
| Field              | Value                                            |
+--------------------+--------------------------------------------------+
| checksum           | 443b7623e27ecf03dc9e01ee93f67afe                 |
| container_format   | bare                                             |
| created_at         | 2022-11-23T09:23:26Z                             |
| disk_format        | qcow2                                            |
| file               | /v2/images/44…7d/file                            |
| id                 | 44943198-59b5-4e49-a6dc-9006f11b327d             |
| min_disk           | 0                                                |
| min_ram            | 0                                                |
| name               | cirros-0.4.0                                     |
| owner              | 62b92970ed2a42cb9ea0e8f013004062                 |
| properties         | os_hash_algo='sha512',…,os_hidden='False'        |
| protected          | False                                            |
| schema             | /v2/schemas/image                                |
| size               | 12716032                                         |
| status             | active                                           |
| tags               |                                                  |
| updated_at         | 2022-11-23T09:23:26Z                             |
| virtual_size       | None                                             |
| visibility         | shared                                           |
+--------------------+--------------------------------------------------+
```

（2）执行 demo-openstack. sh 脚本，切换 myuser 环境，以 cirros-0.4.0-x86_64-disk. img 文件创建镜像 cirros-0.4.1，容器格式为 bare，硬盘格式为 qcow2。

```
[root@ controller ~]# source demo-openstack.sh
[root@ controller ~]# openstack image create --container-format bare --disk-format qcow2 --file cirros-0.4.0-x86_64-disk.img cirros-0.4.1
+--------------------+--------------------------------------------------+
| Field              | Value                                            |
+--------------------+--------------------------------------------------+
| checksum           | 443b7623e27ecf03dc9e01ee93f67afe                 |
| container_format   | bare                                             |
| created_at         | 2022-11-23T09:24:33Z                             |
| disk_format        | qcow2                                            |
| file               | /v2/images/5680ebeb-f740-4d6a-b096-5cb2e0b74e3f/ |
|                    | file                                             |
| id                 | 5680ebeb-f740-4d6a-b096-5cb2e0b74e3f             |
| min_disk           | 0                                                |
| min_ram            | 0                                                |
| name               | cirros-0.4.1                                     |
| owner              | e598a75ab1434d4ea2cfa4e0da5ae9ee                 |
```

```
| properties        | os_hash_algo='sha512',…,os_hidden='False'       |
| protected         | False                                            |
| schema            | /v2/schemas/image                                |
| size              | 12716032                                         |
| status            | active                                           |
| tags              |                                                  |
| updated_at        | 2022-11-23T09:24:33Z                             |
| virtual_size      | None                                             |
| visibility        | shared                                           |
+-------------------+--------------------------------------------------+
```

（3）在 admin 用户下，以 cirros-0.4.0-x86_64-disk.img 文件创建镜像 cirros-0.4.2，容器格式为 bare，硬盘格式为 qcow2，指定镜像在 myproject 项目下，开启保护模式。

```
[root@ controller ~]# source./admin-openstack.sh
[root@ controller ~]# openstack image create --container-format bare --disk-format qcow2 --file cirros-0.4.0-x86_64-disk.img --protected --project myproject cirros-0.4.2
+-------------------+--------------------------------------------------+
| Field             | Value                                            |
+-------------------+--------------------------------------------------+
| checksum          | 443b7623e27ecf03dc9e01ee93f67afe                 |
| container_format  | bare                                             |
| created_at        | 2022-11-23T09:34:20Z                             |
| disk_format       | qcow2                                            |
| file              | /v2/images/f4ee0d50-eefe-466f-9833-4a7ff7a422ff/ |
|                   | file                                             |
| id                | f4ee0d50-eefe-466f-9833-4a7ff7a422ff             |
| min_disk          | 0                                                |
| min_ram           | 0                                                |
| name              | cirros-0.4.2                                     |
| owner             | e598a75ab1434d4ea2cfa4e0da5ae9ee                 |
| properties        | os_hash_algo='sha512',…,os_hidden='False'        |
| protected         | True                                             |
| schema            | /v2/schemas/image                                |
| size              | 12716032                                         |
| status            | active                                           |
| tags              |                                                  |
| updated_at        | 2022-11-23T09:34:20Z                             |
| virtual_size      | None                                             |
| visibility        | shared                                           |
+-------------------+--------------------------------------------------+
```

cirros-0.4.0 镜像和 cirros-0.4.1 镜像在创建时未指定所属的项目，但从创建结果的 owner 属性可知，创建镜像所属项目默认和用户相同，未指定镜像的可见性默认为 shared。

以 admin 用户登录云平台，admin 用户可以看到所有项目下任何可见的镜像，如图 5－1 所示。

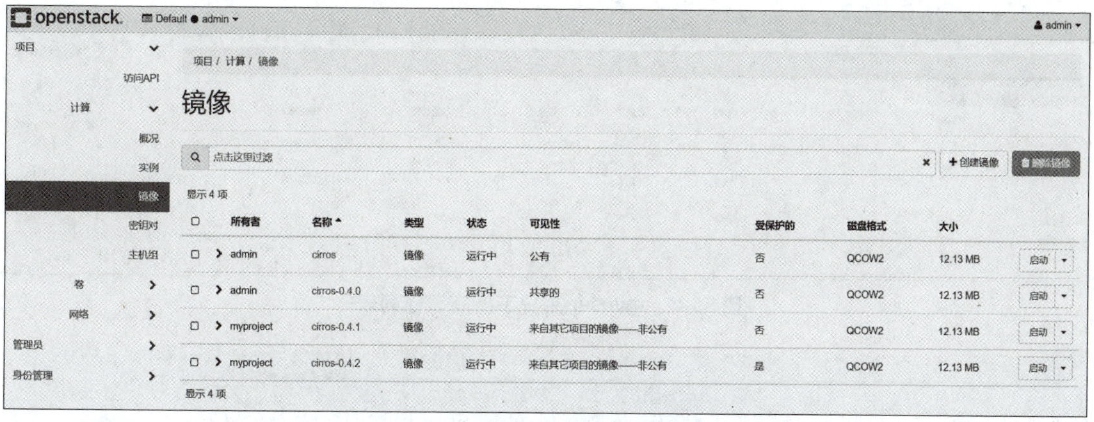

图 5－1 admin 用户下所有镜像列表

cirros－0.4.2 镜像的 protected 属性设置为 true，此时任何用户都无法删除该镜像（包括 admin 用户），在 admin 用户下单击镜像的更多选项，看不到删除选项，如图 5－2 所示。

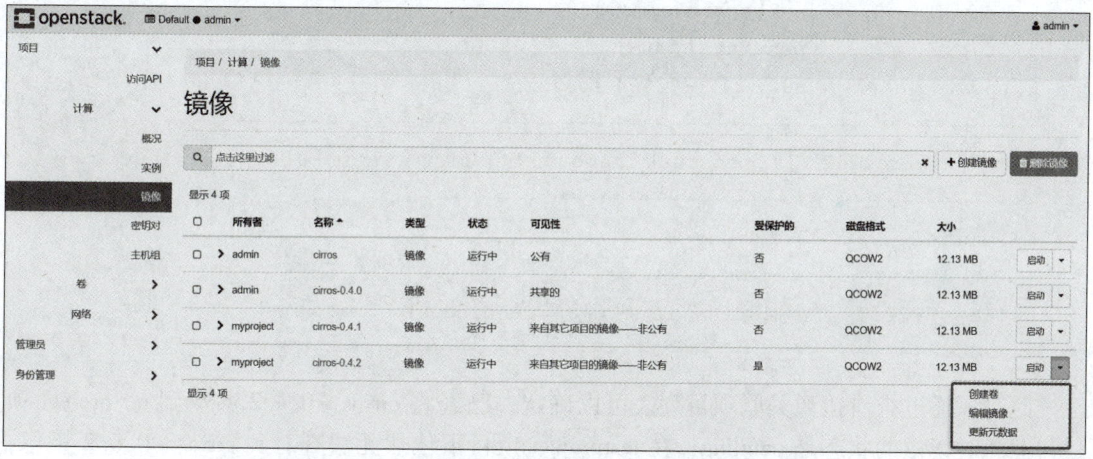

图 5－2 admin 用户下 cirros－0.4.2 镜像的更多选项

在当前系统中包含 4 个镜像：cirros、cirros－0.4.0、cirros－0.4.1、cirros－0.4.2，其中，cirros 和 cirros－0.4.0 在 admin 项目下，cirros－0.4.1 和 cirros－0.4.2 在 myproject 项目下，以 myuser 用户登录云平台，可以看到与 myuser 用户在同一项目下的镜像和其他项目下的公有镜像，如图 5－3 所示。

5）镜像共享到项目

将 myproject 项目下的共享镜像 cirros－0.4.2 共享到 testproject 项目。镜像共享的前提是该镜像的可见性为“shared”。

```
[root@ controller ~]# openstack image add project cirros-0.4.2 testproject
```

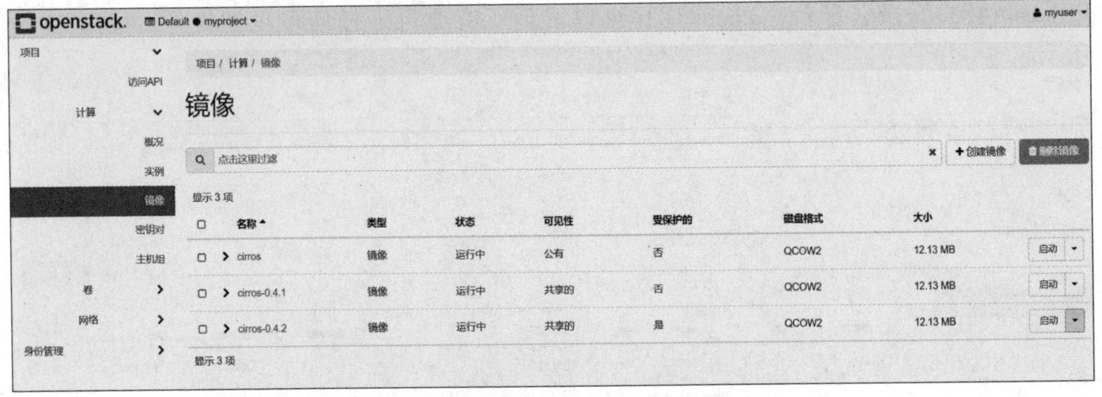

图 5－3　myuser 用户下所有镜像列表

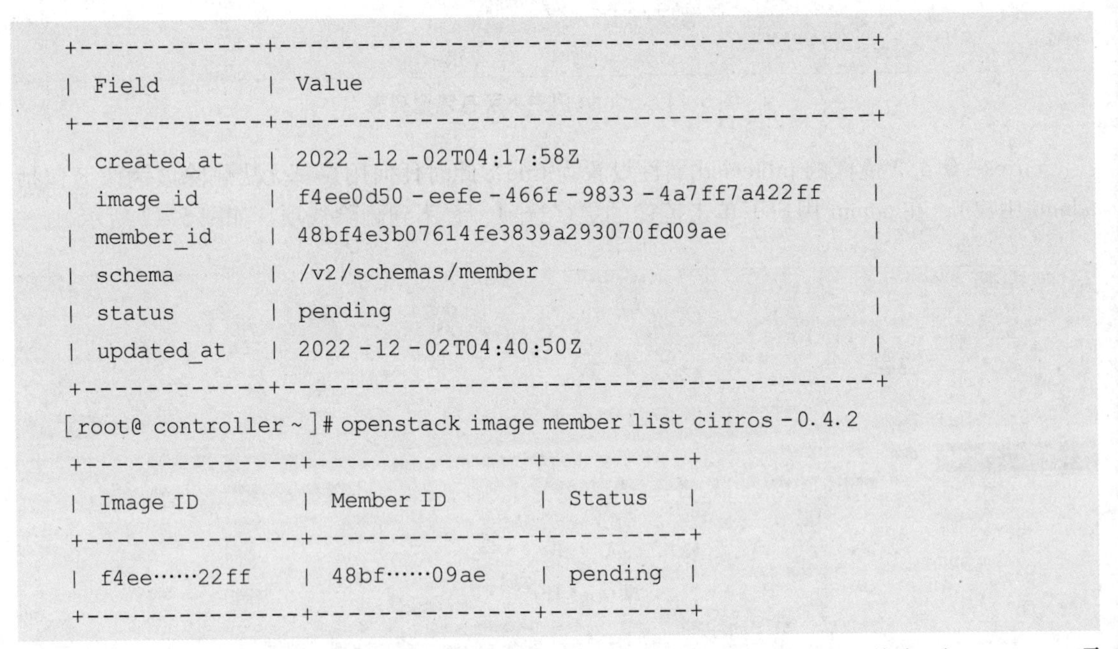

```
+------------+---------------------------------------+
| Field      | Value                                 |
+------------+---------------------------------------+
| created_at | 2022-12-02T04:17:58Z                  |
| image_id   | f4ee0d50-eefe-466f-9833-4a7ff7a422ff  |
| member_id  | 48bf4e3b07614fe3839a293070fd09ae      |
| schema     | /v2/schemas/member                    |
| status     | pending                               |
| updated_at | 2022-12-02T04:40:50Z                  |
+------------+---------------------------------------+
[root@ controller ~]# openstack image member list cirros-0.4.2
+--------------+--------------+---------+
| Image ID     | Member ID    | Status  |
+--------------+--------------+---------+
| f4ee……22ff   | 48bf……09ae   | pending |
+--------------+--------------+---------+
```

此时，通过查看镜像关联项目列表可以确定，已经将 cirros－0.4.2 共享到 testproject 项目，但此时镜像的状态是 pending，在 testproject 项目下依然无法查看到 cirros－0.4.2 镜像，只有公共镜像 cirros，如图 5－4 所示。

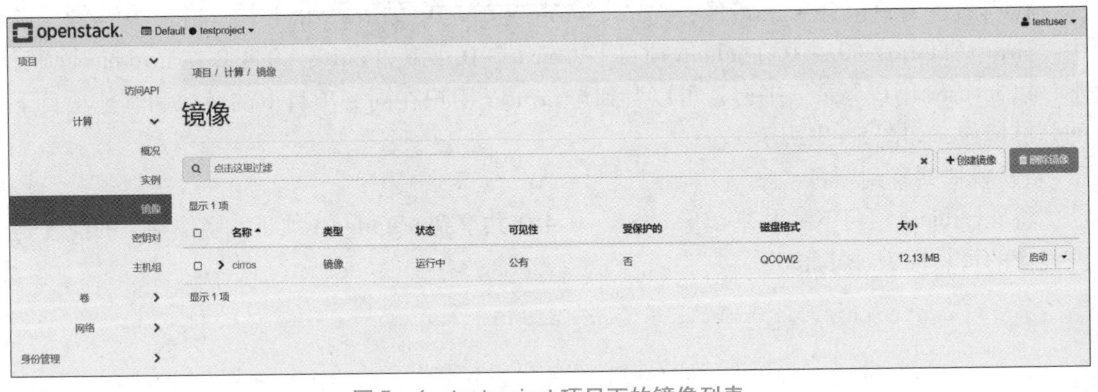

图 5－4　testproject 项目下的镜像列表

想要在 testproject 项目下能够查看到 cirros -0.4.2 镜像，还需要在 test 项目下接收该镜像。接收后，在 testproject 项目下可以看到共享的 cirros -0.4.2 镜像，如图 5-5 所示。

```
[root@ controller ~]# source. /test-openstack.sh
[root@ controller ~]# openstack image set --accept f4ee0d50-eefe-466f-9833-4a7ff7a422ff
[root@ controller ~]# openstack image member list cirros-0.4.2
+--------------+--------------+---------+
| Image ID     | Member ID    | Status  |
+--------------+--------------+---------+
| f4ee……22ff   | 48bf……09ae   | pending |
+--------------+--------------+---------+
```

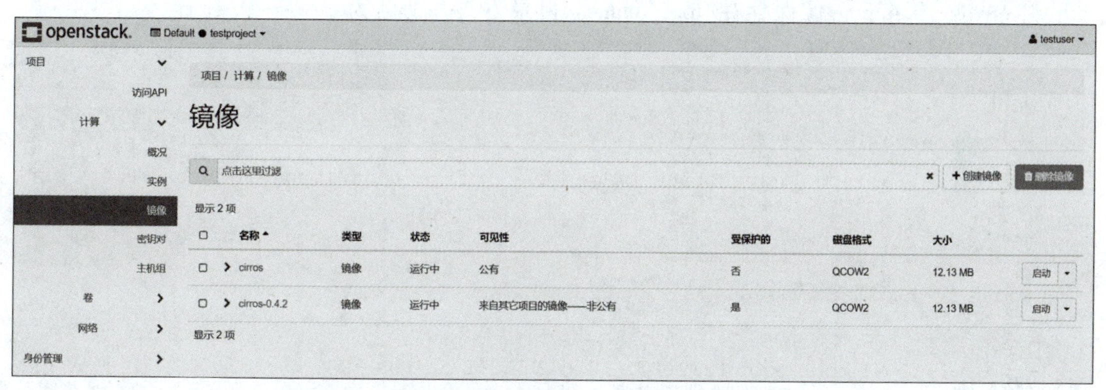

图 5-5　接收镜像后 testproject 项目下的共享镜像列表

6）显示镜像关联项目

查看 cirros -0.4.2 镜像关联的项目列表，显示结果中，仅显示共享给其他项目的信息，镜像本身所属的项目信息不显示。

```
[root@ controller ~]# openstack image member list cirros-0.4.2
+--------------+--------------+---------+
| Image ID     | Member ID    | Status  |
+--------------+--------------+---------+
| f4ee……22ff   | 48bf……09ae   | pending |
+--------------+--------------+---------+
```

7）解除镜像到项目的共享

解除 cirros -0.4.2 镜像到 testproject 项目的共享，解除共享关联之后，在 testproject 项目下看不到 cirros -0.4.2 镜像，如图 5-6 所示。

```
[root@ controller ~]# openstack image remove project cirros-0.4.2 testproject
```

8）设置镜像属性

将 cirros -0.4.2 镜像的名字改为 cirros -0.4.3，镜像改为公共镜像，受保护。

```
[root@ controller ~]# openstack image set --name cirros-0.4.3 --unprotected --public cirros-0.4.2
```

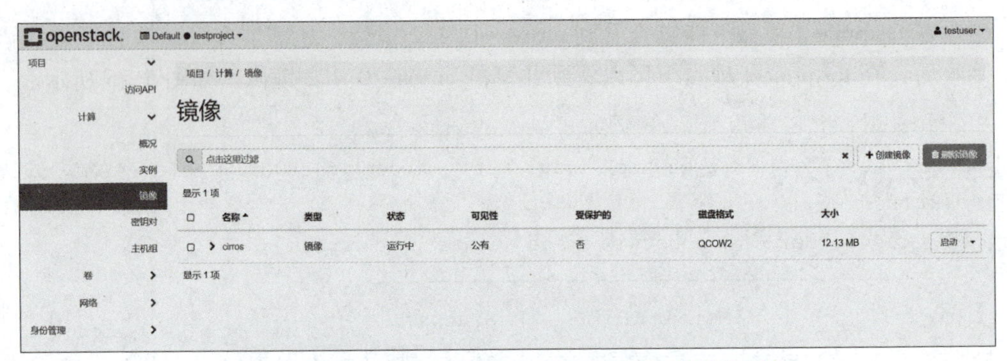

图 5-6　解除镜像共享后 testproject 项目下的镜像列表

9）保存镜像

将 cirros-0.4.3 镜像保存在/root/images/目录下，并命名为 cirros-0.4.4。

```
[root@ controller ~]# pwd
/root
[root@ controller ~]# mkdir images
[root@ controller ~]# openstack image save --file/root/images/cirros-0.4.4 cir-
ros-0.4.3
[root@ controller ~]# cd images/
[root@ controller images]# ll
总用量 12420
-rw-r--r--1 root root 12716032 12 月 2 13:51 cirros-0.4.4
```

2. 制作 CentOS 7 镜像

1）下载 CentOS 7 的 ISO 文件

（1）链接地址：https://www.centos.org/download/mirrors/。

（2）在该链接下按区域过滤，如图 5-7 所示。

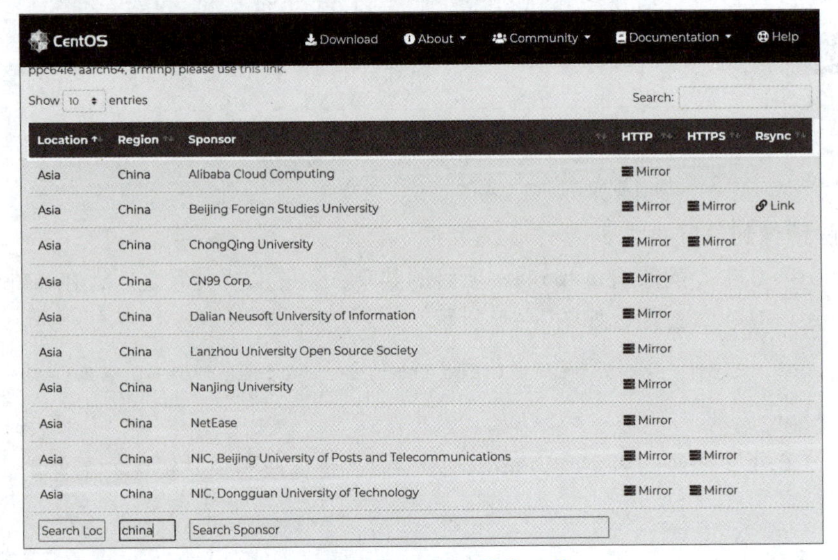

图 5-7　CentOS 7 镜像下载

（3）选择国内某一镜像地址，单击 http 下的 mirror 链接（例如阿里云的镜像）。

（4）依次单击 7/→isos/→x86_64。

（5）选择某个镜像，并通过 wget 在 controller 中下载，例如：CentOS-7-x86_64-NetInstall-2009.iso 就是个不错的选择，因为这个镜像更小，许多软件包可以在安装过程中通过网络下载。鼠标右键复制链接地址，获取 wget 下载地址：http://mirrors.aliyun.com/centos/7/isos/x86_64/CentOS-7-x86_64-NetInstall-2009.iso。

```
[root@ controller ~]# cd/opt
[root@ controller opt]# wget http://mirrors.aliyun.com/centos/7/isos/x86_64/CentOS-7-x86_64-NetInstall-2009.iso
--2021-11-21 14:32:30-- http://mirrors.aliyun.com/centos/7/isos/x86_64/CentOS-7-x86_64-NetInstall-2009.iso
正在解析主机 mirrors.aliyun.com (mirrors.aliyun.com)...220.194.69.120,125.39.43.236,125.36.180.239,...
正在连接 mirrors.aliyun.com(mirrors.aliyun.com)|220.194.69.120|:80... 已连接。
已发出 HTTP 请求，正在等待回应...200 OK
长度：602931200(575M)[application/octet-stream]
正在保存至："CentOS-7-x86_64-NetInstall-2009.iso"
100%[==================================================================================================================================================================================>] 602,931,200 1.39MB/s 用时 6m 40s
2021-11-21 14:39:10 (1.44 MB/s) - 已保存 "CentOS-7-x86_64-NetInstall-2009.iso"[602931200/602931200])
```

2）安装虚拟化工具软件包

（1）安装工具软件包，安装完成后重启 controller 节点。

```
[root@ controller ~]# yum install virt-manager qemu-kvm libvirt qemu-img virt-viewer virt-install
```

软件包说明如下：

virt-manager：KVM 图形化管理工具。

qemu-kvm：KVM 的基本包，包括 KVM 内核模块和 QEMU 模拟器。

libvirt：提供 hypervisor 及虚机管理的 API。

qemu-img：QEMU 磁盘镜像管理工具。

virt-viewer：显示虚拟化客户机的图形界面的工具。

virt-install：创建和克隆虚拟机的命令行工具。

（2）修改/etc/libvirt/qemu.conf 配置文件，在该文件中添加 vnc_listen = "0.0.0.0" 的配置并重启 libvirtd 服务。

```
[root@ controller ~]# systemctl restart libvirtd
```

（3）确认 libvirt 默认网络已开启，如果未开启，执行 virsh net-start default 开启，如图 5-8 所示。

```
[root@ controller ~]# virsh net-list
```

图 5－8　查看 libvirt 网络状态

（4）切换到/tmp/images 目录下，创建一个大小为 10 GB 的镜像文件，如图 5－9 所示。

```
[root@ controller images]# qemu - img create - f qcow2 CentOS7. qcow2 10G
```

```
[root@controller tmp]# qemu-img create -f qcow2 centos7.qcow2 10G
Formatting 'centos7.qcow2', fmt=qcow2 size=10737418240 cluster_size=65536 lazy_refcounts=off refcount_bits=16
```

图 5－9　创建镜像文件

（5）启动部署虚拟机。

```
[root@ controller ~]# virt - install -- name CentOS7 -- ram 1024 \
-- disk/tmp/CentOS7. qcow2,format = qcow2 \
-- network network = default -- graphics vnc,listen = 0. 0. 0. 0,port = 5901 \
-- noautoconsole -- os - type = linux -- os - variant = CentOS7. 0 \
-- location = /opt/CentOS - 7 - x86_64 - NetInstall - 2009. iso
```

（6）执行 virsh vncdisplay <CentOS 7 虚拟机名字>查看 VNC 端口号，如图 5－10 所示。

```
[root@controller ~]# virsh vncdisplay cloud_centos7_64bit
:0
```

图 5－10　查看虚拟机的 VNC 端口号

3. 安装虚拟机操作系统

（1）通过 VNC Viewer 连接到虚拟桌面，如图 5－11 所示。

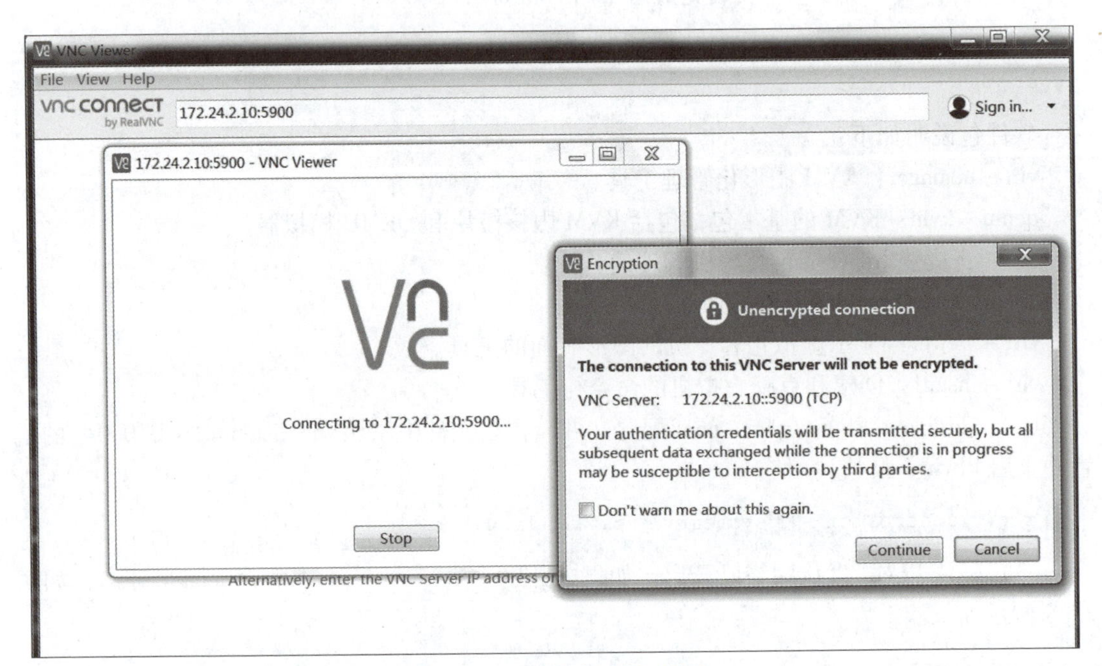

图 5－11　连接到虚拟机桌面

（2）进入安装界面后，首先选择语言，如图 5－12 所示。

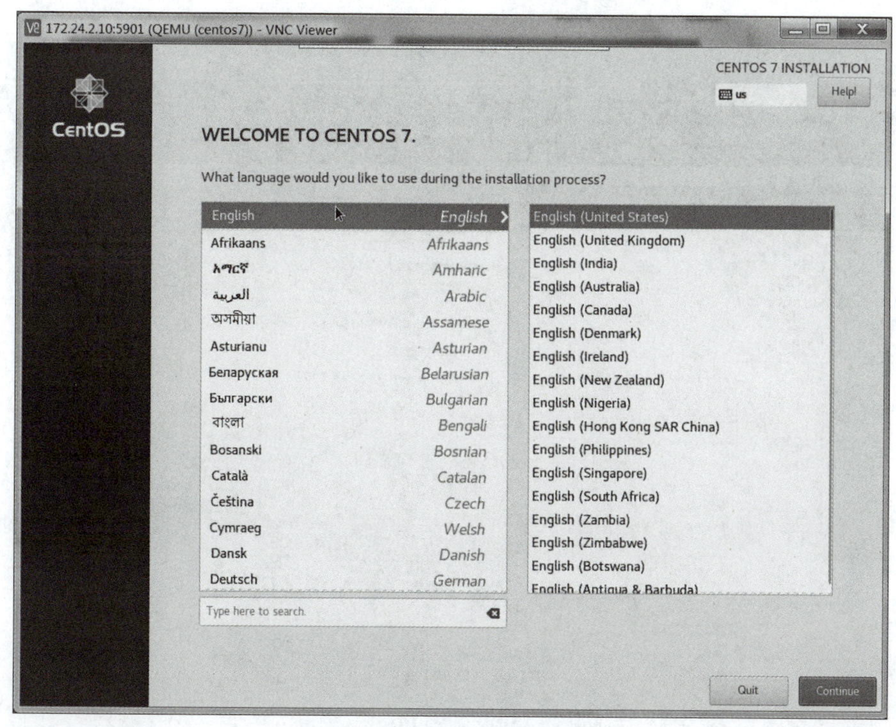

图 5－12　选择虚拟机语言

（3）完成相应的安装设置，主要包括网络设置、安装源、安装软件选择、分区等，如图 5－13 所示。

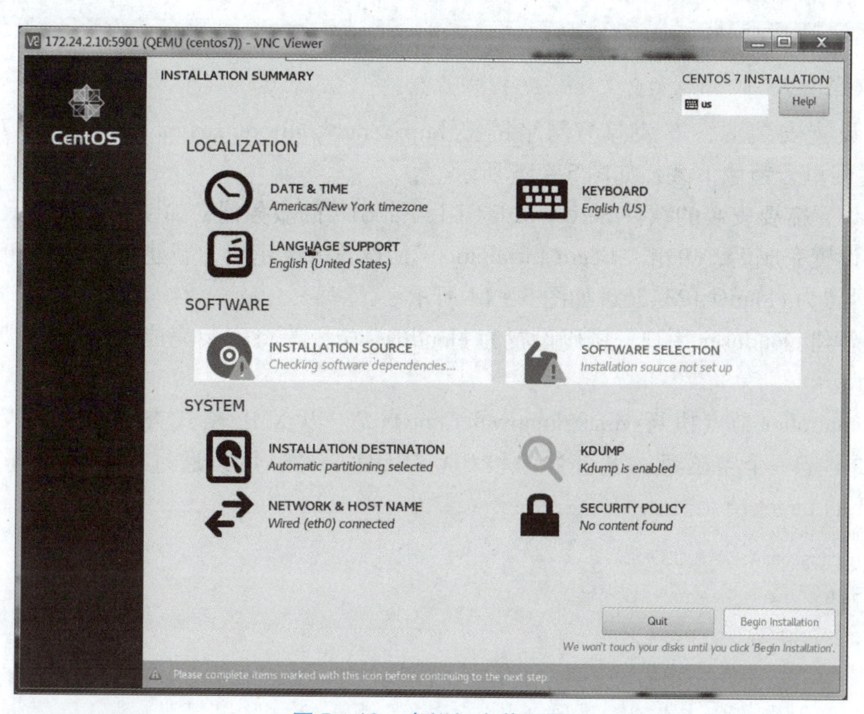

图 5－13　虚拟机安装设置

（4）由于需要用到网络安装，所以先设置网络。打开网络，如图 5－14 所示。

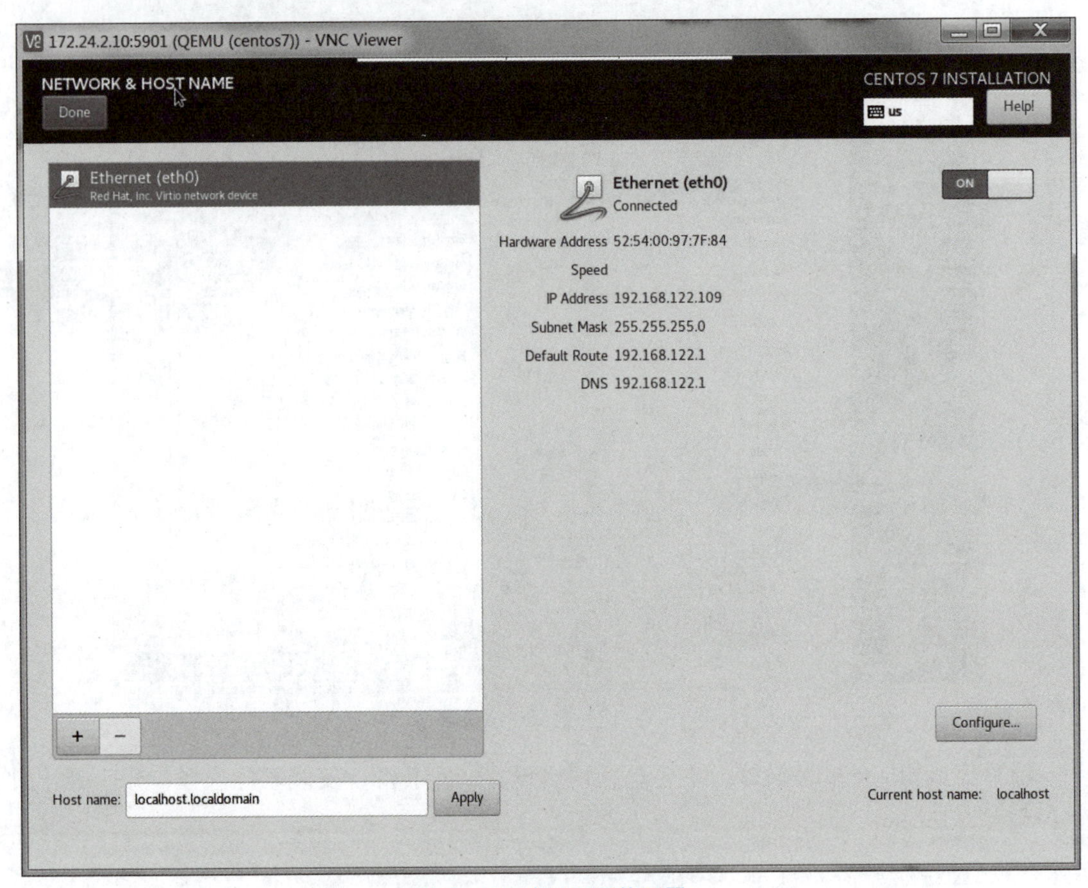

图 5－14　虚拟机网络配置

（5）设置安装源，安装源为网络链接 http://pub. mirrors. aliyun. com/centos/7/os/x86_64/，通过阿里云镜像下载，如图 5－15 所示。

（6）设置需要安装的软件 Server with GUI，带 GUI 的服务器，如图 5－16 所示。

（7）设置完成后，单击"Begin Installation"，设置 root 密码，新建登录用户和密码，将 root 密码设置为 cloud@123456，如图 5－17 所示。

（8）新建 clouduser 用户，密码设置为 cloudpasswd，等待安装完成后重启虚拟机，如图 5－18 所示。

（9）controller 节点执行 virsh dumpxml CentOS 7，以 XML 格式查看 CentOS 7 的信息，libvirt 需要绑定一个空磁盘，磁盘目标和 CD－ROM 一致，可以通过以下命令查看正确的 CD－ROM 的 target 值。

```
#virsh dumpxml CentOS7
<domain type='qemu' id='1'>
<name>CentOS7</name>
<uuid>09c89eaa-b668-4d2a-90f0-345cec033cdd</uuid>
......
```

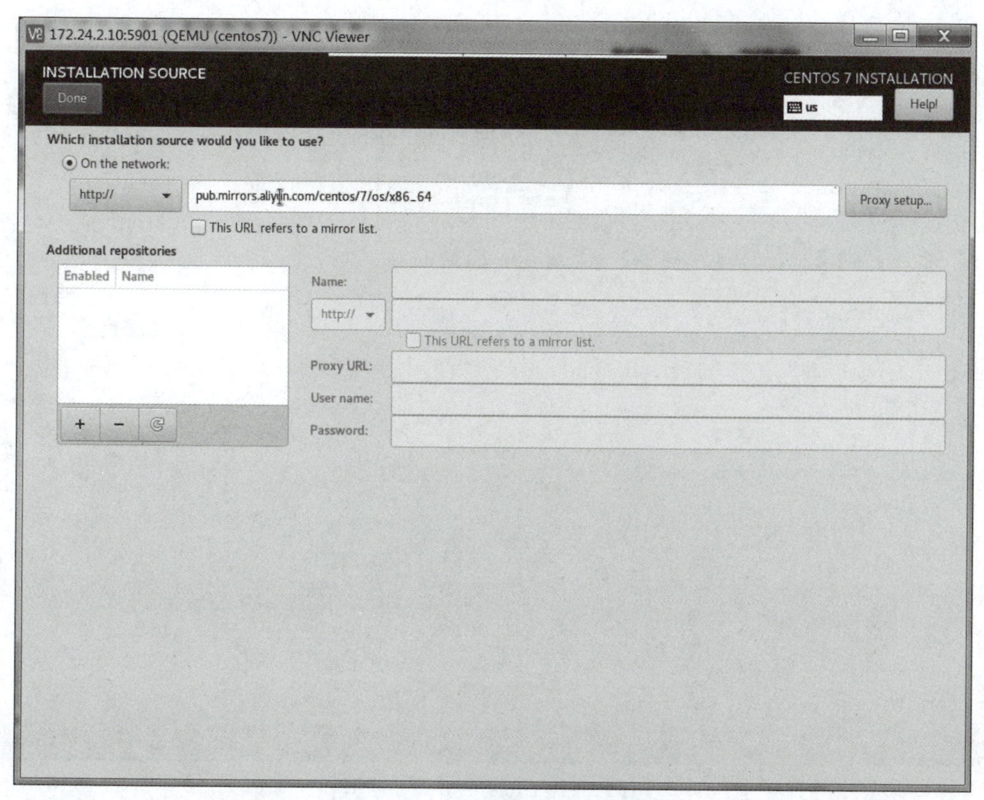

图 5-15　设置操作系统安装源

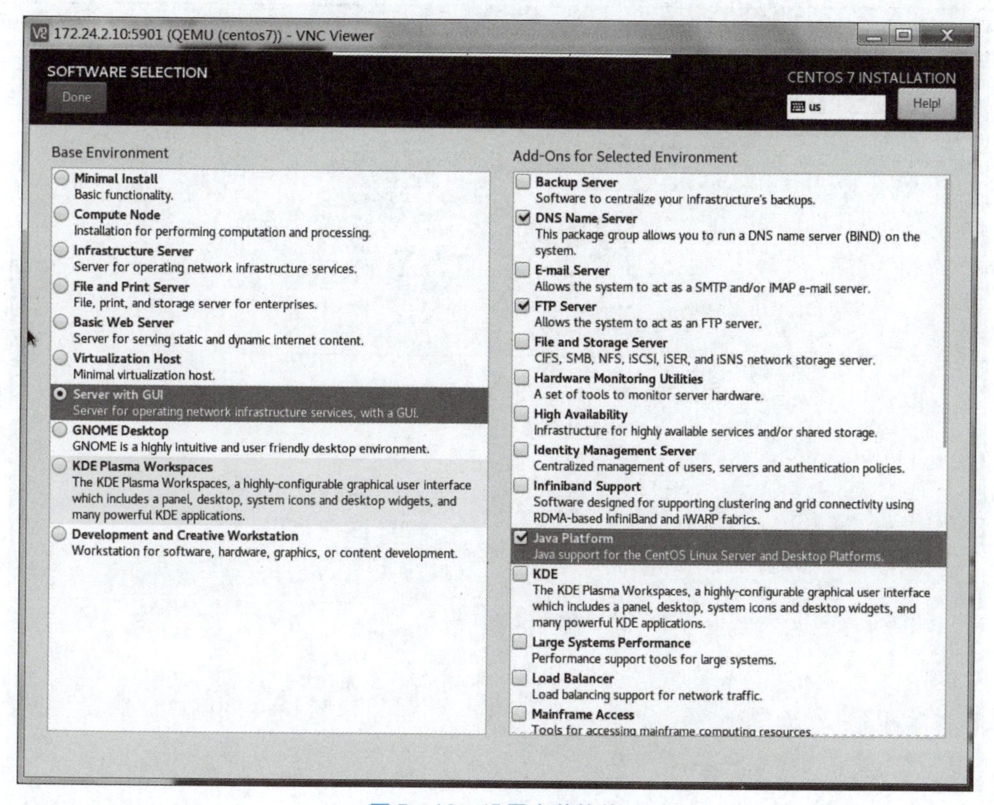

图 5-16　设置安装软件

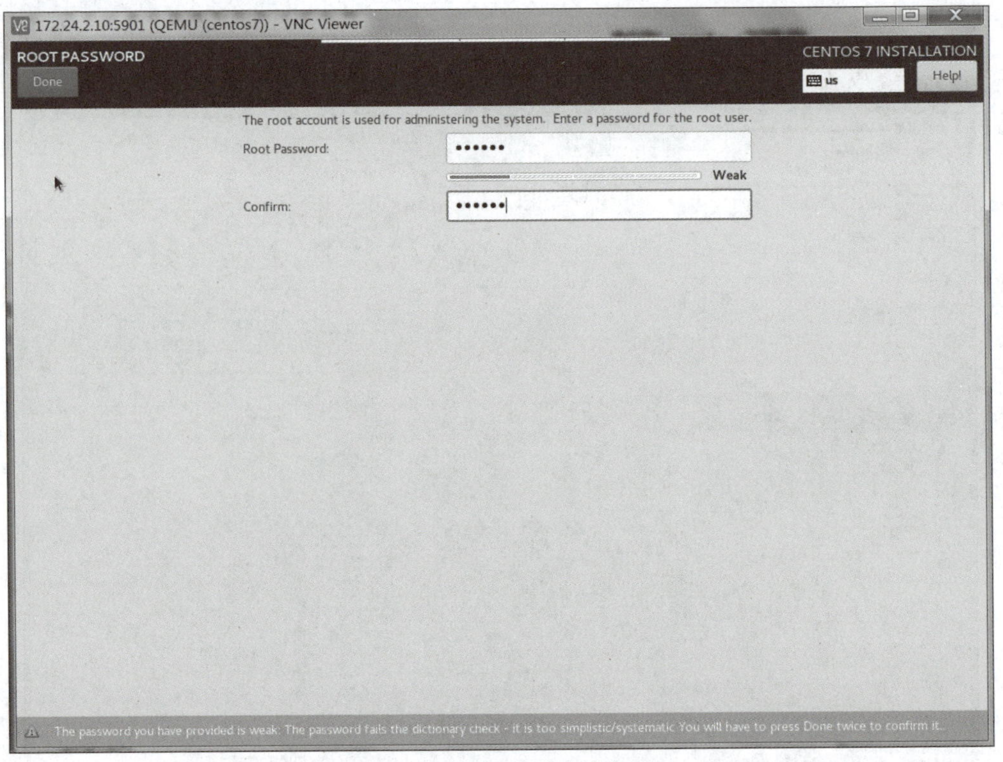

图 5-17 设置 root 用户密码

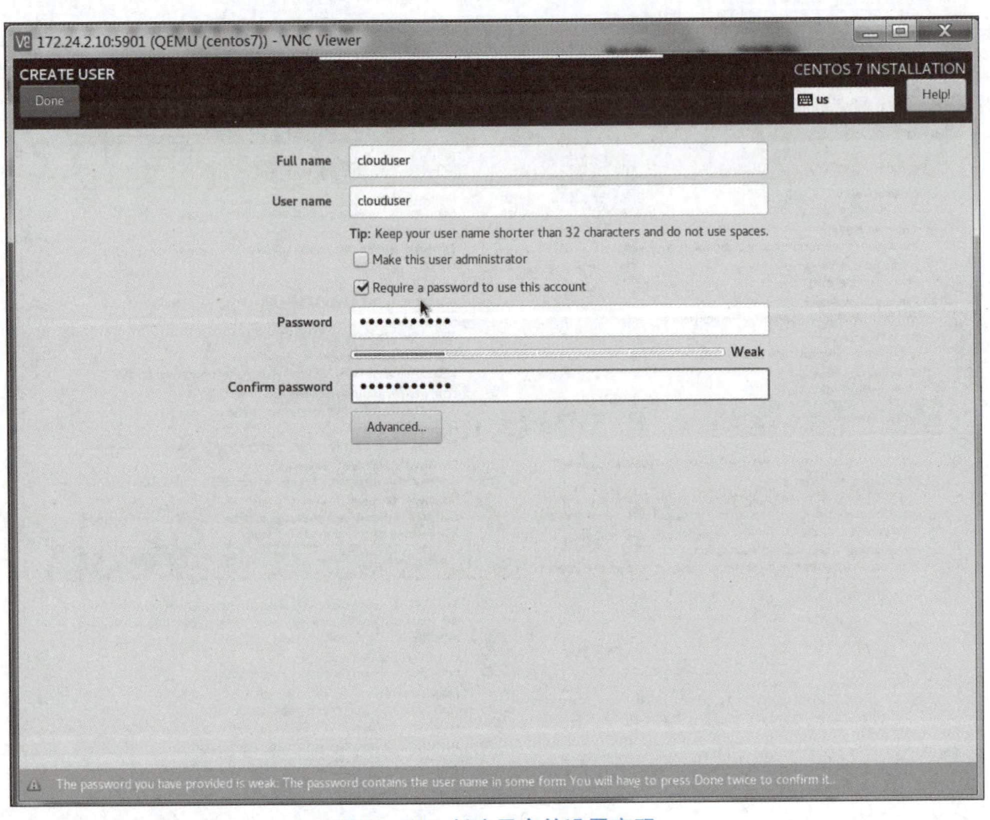

图 5-18 创建用户并设置密码

```
<devices>
<emulator>/usr/libexec/qemu-kvm</emulator>
<disk type='file' device='disk'>
<driver name='qemu' type='qcow2'/>
<source file='/tmp/CentOS7.qcow2'/>
<backingStore/>
<target dev='vda' bus='virtio'/>
<alias name='virtio-disk0'/>
<address type='pci' domain='0x0000' bus='0x00' slot='0x06' function='0x0'/>
</disk>
<disk type='block' device='cdrom'>
   <driver name='qemu' type='raw'/>
   <target dev='hda' bus='ide' tray='open'/>
   <readonly/>
   <alias name='ide0-0-0'/>
   <address type='drive' controller='0' bus='0' target='0' unit='0'/>
 </disk>
......
</domain>
```

（10）controller 运行以下命令，以弹出磁盘并使用 virsh 重新启动 CentOS 7。

```
[root@ controller ~]# virsh attach-disk --type cdrom --mode readonly centos
"" hda
[root@ controller ~]# virsh reboot CentOS7
```

（11）VNC 连接虚拟机，重新启动，设置 license 信息，并勾选接受协议，然后启动虚拟机。跟随导航配置环境界面，进入 CentOS 7 虚拟机桌面。

【任务工单】

工单号：5－2

<table>
<tr><td colspan="2">项目名称：OpenStack 云平台运维</td><td colspan="2">任务名称：镜像服务 Glance 管理</td></tr>
<tr><td colspan="2">班级：</td><td>学号：</td><td>姓名：</td></tr>
<tr><td>任务安排</td><td colspan="3">□创建 cirros 的镜像
□镜像在项目间共享，并通过命令行和 Web 端查看
□显示镜像关联项目
□解除镜像到项目的共享
□设置镜像属性
□制作一个 CentOS 7 镜像</td></tr>
<tr><td>成果交付</td><td colspan="3">实验案例整理成操作指导交付文档</td></tr>
<tr><td>任务实施总结</td><td colspan="3">任务自评（0～10 分）：
任务收获：

改进点：</td></tr>
<tr><td>成果验收</td><td colspan="3">□完全满足任务要求
□基本满足任务要求
要求全部完成，能够对镜像管理，但独立制作 CentOS 7 镜像存在一些问题：

□不能满足需求
要求不能独立完成，无法独立制作 CentOS 7 镜像，对镜像的管理操作不熟练：</td></tr>
</table>

【知识巩固】

1. OpenStack 创建镜像时，--disk-format 参数指定的是（　　）。

A. 镜像磁盘格式　B. 镜像容器格式　C. 镜像内存格式　D. 镜像缓存格式

2. virsh vncdisplay <CentOS 7 虚拟机名字>命令显示的结果是“:1”，那么 VNC 连接到虚拟机时，通过（　　）端口连接。

A. 1　B. 5900　C. 5901　D. 5911

【小李的反思】

合抱之木，生于毫末；九层之台，起于累土；千里之行，始于足下。

语出春秋·楚·李耳《老子》，意思是做事要脚踏实地，一步一个脚印，夯实基础非常重要。在镜像服务 Glance 管理中，通过 Glance 组件制作自定义的 CentOS 7 镜像作为基础工作，关乎后期镜像文件的使用，因此要夯实基础工作，才能在后期工作中顺利使用镜像文件。

衣服穿得好，扣好第一粒扣子很重要。习近平总书记多次使用扣好“第一粒扣子”来比喻引导青少年价值观、帮助青少年迈好人生第一个台阶的重要性。党的十九届六中全会通过的《中共中央关于党的百年奋斗重大成就和历史经验的决议》强调，“党和人民事业发展需要一代代中国共产党人接续奋斗，必须抓好后继有人这个根本大计”，这就更需要帮助青少年扣好人生第一粒扣子。

党的二十大报告指出，全面建成社会主义现代化强国，总的战略安排是分两步走：从二〇二〇年到二〇三五年基本实现社会主义现代化；从二〇三五年到 21 世纪中叶把我国建成富强民主文明和谐美丽的社会主义现代化强国。未来五年是全面建设社会主义现代化国家开局起步的关键时期，作为新时代的青年学子，要扣好人生的第一粒扣子，发挥主观能动性，发奋图强，将来学有所成，为祖国的现代化贡献力量！

任务3 网络服务 Neutron 管理

【任务描述】

小李已经完成学会了在 OpenStack 云平台创建镜像，并通过 VNC 连接使用虚拟机，但虚拟机需要能够远程连接，那么就需要进行 OpenStack 的网络管理，小李决定通过 Neutron 组件创建自服务网络给云主机实例使用。

【知识要点】

1. 网络管理

网络是独立的二层网段，包含两种类型：项目网络和供应商网络。项目网络完全隔离，不与其他项目共享。供应商网络映射到数据中心物理网络，并为服务器和其他资源提供外部网络访问。只有 OpenStack 管理员才能创建供应商网络。网络间可以通过路由器连接。

1）网络创建

```
openstack network create
[ --project <project > ]
[ --enable | --disable]
[ --share | --no - share]
[ --description <description > ]
[ --enable - port - security | --disable - port - security]
[ --external | --internal]
[ --provider - network - type <provider - network - type > ]
[ --provider - physical - network <provider - physical - network > ]
<name >
```

-- project <project>：指定该网络所属的项目。

-- enable | -- disable：启用网络 | 不启用网络。

-- share | -- no - share：项目间共享网络 | 项目间不共享网络。

-- description <description>：添加网络描述信息。

-- enable - port - security | -- disable - port - security：在该网络上创建的端口默认开启端口安全 | 在该网络上创建的端口默认不开启端口安全。

-- external | -- internal：创建的网络为外部网络 | 创建的网络为内部网络。

-- provider - network - type <provider - network - type>：网络虚拟化的物理机制，可以是 flat、geneve、gre、local、vlan、vxlan。

-- provider - physical - network <provider - physical - network>：实现虚拟网络的物理网络。

2）查看网络列表

```
openstack network list
[ --external | --internal]
[ --long]
[ --name <name> ]
[ --enable | --disable]
[ --project <project> ]
[ --share | --no-share]
[ --status <status> ]
[ --provider-network-type <provider-network-type> ]
[ --provider-physical-network <provider-physical-network> ]
```

--external | --internal：根据是内部网络还是外部网络过滤。

--long：显示更多字段。

--name <name>：根据网络名字过滤。

--enable | --disable：根据网络是否启用过滤。

--project <project>：根据项目过滤。

--share | --no-share：根据是否共享过滤。

--status <status>：根据网络状态过滤。

--provider-network-type：根据虚拟化类型过滤。

--provider-physical-network：根据虚拟的物理网络过滤。

3）查看网络详细信息

```
openstack network show
<network>
```

4）设置网络信息

```
openstack network set
[ --name <name> ]
[ --enable | --disable]
[ --share | --no-share]
[ --description <description> ]
[ --enable-port-security | --disable-port-security]
[ --external | --internal]
[ --provider-network-type <provider-network-type> ]
[ --provider-physical-network <provider-physical-network> ]
<network>
```

各项选项参数与 openstack network create 指令相同。

5）删除网络

```
openstack network delete
<network>
```

2. 子网管理

1）子网创建

```
openstack subnet create
[ --project <project> ]
[ --subnet-range <subnet-range> ]
[ --allocation-pool start=<ip-address>,end=<ip-address> ]
[ --dhcp | --no-dhcp]
[ --dns-nameserver <dns-nameserver> ]
[ --gateway <gateway> ]
[ --host-route destination=<subnet>,gateway=<ip-address> ]
[ --ip-version {4,6}]
[ --description <description> ]
 --network <network>
 <name>
```

--project <project>：指定子网所属的项目。

--subnet-range <subnet-range>：以 CIDR 形式指定子网段。

--allocation-pool start=<ip-address>,end=<ip-address>：指定子网段内可分配的 IP 地址池。

--dhcp | --no-dhcp：开启 DHCP | 不开启 DHCP。

--dns-nameserver <dns-nameserver>：设置子网的 DNS 服务器。

--gateway <gateway>：设置子网的网关，可用选项包含三个：IP 地址，表示以某个具体的 IP 地址作为网关地址；auto，表示从子网内自动选择一个作为子网的网关；none，表示不使用网关。

--host-route destination=<subnet>,gateway=<ip-address>：添加路由，例如：destination=10.10.0.0/16，gateway=192.168.71.254，destination 表示目的网段，gateway 表示下一跳 IP 地址。

--ip-version {4,6}：指定 IP 地址版本，当子网地址池指定后，IP 地址版本由 IP 地址池确定，此选项失效。

--description <description>：添加子网描述。

--network <network>：指定子网所属的网络。

2）查看子网列表

```
openstack subnet list
[ --long]
[ --ip-version {4,6}]
[ --dhcp | --no-dhcp]
[ --project <project> ]
[ --network <network> ]
[ --gateway <gateway> ]
[ --name <name> ]
```

```
[--subnet-range <subnet-range>]
```

各选项含义与 openstack subnet create 命令相同。

3）查看子网详细信息

```
openstack subnet show
<subnet>
```

4）设置子网信息

```
openstack subnet set
[--allocation-pool start=<ip-address>,end=<ip-address>]
[--no-allocation-pool]
[--dhcp|--no-dhcp]
[--dns-nameserver <dns-nameserver>]
[--no-dns-nameserver]
[--gateway <gateway-ip>]
[--host-route destination=<subnet>,gateway=<ip-address>]
[--no-host-route]
[--name <new-name>]
[--description <description>]
<subnet>
```

--no-allocation-pool：清除子网的 IP 地址池，一般将 --allocation-pool 和 --no-allocation-pool 选项一起使用，覆盖当前的 IP 地址池。

--no-dns-nameserver：清除当前的 DNS 服务器，一般将 --dns-nameserver 和 --no-dns-nameserver 选项一起使用，覆盖当前的 DNS 服务器。

--no-host-route：清除当前的主机路由信息，一般将 --host-route 和 --no-host-route 选项一起使用，覆盖当前的主机路由。

--name <new-name>：设置子网的名字。

其他选项含义与 openstack subnet create 命令相同。

5）取消子网信息设置

```
openstack subnet unset
[--allocation-pool start=<ip-address>,end=<ip-address>[...]]
[--dns-nameserver <dns-nameserver>[...]]
[--host-route destination=<subnet>,gateway=<ip-address>[...]]
<subnet>
```

各选项含义与 openstack subnet create 命令相同。

6）删除子网

```
openstack subnet delete
<subnet>
```

3. 路由器管理

路由器是一个用来在网络间进行数据转发的逻辑部件，同时也提供三层和 NAT 功能，

为项目网络提供外部网络访问。

1）创建路由器

```
openstack router create
[ --project <project> ]
[ --enable | --disable]
<name>
```

--project <project>：指定路由器所属的项目。

--enable | --disable：指定路由器启用 | 不启用。

2）查看路由器列表

```
openstack router list
[ --name <name> ]
[ --enable | --disable]
[ --long]
[ --project <project> ]
```

--name <name>：根据名字进行过滤。

--enable | --disable：根据是否启用过滤。

--project <project>：根据项目过滤。

--long：查看更多选项。

3）查看路由器详细信息

```
openstack router show
<router>
```

4）设置路由器信息

```
openstack router set
[ --name <name> ]
[ --enable | --disable]
[ --route destination = <subnet>,gateway = <ip-address> | --no-route]
[ --external-gateway <network> ]
<router>
```

--name <name>：设置路由器名字。

--enable | --disable：设置路由器是否启用。

--route destination = <subnet>, gateway = <ip-address>：为路由器添加路由信息，destination 表示目的地址，gateway 表示下一跳。

--no-route：清除所有路由信息，一般将 --route 和 --no-route 选项一起使用，覆盖当前的路由信息。

--external-gateway <network>：设置外部网络作为路由器网关。

5）取消路由器信息设置

```
openstack router unset
[ --route destination = <subnet>,gateway = <ip-address> ]
<router>
```

--route destination=<subnet>,gateway=<ip-address>：取消具体某条路由信息的设置。

6）向路由器添加子网

```
openstack router add subnet
<router>
<subnet>
```

7）从路由器删除子网

```
openstack router remove subnet
<router>
<subnet>
```

8）删除路由器

```
openstack router delete
<router>[<router>...]
```

4. 安全组管理

在传统网络拓扑中，主机可以访问哪些数据流量，可以通过定义访问控制列表（ACL）对进出网络的流量进行过滤。对于云主机来说，能够访问哪些流量可以通过安全组（Security Group）定义。安全组作为网络中服务器和其他资源的虚拟防火墙，它是由一系列安全组规则组成的，通过安全组规则定义数据流量的过滤规则。

1）创建安全组

```
openstack security group create
[--description<description>]
[--project<project>]
<name>
```

--description <description>：指定安全组的描述信息。

--project <project>：指定安全组所属的项目。

2）查看安全组列表

```
openstack security group list
[--all-projects]
[--project<project>]
```

--project <project>：根据项目进行安全组过滤。

--all-projects：查看所有项目的安全组。

3）查看安全组详细信息

```
openstack security group show
<group>
```

4）设置安全组信息

```
openstack security group set
```

```
[ --name <new-name> ]
[ --description <description> ]
<group>
```

--name <new-name>：修改安全组名字。

--description <description>：修改安全组描述信息。

5）删除安全组

```
openstack security group delete
<group>[ <group>... ]
```

5. 安全组规则管理

1）安全组规则创建

```
openstack security group rule create
[ --remote-ip <ip-address> ]
[ --dst-port <port-range> ]
[ --protocol <protocol> ]
[ --ingress | --egress]
[ --project <project> ]
[ --description <description> ]
<group>
```

--remote-ip <ip-address>：指定远端 IP 地址段（以 CIRD 格式指定，默认是 0.0.0.0/0，表示所有 IP 地址）。

--dst-port <port-range>：目的端口，可以是一个端口（例如：22），也可以是一个端口范围（例如：137:139，表示 137、138 和 139 三个端口），主要用于 TCP 协议和 UDP 协议。

--protocol <protocol>：指定 IP 协议。

--ingress |--egress：指定安全组规则应用于入方向（默认）| 出方向。

--project <project>：指定所属的项目。

--description <description>：添加描述信息。

2）查看安全组规则列表

```
openstack security group rule list
[ --all-projects]
[ --protocol <protocol> ]
[ --ingress | --egress]
[ --long]
[ <group> ]
```

--all-projects：查看所有项目的安全组规则。

--protocol <protocol>：根据协议进行过滤。

--ingress |--egress：根据规则应用在入方向 | 出方向进行过滤。

--long：显示更多字段。

<group>：查看某个安全组内的全部安全组规则。

3）查看安全组规则详细信息

```
openstack security group rule show
<rule>
```

4）删除安全组规则

```
openstack security group rule delete
<rule>[<rule>...]
```

6. 浮动 IP 管理

1）浮动 IP 创建

```
openstack floating ip create
[--subnet <subnet>]
[--port <port>]
[--floating-ip-address <ip-address>]
[--fixed-ip-address <ip-address>]
[--description <description>]
[--project <project>]
<network>
```

--subnet <subnet>：指定浮动 IP 所属的子网。

--port <port>：指定浮动 IP 应用的端口。

--floating-ip-address <ip-address>：设定浮动 IP 地址。

--fixed-ip-address <ip-address>：设定映射到浮动 IP 的固定 IP 地址。

--description <description>：设定浮动 IP 的描述信息。

--project <project>：设定浮动 IP 所属的项目。

2）查看浮动 IP 地址

```
openstack floating ip list
[--network <network>]
[--port <port>]
[--long]
[--project <project>]
[--router <router>]
```

--network <network>：根据网络过滤浮动 IP。

--port <port>：根据端口过滤。

--long：查看更多字段。

--project <project>：根据项目过滤。

--router <router>：根据路由器过滤。

3）查看浮动 IP 具体信息

```
openstack floating ip show <floating-ip>
```

4）设置浮动 IP 信息

```
openstack floating ip set
[ --port <port> ]
[ --fixed-ip-address <ip-address> ]
<floating-ip>
```

--port <port>：设定浮动 IP 绑定的端口。

--fixed-ip-address <ip-address>：设定浮动 IP 映射的固定 IP 地址。

5）取消浮动 IP 信息设置

```
openstack floating ip unset
[ --port]
<floating-ip>
```

--port：取消浮动 IP 与端口的绑定。

6）删除浮动 IP

```
openstack floating ip delete <floating-ip>[ <floating-ip>... ]
```

【任务实施】

1. 网络管理

1）网络创建

（1）创建一个 int-net 网络，该网络用于内部连接云主机。

```
[root@ controller ~]# openstack network create int-net
+---------------------------+--------------------------------------+
| Field                     | Value                                |
+---------------------------+--------------------------------------+
| admin_state_up            | UP                                   |
| availability_zone_hints   |                                      |
| availability_zones        |                                      |
| created_at                | 2022-12-08T08:58:01Z                 |
| description               |                                      |
| dns_domain                | None                                 |
| id                        | cbefdbc8-50cb-497d-ae51-665dfbd9c50b |
| ipv4_address_scope        | None                                 |
| ipv6_address_scope        | None                                 |
| is_default                | False                                |
| is_vlan_transparent       | None                                 |
| mtu                       | 1450                                 |
| name                      | int-net                              |
| port_security_enabled     | True                                 |
| project_id                | 62b92970ed2a42cb9ea0e8f013004062     |
```

```
| provider:network_type        | vxlan                                 |
| provider:physical_network    | None                                  |
| provider:segmentation_id     | 17                                    |
| qos_policy_id                | None                                  |
| revision_number              | 1                                     |
| router:external              | Internal                              |
| segments                     | None                                  |
| shared                       | False                                 |
| status                       | ACTIVE                                |
| subnets                      |                                       |
| tags                         |                                       |
| updated_at                   | 2022-12-08T08:58:01Z                  |
+------------------------------+---------------------------------------+
```

（2）创建一个 ext-net 网络用于与外网连接，所有项目共享，在 provider 网络上通过 flat 方式实现虚拟化。

```
[root@ controller ~]# openstack network create --share --external \
>--provider-physical-network provider \
>--provider-network-type flat ext-net
+------------------------------+---------------------------------------+
| Field                        | Value                                 |
+------------------------------+---------------------------------------+
| admin_state_up               | UP                                    |
| availability_zone_hints      |                                       |
| availability_zones           |                                       |
| created_at                   | 2022-12-08T08:52:36Z                  |
| description                  |                                       |
| dns_domain                   | None                                  |
| id                           | 500a566c-cffe-4700-929e-5947040e49b5 |
| ipv4_address_scope           | None                                  |
| ipv6_address_scope           | None                                  |
| is_default                   | False                                 |
| is_vlan_transparent          | None                                  |
| mtu                          | 1500                                  |
| name                         | ext-net                               |
| port_security_enabled        | True                                  |
| project_id                   | 62b92970ed2a42cb9ea0e8f013004062      |
| provider:network_type        | flat                                  |
| provider:physical_network    | provider                              |
| provider:segmentation_id     | None                                  |
| qos_policy_id                | None                                  |
| revision_number              | 1                                     |
| router:external              | External                              |
```

```
| segments                  | None                                  |
| shared                    | True                                  |
| status                    | ACTIVE                                |
| subnets                   |                                       |
| tags                      |                                       |
| updated_at                | 2022-12-08T08:52:36Z                  |
+---------------------------+---------------------------------------+
```

2）查看网络列表

```
[root@ controller ~]# openstack network list
+--------------------------------------+---------+---------+
| ID                                   | Name    | Subnets |
+--------------------------------------+---------+---------+
| 500a566c-cffe-4700-929e-5947040e49b5 | ext-net |         |
| cbefdbc8-50cb-497d-ae51-665dfbd9c50b | int-net |         |
+--------------------------------------+---------+---------+
```

3）查看网络详细信息

查看 ext-net 网络的详细信息。

```
[root@ controller ~]# openstack network show ext-net
+---------------------------+---------------------------------------+
| Field                     | Value                                 |
+---------------------------+---------------------------------------+
| admin_state_up            | UP                                    |
| availability_zone_hints   |                                       |
| availability_zones        |                                       |
| created_at                | 2022-12-08T08:52:36Z                  |
| description               |                                       |
| dns_domain                | None                                  |
| id                        | 500a566c-cffe-4700-929e-5947040e49b5  |
| ipv4_address_scope        | None                                  |
| ipv6_address_scope        | None                                  |
| is_default                | False                                 |
| is_vlan_transparent       | None                                  |
| mtu                       | 1500                                  |
| name                      | ext-net                               |
| port_security_enabled     | True                                  |
| project_id                | 62b92970ed2a42cb9ea0e8f013004062      |
| provider:network_type     | flat                                  |
| provider:physical_network | provider                              |
| provider:segmentation_id  | None                                  |
| qos_policy_id             | None                                  |
| revision_number           | 1                                     |
```

```
| router:external              | External                              |
| segments                     | None                                  |
| shared                       | True                                  |
| status                       | ACTIVE                                |
| subnets                      |                                       |
| tags                         |                                       |
| updated_at                   | 2022-12-08T08:52:36Z                  |
+------------------------------+---------------------------------------+
```

4）设置网络信息

修改 int-net 网络信息，将名字改为 selfservice，非共享，描述信息为“selfservice network”。

```
[root@ controller ~]# openstack network set --name selfservice --no-share --description "selfservice network" int-net
[root@ controller ~]# openstack network show selfservice
+------------------------------+---------------------------------------+
| Field                        | Value                                 |
+------------------------------+---------------------------------------+
| admin_state_up               | UP                                    |
| availability_zone_hints      |                                       |
| availability_zones           |                                       |
| created_at                   | 2022-12-08T08:58:01Z                  |
| description                  | selfservice network                   |
| dns_domain                   | None                                  |
| id                           | cbefdbc8-50cb-497d-ae51-665dfbd9c50b |
| ipv4_address_scope           | None                                  |
| ipv6_address_scope           | None                                  |
| is_default                   | None                                  |
| is_vlan_transparent          | None                                  |
| mtu                          | 1450                                  |
| name                         | selfservice                           |
| port_security_enabled        | True                                  |
| project_id                   | 62b92970ed2a42cb9ea0e8f013004062      |
| provider:network_type        | vxlan                                 |
| provider:physical_network    | None                                  |
| provider:segmentation_id     | 17                                    |
| qos_policy_id                | None                                  |
| revision_number              | 4                                     |
| router:external              | Internal                              |
| segments                     | None                                  |
| shared                       | False                                 |
| status                       | ACTIVE                                |
| subnets                      |                                       |
```

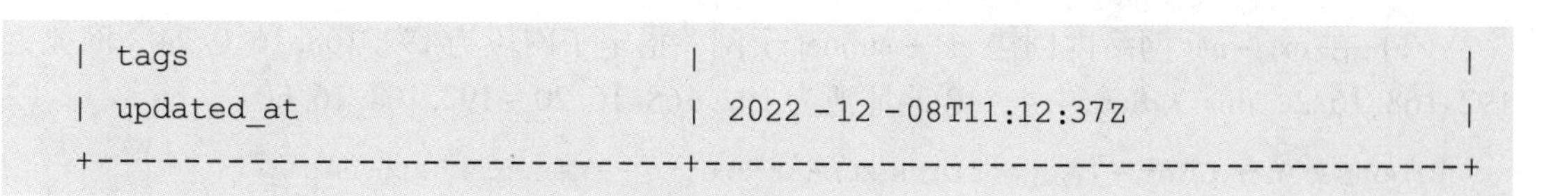

```
| tags                 |                                        |
| updated_at           | 2022-12-08T11:12:37Z                   |
+----------------------+----------------------------------------+
```

从 openstack network show 命令的显示结果可以看出，openstack network set 命令设置的参数已经生效。

5）删除网络

删除 selfservice 网络。

```
[root@ controller ~]# openstack network delete selfservice
```

2. 子网管理

1）创建子网

（1）在 int-net 网络下创建 int-subnet 子网，设定 dns 为 8.8.4.4，子网段为 172.16.1.0/24，网关为 172.16.1.1。

```
[root@ controller ~]# openstack subnet create --network int-net \
>--dns-nameserver 8.8.4.4 --gateway 172.16.1.1 \
>--subnet-range 172.16.1.0/24 int-subnet
+-------------------+--------------------------------------+
| Field             | Value                                |
+-------------------+--------------------------------------+
| allocation_pools  | 172.16.1.2-172.16.1.254              |
| cidr              | 172.16.1.0/24                        |
| created_at        | 2022-12-08T13:45:14Z                 |
| description       |                                      |
| dns_nameservers   | 8.8.4.4                              |
| enable_dhcp       | True                                 |
| gateway_ip        | 172.16.1.1                           |
| host_routes       |                                      |
| id                | 38bc7e6e-9696-4e66-a1cc-827fdabd76e4 |
| ip_version        | 4                                    |
| ipv6_address_mode | None                                 |
| ipv6_ra_mode      | None                                 |
| name              | int-subnet                           |
| network_id        | 141f4049-d056-4139-b8a1-fc8428b66751 |
| project_id        | 62b92970ed2a42cb9ea0e8f013004062     |
| revision_number   | 0                                    |
| segment_id        | None                                 |
| service_types     |                                      |
| subnetpool_id     | None                                 |
| tags              |                                      |
| updated_at        | 2022-12-08T13:45:14Z                 |
+-------------------+--------------------------------------+
```

（2）在 ext - net 网络内创建 ext - subnet 子网，指定子网段为 192.168.16.0/24，网关为 192.168.16.2，dns 为 8.8.4.4，IP 地址池为 192.168.16.20～192.168.16.60。

```
[root@ controller ~]# openstack subnet create --network ext-net\
>--subnet-range 192.168.16.0/24 --gateway 192.168.16.2\
>--dns-nameserver 8.8.4.4 --allocation-pool start=192.168.16.20,end=
192.168.16.60\
>ext-subnet
+-------------------+--------------------------------------+
| Field             | Value                                |
+-------------------+--------------------------------------+
| allocation_pools  | 192.168.16.20-192.168.16.60          |
| cidr              | 192.168.16.0/24                      |
| created_at        | 2022-12-08T13:37:54Z                 |
| description       |                                      |
| dns_nameservers   | 8.8.4.4                              |
| enable_dhcp       | True                                 |
| gateway_ip        | 192.168.16.2                         |
| host_routes       |                                      |
| id                | 9d95d7ff-c1b2-4435-a70f-e4c22c9fc85b |
| ip_version        | 4                                    |
| ipv6_address_mode | None                                 |
| ipv6_ra_mode      | None                                 |
| name              | ext-subnet                           |
| network_id        | 500a566c-cffe-4700-929e-5947040e49b5 |
| project_id        | 62b92970ed2a42cb9ea0e8f013004062     |
| revision_number   | 0                                    |
| segment_id        | None                                 |
| service_types     |                                      |
| subnetpool_id     | None                                 |
| tags              |                                      |
| updated_at        | 2022-12-08T13:37:54Z                 |
+-------------------+--------------------------------------+
```

2）查看子网列表

（1）查看所有子网列表。

```
[root@ controller ~]# openstack subnet list
+-----------+------------+-----------+----------------+
| ID        | Name       | Network   | Subnet         |
+-----------+------------+-----------+----------------+
| 38bc…76e4 | int-subnet | 141f…6751 | 172.16.1.0/24  |
| 9d95…c85b | ext-subnet | 500a…49b5 | 192.168.16.0/24 |
+-----------+------------+-----------+----------------+
```

（2）查看 ext - net 网络下的子网列表。

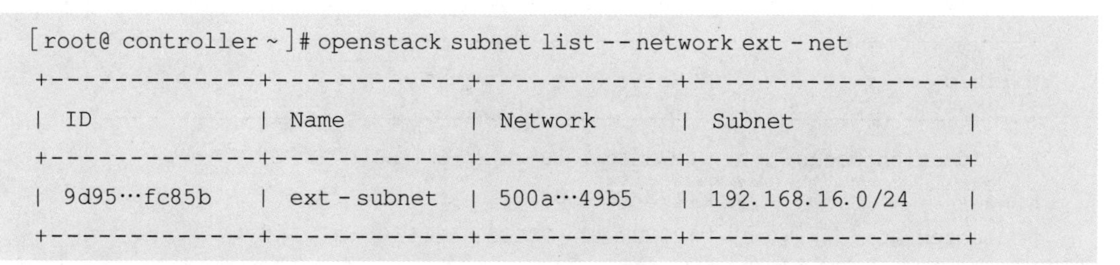

```
[root@ controller ~]# openstack subnet list --network ext-net
+------------+------------+------------+---------------+
| ID         | Name       | Network    | Subnet        |
+------------+------------+------------+---------------+
| 9d95…fc85b | ext-subnet | 500a…49b5  | 192.168.16.0/24 |
+------------+------------+------------+---------------+
```

3）查看子网详细信息

查看 ext-subnet 子网的详细信息。

```
[root@ controller ~]# openstack subnet show ext-subnet
+-------------------+--------------------------------------+
| Field             | Value                                |
+-------------------+--------------------------------------+
| allocation_pools  | 192.168.16.20-192.168.16.60          |
| cidr              | 192.168.16.0/24                      |
| created_at        | 2022-12-08T13:37:54Z                 |
| description       |                                      |
| dns_nameservers   | 8.8.4.4                              |
| enable_dhcp       | True                                 |
| gateway_ip        | 192.168.16.2                         |
| host_routes       |                                      |
| id                | 9d95d7ff-c1b2-4435-a70f-e4c22c9fc85b |
| ip_version        | 4                                    |
| ipv6_address_mode | None                                 |
| ipv6_ra_mode      | None                                 |
| name              | ext-subnet                           |
| network_id        | 500a566c-cffe-4700-929e-5947040e49b5 |
| project_id        | 62b92970ed2a42cb9ea0e8f013004062     |
| revision_number   | 0                                    |
| segment_id        | None                                 |
| service_types     |                                      |
| subnetpool_id     | None                                 |
| tags              |                                      |
| updated_at        | 2022-12-08T13:37:54Z                 |
+-------------------+--------------------------------------+
```

4）设置子网信息

设置 int-subnet 子网的子网段：

```
[root@ controller ~]# openstack subnet set int-subnet \
>--no-allocation-pool --allocation-pool start=172.16.1.10,end=172.16.1.20 \
>--no-dns-nameserver --dns-nameserver 8.8.8.8 \
>--gateway 172.16.1.2 --name subint-net
[root@ controller ~]# openstack subnet show subint-net
```

```
+-------------------+---------------------------------------+
| Field             | Value                                 |
+-------------------+---------------------------------------+
| allocation_pools  | 172.16.1.10-172.16.1.20               |
| cidr              | 172.16.1.0/24                         |
| created_at        | 2022-12-08T13:45:14Z                  |
| description       |                                       |
| dns_nameservers   | 8.8.8.8                               |
| enable_dhcp       | True                                  |
| gateway_ip        | 172.16.1.2                            |
| host_routes       |                                       |
| id                | 38bc7e6e-9696-4e66-a1cc-827fdabd76e4  |
| ip_version        | 4                                     |
| ipv6_address_mode | None                                  |
| ipv6_ra_mode      | None                                  |
| name              | subint-net                            |
| network_id        | 141f4049-d056-4139-b8a1-fc8428b66751  |
| project_id        | 62b92970ed2a42cb9ea0e8f013004062      |
| revision_number   | 1                                     |
| segment_id        | None                                  |
| service_types     |                                       |
| subnetpool_id     | None                                  |
| tags              |                                       |
| updated_at        | 2022-12-08T13:59:40Z                  |
+-------------------+---------------------------------------+
```

从 openstack subnet show 命令的显示结果可以看出，openstack subnet set 命令设置的参数已经生效。

5）取消子网信息设置

取消 subint-net 子网的 IP 地址池和 DNS 设置：

```
[root@ controller ~]# openstack subnet unset \
>--allocation-pool start=172.16.1.10,end=172.16.1.20 \
>--dns-nameserver 8.8.8.8 subint-net
[root@ controller ~]# openstack subnet show subint-net
+-------------------+---------------------------------------+
| Field             | Value                                 |
+-------------------+---------------------------------------+
| allocation_pools  |                                       |
| cidr              | 172.16.1.0/24                         |
| created_at        | 2022-12-08T13:45:14Z                  |
| description       |                                       |
| dns_nameservers   |                                       |
| enable_dhcp       | True                                  |
```

```
| gateway_ip           | 172.16.1.2                            |
| host_routes          |                                       |
| id                   | 38bc7e6e-9696-4e66-a1cc-827fdabd76e4  |
| ip_version           | 4                                     |
| ipv6_address_mode    | None                                  |
| ipv6_ra_mode         | None                                  |
| name                 | subint-net                            |
| network_id           | 141f4049-d056-4139-b8a1-fc8428b66751  |
| project_id           | 62b92970ed2a42cb9ea0e8f013004062      |
| revision_number      | 2                                     |
| segment_id           | None                                  |
| service_types        |                                       |
| subnetpool_id        | None                                  |
| tags                 |                                       |
| updated_at           | 2022-12-08T14:03:46Z                  |
+----------------------+---------------------------------------+
```

从 openstack subnet show 命令的显示结果可以看出，openstack subnet unset 命令取消的参数设置已经生效。

6）删除子网

删除 subint-net 子网：

```
[root@ controller ~]# openstack subnet delete subint-net
[root@ controller ~]# openstack subnet list
+------------+------------+------------+-----------------+
| ID         | Name       | Network    | Subnet          |
+------------+------------+------------+-----------------+
| 9d95…c85b  | ext-subnet | 500a…49b5  | 192.168.16.0/24 |
+------------+------------+------------+-----------------+
```

从查看子网列表的结果中可以看出，subint-net 子网已经被删除不存在了。

3. 路由器管理

1）路由器创建

创建一个路由器，命名为 router：

```
[root@ controller ~]# openstack router create router
+-------------------------+--------------------------------------+
| Field                   | Value                                |
+-------------------------+--------------------------------------+
| admin_state_up          | UP                                   |
| availability_zone_hints |                                      |
| availability_zones      |                                      |
| created_at              | 2022-12-09T06:16:13Z                 |
| description             |                                      |
```

```
| distributed               | False                                   |
| external_gateway_info     | None                                    |
| flavor_id                 | None                                    |
| ha                        | False                                   |
| id                        | 90bcb531-23d2-44df-b10b-1f766fa0395f    |
| name                      | router                                  |
| project_id                | 62b92970ed2a42cb9ea0e8f013004062        |
| revision_number           | 1                                       |
| routes                    |                                         |
| status                    | ACTIVE                                  |
| tags                      |                                         |
| updated_at                | 2022-12-09T06:16:13Z                    |
+---------------------------+-----------------------------------------+
```

2）查看路由器列表

```
[root@ controller ~]# openstack router list
+-------------+------+--------+------+-----------+-------------+----+
| ID          | Name | Status | State | Project  | Distributed | HA |
+-------------+------+--------+------+-----------+-------------+----+
| 90bc…395f   | router| ACTIVE | UP  | 62b9…4062 | False       | False
+-------------+------+--------+------+-----------+-------------+----+
```

3）查看路由器详细信息

查看路由器 router 的详细信息：

```
[root@ controller ~]# openstack router show router
+---------------------------+-----------------------------------------+
| Field                     | Value                                   |
+---------------------------+-----------------------------------------+
| admin_state_up            | UP                                      |
| availability_zone_hints   |                                         |
| availability_zones        |                                         |
| created_at                | 2022-12-09T06:16:13Z                    |
| description               |                                         |
| distributed               | False                                   |
| external_gateway_info     | None                                    |
| flavor_id                 | None                                    |
| ha                        | False                                   |
| id                        | 90bcb531-23d2-44df-b10b-1f766fa0395f    |
| interfaces_info           | []                                      |
| name                      | router                                  |
| project_id                | 62b92970ed2a42cb9ea0e8f013004062        |
| revision_number           | 1                                       |
| routes                    |                                         |
```

```
| status                       | ACTIVE                                    |
| tags                         |                                           |
| updated_at                   | 2022-12-09T06:16:13Z                      |
+------------------------------+-------------------------------------------+
```

4）设置路由器信息

（1）设置 ext-net 网络作为路由器 router 连接外网的网关。

```
[root@ controller ~]# openstack router set router --external-gateway ext-net
```

查看路由器详细信息，可以看到，router 路由器的 external_gateway_info 的值为 {"network_id":"a70e14a6-2567-4de3-9d4f-1bca3cdcba13","enable_snat":true,"external_fixed_ips":[{"subnet_id":"1886f56c-fbe6-43f5-8def-5a22aa8fe175","ip_address":"192.168.16.24"}]}，可以看出路由器连接外网的网关端口 IP 为 192.168.16.24。

（2）从 controller 节点 ping 该 IP 地址。

```
[root@ controller ~]# ping 192.168.16.24 -c 5
PING 192.168.16.24(192.168.16.24)56(84)bytes of data.
64bytes from 192.168.16.24:icmp_seq=1 ttl=64 time=0.099ms
64bytes from 192.168.16.24:icmp_seq=2 ttl=64 time=0.065ms
64bytes from 192.168.16.24:icmp_seq=3 ttl=64 time=0.060ms
64bytes from 192.168.16.24:icmp_seq=4 ttl=64 time=0.060ms
64bytes from 192.168.16.24:icmp_seq=5 ttl=64 time=0.055ms

---192.168.16.24 ping statistics---
5 packets transmitted,5 received,0% packet loss,time 3999ms
rtt min/avg/max/mdev=0.055/0.067/0.099/0.019 m
```

（3）从物理主机 ping 该 IP 地址。

```
C:\Users\ly>ping 192.168.16.24
```

正在 ping 192.168.16.24 具有 32 字节的数据：
来自 192.168.16.24 的回复：字节 =32 时间 <1 ms TTL=64。
来自 192.168.16.24 的回复：字节 =32 时间 <1 ms TTL=64。
来自 192.168.16.24 的回复：字节 =32 时间 <1 ms TTL=64。
来自 192.168.16.24 的回复：字节 =32 时间 <1 ms TTL=64。
192.168.16.24 的 ping 统计信息：
数据包：已发送 =4，已接收 =4，丢失 =0（0% 丢失）
往返行程的估计时间（以毫秒为单位）：
最短 =0 ms，最长 =0 ms，平均 =0 ms

5）向路由器添加子网

将 int-subnet 子网添加到路由器 router：

```
[root@ controller ~]# openstack router add subnet router int-subnet
```

查看路由器详细信息，可以看到，router 路由器的 interfaces_info 值为 [{"subnet_id":

"bd8670d7 - aea3 - 44e8 - 83ee - 82714e0ce644"," ip_address":"172.16.1.1"," port_id":"64c625b4 - 0a7a - 4ad0 - 8632 - 2eebd43a5558"}]

6）从路由器删除子网

将 int - subnet 子网从路由器 router 移除：

```
[root@ controller ~]# openstack router remove subnet router int - subnet
```

查看路由器详细信息可以看到，router 路由器的 interfaces_ info 信息里已经移除了 int - subnet 的子网信息。

7）删除路由器

删除 router 路由器：

```
[root@ controller ~]# openstack router delete router
```

查看路由器列表，发现 router 路由器已经不存在了。

4. 安全组管理

1）创建安全组

创建安全组 sg - test，描述信息为“security group for test”，所属项目为 myproject。

```
[root@ controller ~]# openstack security group create -- description "security group for test" -- project myproject sg - test
+-----------------+-----------------------------------------------------+
| Field           | Value                                               |
+-----------------+-----------------------------------------------------+
| created_at      | 2022 -12 -15T05:55:33Z                              |
| description     | security group for test                             |
| id              | 439ae5ea -50f9 -44f6 -90aa -0645f21f87aa            |
| name            | sg - test                                           |
| project_id      | e598a75ab1434d4ea2cfa4e0da5ae9ee                    |
| revision_number | 1                                                   |
| rules           | created_at ='…',…,id =' 64…c1',updated_at ='…'      |
|                 | created_at ='…',…,id ='c7…e1',updated_at ='…'       |
| tags            | []                                                  |
| updated_at      | 2022 -12 -15T05:55:33Z                              |
+-----------------+-----------------------------------------------------+
```

2）查看安全组列表

（1）查看安全组列表，默认查看所有项目。

```
[root@ controller ~]# openstack security group list
+---------+---------+-------------------------+---------+------+
| ID      | Name    | Description             | Project | Tags |
+---------+---------+-------------------------+---------+------+
| 439…7aa | sg - test | security group for test | e59…9ee | []   |
| 543…220 | default | Default security group  |         | []   |
```

```
| 55b…731    | default   | Default security group         | e59…9ee   | []  |
| cfa…e27    | default   | Default security group         | 62b…062   | []  |
| cff…f07    | default   | Default security group         | 836…b97   | []  |
+----------+---------+---------------------------+----------+-----+
```

(2) 查看 myproject 项目下的安全组列表。

```
[root@ controller ~]# openstack security group list --project myproject
+----------+---------+---------------------------+----------+-----+
| ID         | Name      | Description                    | Project   | Tags |
+----------+---------+---------------------------+----------+-----+
| 439…7aa    | sg-test   | security group for test        | e59…9ee   | []  |
| 55b…731    | default   | Default security group         | e59…9ee   | []  |
+----------+---------+---------------------------+----------+-----+
```

3) 查看安全组详细信息

查看 sg-test 安全组的详细信息：

```
[root@ controller ~]# openstack security group show sg-test
+-----------------+---------------------------------------------+
| Field            | Value                                                  |
+-----------------+---------------------------------------------+
| created_at       | 2022-12-15T05:55:33Z                                   |
| description      | security group for test                                |
| id               | 439ae5ea-50f9-44f6-90aa-0645f21f87aa                   |
| name             | sg-test                                                |
| project_id       | e598a75ab1434d4ea2cfa4e0da5ae9ee                       |
| revision_number  | 1                                                      |
| rules            | created_at='…',…,id=' 64…c1',updated_at='…'            |
|                  | created_at='…',…,id='c7…e1',updated_at='…'             |
| tags             | []                                                     |
| updated_at       | 2022-12-15T05:55:33Z                                   |
+-----------------+---------------------------------------------+
```

4) 设置安全组信息

修改 sg-test 安全组名称为 sg-new，描述信息为“sg-test changed to sg-new”。

```
[root@ controller ~]# openstack security group set --name sg-new --description "
sg-test changed to sg-new" sg-test
[root@ controller ~]# openstack security group show sg-new
+-----------------+---------------------------------------------+
| Field            | Value                                                  |
+-----------------+---------------------------------------------+
| created_at       | 2022-12-15T05:55:33Z                                   |
| description      | sg-test changed to sg-new                              |
| id               | 439a…7aa                                               |
```

```
| name              | sg-new                                          |
| project_id        | e598a75ab1434d4ea2cfa4e0da5ae9ee                |
| revision_number   | 2                                               |
| rules             | created_at='…',…,id='64…c1',updated_at='…'      |
|                   | created_at='…',…,id='c7…e1',updated_at='…'      |
| tags              | []                                              |
| updated_at        | 2022-12-15T06:19:09Z                            |
+-------------------+-------------------------------------------------+
```

5）删除安全组

删除 sg-new 安全组：

```
[root@ controller ~]# openstack security group delete sg-new
```

5. 安全组规则管理

1）安全组规则创建

在 sg-test 安全组下添加安全组规则，放行 TCP 协议的 22 端口，描述信息为“allow tcp:22 for ssh”。

```
[root@ controller ~]# openstack security group rule create sg-test --protocol tcp --dst-port 22 --description "allow tcp:22 for ssh"
+-------------------+------------------------------------------+
| Field             | Value                                    |
+-------------------+------------------------------------------+
| created_at        | 2022-12-15T06:37:00Z                     |
| description       | allow tcp:22 for ssh                     |
| direction         | ingress                                  |
| ether_type        | IPv4                                     |
| id                | 642d6ebe-3fb6-4a52-bd1a-889d196351fa     |
| name              | None                                     |
| port_range_max    | 22                                       |
| port_range_min    | 22                                       |
| project_id        | 62b92970ed2a42cb9ea0e8f013004062         |
| protocol          | tcp                                      |
| remote_group_id   | None                                     |
| remote_ip_prefix  | 0.0.0.0/0                                |
| revision_number   | 0                                        |
| security_group_id | a4b65b0d-2e48-47b9-8c12-9693e9176dec     |
| updated_at        | 2022-12-15T06:37:00Z                     |
+-------------------+------------------------------------------+
```

2）查看安全组规则列表

（1）查看开放 TCP 协议的安全组规则。

```
[root@ controller ~]# openstack security group rule list --protocol tcp
```

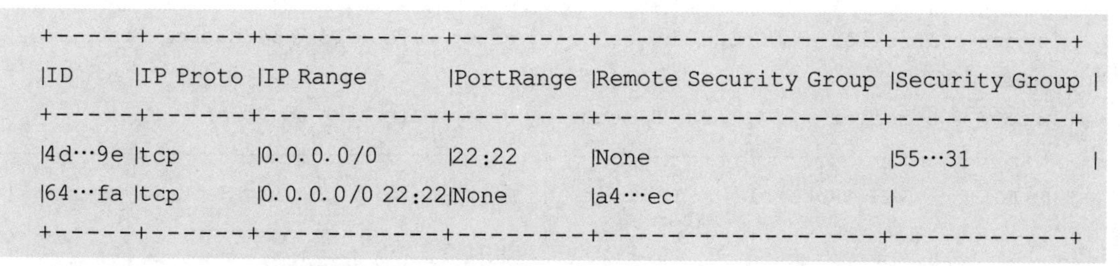

```
+-----+------+----------+--------+-------------------+-----------+
|ID    |IP Proto |IP Range       |PortRange |Remote Security Group |Security Group |
+-----+------+----------+--------+-------------------+-----------+
|4d…9e |tcp      |0.0.0.0/0      |22:22     |None                  |55…31          |
|64…fa |tcp      |0.0.0.0/0 22:22|None      |a4…ec                 |
+-----+------+----------+--------+-------------------+-----------+
```

（2）查看 sg－test 安全组下的安全组规则。

```
[root@ controller ~]# openstack security group rule list sg-test
+---------+-----------+---------+---------+------------------+
| ID       | IP Protocol  | IP Range    | Port Range | Remote Security Group |
+---------+-----------+---------+---------+------------------+
| 01…5e   | None         | None        |            | None                  |
| 51…18   | None         | None        |            | None                  |
| 64…fa   | tcp          | 0.0.0.0/0   | 22:22      | None                  |
+---------+-----------+---------+---------+------------------+
```

3）查看安全组规则详细信息

查看 id 为 642d6ebe－3fb6－4a52－bd1a－889d196351fa 的安全组规则的详细信息。

```
[root@ controller ~]# openstack security group rule show 642d6ebe-3fb6-4a52-
bd1a-889d196351fa
+------------------+-------------------------------------+
| Field             | Value                                  |
+------------------+-------------------------------------+
| created_at        | 2022-12-15T06:37:00Z                   |
| description       | allow tcp:22 for ssh                   |
| direction         | ingress                                |
| ether_type        | IPv4                                   |
| id                | 642d6ebe-3fb6-4a52-bd1a-889d196351fa   |
| name              | None                                   |
| port_range_max    | 22                                     |
| port_range_min    | 22                                     |
| project_id        | 62b92970ed2a42cb9ea0e8f013004062       |
| protocol          | tcp                                    |
| remote_group_id   | None                                   |
| remote_ip_prefix  | 0.0.0.0/0                              |
| revision_number   | 0                                      |
| security_group_id | a4b65b0d-2e48-47b9-8c12-9693e9176dec   |
| updated_at        | 2022-12-15T06:37:00Z                   |
+------------------+-------------------------------------+
```

4）删除安全组规则

删除 sg－test 安全组下 id 为 642d6ebe－3fb6－4a52－bd1a－889d196351fa 的安全组规则。

```
[root@ controller ~]# openstack security group rule delete 642d6ebe-3fb6-4a52-bd1a-889d196351fa
[root@ controller ~]# openstack security group rule list sg-test
+-------+------------+----------+------------+----------------------+
| ID    | IP Protocol | IP Range | Port Range | Remote Security Group |
+-------+------------+----------+------------+----------------------+
| 01…5e | None       | None     |            | None                 |
| 51…18 | None       | None     |            | None                 |
+-------+------------+----------+------------+----------------------+
```

从显示结果可以看出，id 为 642d6ebe-3fb6-4a52-bd1a-889d196351fa 的安全组规则已经被删除了。

6. 浮动 IP 管理

1）浮动 IP 创建

创建 ext-net 网络内 ext-subnet 子网的浮动 IP 地址 192.168.16.26，设定描述信息为"floating ip"，所属项目为 myproject。

```
[root@ controller ~]# openstack floating ip create --subnet ext-subnet --floating-ip-address 192.168.16.26 --description "floating ip" --project myproject ext-net
+---------------------+----------------------------------------+
| Field               | Value                                  |
+---------------------+----------------------------------------+
| created_at          | 2022-12-15T07:07:29Z                   |
| description         | floating ip                            |
| dns_domain          | None                                   |
| dns_name            | None                                   |
| fixed_ip_address    | None                                   |
| floating_ip_address | 192.168.16.26                          |
| floating_network_id | f6617eb3-fac0-44e9-9eae-82a4c94263dd   |
| id                  | 7d25b141-5f3d-4b8f-9d37-f7ed89865c42   |
| name                | 192.168.16.26                          |
| port_details        | None                                   |
| port_id             | None                                   |
| project_id          | e598a75ab1434d4ea2cfa4e0da5ae9ee       |
| qos_policy_id       | None                                   |
| revision_number     | 0                                      |
| router_id           | None                                   |
| status              | DOWN                                   |
| subnet_id           | d3a9cb74-9a07-4bbb-b642-17d83f06f18e   |
| tags                | []                                     |
| updated_at          | 2022-12-15T07:07:29Z                   |
+---------------------+----------------------------------------+
```

2）查看浮动 IP 地址

查看 router 路由器的浮动 IP 地址：

```
[root@ controller ~]# openstack floating ip list --router router
+-----+-----------------+------------+------+-------------------+------+
|ID    |FloatingIPAddress |FixedIPAddress |Port    |FloatingNetwork   |Project |
+-----+-----------------+------------+------+-------------------+------+
|5a…64 |192.168.16.24     |172.16.1.19    |d1…a7  |f6…3dd            |e5…ee   |
+-----+-----------------+------------+------+-------------------+------+
```

3）查看浮动 IP 地址详细信息

查看浮动 IP 192.168.16.26 的详细信息：

```
[root@ controller ~]# openstack floating ip show 192.168.16.26
+----------------------+--------------------------------------+
| Field                | Value                                |
+----------------------+--------------------------------------+
| created_at           | 2022-12-15T07:07:29Z                 |
| description          | floating ip                          |
| dns_domain           | None                                 |
| dns_name             | None                                 |
| fixed_ip_address     | None                                 |
| floating_ip_address  | 192.168.16.26                        |
| floating_network_id  | f6617eb3-fac0-44e9-9eae-82a4c94263dd |
| id                   | 7d25b141-5f3d-4b8f-9d37-f7ed89865c42 |
| name                 | 192.168.16.26                        |
| port_details         | None                                 |
| port_id              | None                                 |
| project_id           | e598a75ab1434d4ea2cfa4e0da5ae9ee     |
| qos_policy_id        | None                                 |
| revision_number      | 0                                    |
| router_id            | None                                 |
| status               | DOWN                                 |
| subnet_id            | None                                 |
| tags                 | []                                   |
| updated_at           | 2022-12-15T07:07:29Z                 |
+----------------------+--------------------------------------+
```

4）删除浮动 IP

删除浮动 IP 192.168.16.26：

```
[root@ controller ~]# openstack floating ip delete 192.168.16.26
```

【任务工单】

工单号：5－3

<table>
<tr><td colspan="2">项目名称：OpenStack 云平台运维</td><td colspan="2">任务名称：网络服务 Neutron 管理</td></tr>
<tr><td colspan="2">班级：</td><td>学号：</td><td>姓名：</td></tr>
<tr><td>任务安排</td><td colspan="3">□完成网络的管理，包括网络的创建、修改、删除等
□完成子网的管理，包括在网络内创建子网、对子网修改和查看等
□完成路由器的管理，包括路由器的创建、信息的设置与修改等，通过路由器实现网络连通
□完成安全组的管理，包括安全组的创建、安全组规则的添加与移除等
□完成浮动 IP 的管理，包括 IP 的创建、修改等</td></tr>
<tr><td>成果交付</td><td colspan="3">云平台创建一台云主机实例，并通过网络管理实现正常访问外网</td></tr>
<tr><td>任务实施总结</td><td colspan="3">任务自评（0～10 分）：
任务收获：

改进点：</td></tr>
<tr><td>成果验收</td><td colspan="3">□完全满足任务要求
□基本满足任务要求
要求全部完成，但对网络的管理操作不够熟练：

□不能满足需求
要求不能独立完成，无法独立完成云主机实例外网访问的网络管理：</td></tr>
</table>

【知识巩固】

openstack subnet create 创建子网时，可以作为 --gateway 指定的参数的是（　　）。

A. --gateway 192.168.9.1　　B. --gateway auto

C. --gateway none　　D. --gateway dhcp

【小李的反思】

凡事预则立，不预则废。

语出《礼记·中庸》，意思是任何事情，事前有规划就可以成功，没有规划就要失败。在 OpenStack 云平台运维中，网络服务 Neutron 管理从计算机网络技术层面对平台的运维作出了规划，需要结合子网、VLAN、GRE 等网络技术，认真设计网络，保证平台安全以及减少网络堵塞。

美国危害全球网络安全，2013 年“棱镜门”事件揭露美国对包括中国在内的全球各国进行网络窃密，美国甚至要求微软、雅虎、谷歌、苹果等在内的 9 家国际网络巨头配合美国政府秘密监听，并入侵德国、韩国等多个国家的网络设备。另外，2018 年至今，美国在没有任何证据的情况下，以数据安全为由无理打压中国企业，还胁迫遭受美国窃密之害的盟友加入这一行列。

党的二十大报告指出，建设现代化产业体系，坚持把发展经济的着力点放在实体经济上，推进新型工业化，加快建设网络强国，网络强国成为当前我国网信事业发展的重要目标。随着数字技术发展进步与应用范围不断扩大，网络已经深度融入百姓生活、经济发展、科技进步、国家治理等各个环节。面向新时代新征程，我们要坚持以全面学习、全面把握、全面落实党的二十大精神为统领，深入贯彻习近平新时代中国特色社会主义思想特别是习近平总书记关于网络强国的重要思想，始终站在“没有网络安全就没有国家安全，没有信息化就没有现代化”的全局和战略高度，牢牢把握信息革命时代潮流与中华民族伟大复兴发展大势、信息化与中国式现代化、网络强国与社会主义现代化强国的内在关系，抓住战略机遇期，抢占发展制高点，赢得时代主动权，以网络强国建设新成效为全面建设社会主义现代化国家、全面推进中华民族伟大复兴提供有力服务、支撑和保障。

任务 4 计算服务 Nova 管理

【任务描述】

小李已经可以在 OpenStack 云平台对云主机进行身份认证、镜像和网络等的管理，按照部署云平台的组件顺序，小李需要了解计算组件的使用，小李决定通过 Nova 组件对云主机的实例类型模板和服务器进行管理。

【知识要点】

1. 实例类型模板（flavor）管理

1）创建实例类型模板

```
openstack flavor create
[--ram <size-mb>]
[--disk <size-gb>]
[--swap <size-mb>]
[--vcpus <num-cpu>]
[--public|--private]
[--project <project>]
[--description <description>]
<flavor-name>
```

--ram <size-mb>：指定内存大小，默认单位是 MB。

--disk <size-gb>：指定磁盘大小，默认单位是 GB。

--swap <size-mb>：指定交换空间大小，默认单位是 MB。

--vcpus <num-cpu>：指定虚拟 CPU 数量。

--public|--private：指定实例类型模板是公有的还是私有的。

--project <project>：允许访问私有实例类型模板的项目，由于只能用于私有实例类型模板，所以必须和 --private 一起使用。

--description <description>：对实例类型模板添加描述信息，--os-compute-api 在版本 2.55 以上生效。

2）查看实例类型模板列表

```
openstack flavor list
[--public|--private|--all]
[--long]
[--limit <num-flavors>]
```

--public：查看公共实例类型模板。

--private：查看私有实例类型模板。

--all：查看所有实例类型模板。

--long：列出所有字段。

--limit <num-flavors>：列出实例类型模板的最大数量。

3）设置实例类型模板信息

```
openstack flavor set
[--no-property]
[--property <key=value>[…]]
[--project <project>]
[--description <description>]
<flavor>
```

--property <key=value>：添加或设置实例类型模板的属性，可以通过设置多个 key、value 对来设置多个属性。

--no-property：移除该实例类型模板的所有属性，一般和 --property 属性一起使用。在设置新属性前，通过 --no-proterty 移除当前所有属性。

--project <project>：设置实例类型模板可访问的项目。

--description <description>：设置描述信息。

4）查看实例类型模板详细信息

```
openstack flavor show
<flavor>
```

5）取消实例类型模板信息设置

```
openstack flavor unset
[--property <key>[…]]
[--project <project>]
<flavor>
```

--property <key>：取消属性值设置。

--project <project>：取消对该项目的可见。

6）删除实例类型模板

```
openstack flavor delete
<flavor>[ <flavor>...]
```

2. 服务器管理

1）服务器创建

```
openstack server create
--image <image>
--flavor <flavor>
[--security-group <security-group>]
[--key-name <key-name>]
[--property <key=value>]
```

```
[--nic net-id=net-uuid]
<server-name>
```

--image <image>：指定服务器的系统镜像。

--flavor <flavor>：指定服务器的规格。

--security-group <security-group>：指定服务器应用的安全组。

--key-name <key-name>：注册到该服务器的密钥对。

--nic net-id=net-uuid：为该服务器创建一张网卡，指定该网卡连接的网络。

2）查看服务器列表

```
openstack server list
```

3）查看服务器详细信息

```
openstack server show <server-name>
```

4）设置服务器信息

```
openstack server set
[--name <new-name>]
[--root-password]
[--property <key=value>]
[--state <state>]
<server>
```

--name <new-name>：设置服务器名字。

--root-password：设置服务器的 root 密码（仅交互式）。

--property <key=value>：设置服务器属性信息。

--state <state>：设置服务器状态（有效状态：active、error）。

5）为服务器添加浮动 IP

```
openstack server add floating ip
<server>
<ip-address>
```

6）取消服务器浮动 IP

```
openstack server remove floating ip <server> <ip-address>
```

7）删除服务器

```
openstack server delete <server>
```

【任务实施】

1. 实例类型模板管理

1）创建实例类型模板

（1）创建实例类型模板 tinyflavor，设置内存为 512 MB、硬盘大小为 4 GB、虚拟 CPU 个数为 1。

```
[root@ controller ~]# openstack flavor create --ram 512 --disk 4 --vcpus 1 tiny-
flavor
+----------------------------+--------------------------------------+
| Field                      | Value                                |
+----------------------------+--------------------------------------+
| OS-FLV-DISABLED:disabled   | False                                |
| OS-FLV-EXT-DATA:ephemeral  | 0                                    |
| disk                       | 4                                    |
| id                         | 582bfda4-68fa-4a34-8f73-2ae79cc30f29 |
| name                       | tinyflavor                           |
| os-flavor-access:is_public | True                                 |
| properties                 |                                      |
| ram                        | 512                                  |
| rxtx_factor                | 1.0                                  |
| swap                       |                                      |
| vcpus                      | 1                                    |
+----------------------------+--------------------------------------+
```

查看 flavor 列表可发现，在未指定该实例类型模板是公共的还是私有的情况下，默认设置为公有的。

```
[root@ controller ~]# openstack flavor list
+-----------+------------+-----+------+-----------+-------+-----------+
| ID        | Name       | RAM | Disk | Ephemeral | VCPUs | Is Public |
+-----------+------------+-----+------+-----------+-------+-----------+
| 582b…0f29 | tinyflavor | 512 | 4    | 0         | 1     | True      |
+-----------+------------+-----+------+-----------+-------+-----------+
```

（2）创建实例类型模板 mediumflavor，设置内存为 1 024 MB、硬盘大小为 8 GB、虚拟 CPU 个数为 2。

```
[root@ controller ~]# openstack flavor create --ram 1024 --disk 8 --vcpus 2 --
private mediumflavor
+----------------------------+--------------------------------------+
| Field                      | Value                                |
+----------------------------+--------------------------------------+
| OS-FLV-DISABLED:disabled   | False                                |
| OS-FLV-EXT-DATA:ephemeral  | 0                                    |
| disk                       | 8                                    |
| id                         | f599e498-297e-4765-a35c-56be72aa41e8 |
| name                       | mediumflavor                         |
| os-flavor-access:is_public | False                                |
| properties                 |                                      |
| ram                        | 1024                                 |
```

```
| rxtx_factor                  | 1.0                                   |
| swap                         |                                       |
| vcpus                        | 2                                     |
+------------------------------+---------------------------------------+
```

2）多维度查看实例类型模板

（1）查看实例类型模板列表。

```
[root@ controller ~]# openstack flavor list
+----------+-----------+----+-----+----------+------+---------+
| ID       | Name      | RAM | Disk | Ephemeral | VCPUs | Is Public |
+----------+-----------+----+-----+----------+------+---------+
| 582b…0f29 | tinyflavor | 512 | 4   | 0        | 1    | True    |
+----------+-----------+----+-----+----------+------+---------+
```

默认情况下，查看的是公共实例类型模板。

（2）查看私有实例类型模板类型列表。

```
[root@ controller ~]# openstack flavor list --private
+----------+------------+----+-----+---------+------+----------+
| ID       | Name       | RAM | Disk | Ephemeral | VCPUs | Is Public |
+----------+------------+----+-----+---------+------+----------+
| f599…41e8 | mediumflavor | 1024| 8   | 0       | 2    | False    |
+----------+------------+----+-----+---------+------+----------+
```

（3）查看所有实例类型模板列表。

```
[root@ controller ~]# openstack flavor list --all
+----------+------------+------+------+---------+------+--------+
| ID       | Name       | RAM  | Disk | Ephemeral | VCPUs |Is Public |
+----------+------------+------+------+---------+------+--------+
| 582b…0f29 | tinyflavor | 512  | 4    | 0       | 1    |True    |
| f599…41e8 | mediumflavor | 1024 | 8    | 0       | 2    |False   |
+----------+------------+------+------+---------+------+--------+
```

（4）登录 dashboard，在管理员下的实例类型中，可以看到两个实例类型模板，如图 5-19 所示。

3）设置实例类型模板信息

（1）设置 tinyflavor 实例类型模板对 myproject 项目可见。

```
[root@ controller ~]# openstack flavor set --project myproject tinyflavor
Failed to set flavor access to project:Cannot set access for a public flavor
Command Failed:One or more of the operations failed
```

设置结果报错，提示无法对公共的实例类型模板设置可见性，由此可知，--project 只能对私有实例类型模板使用。

（2）设置 mediumflavor 实例类型模板对 myproject 项目可见。

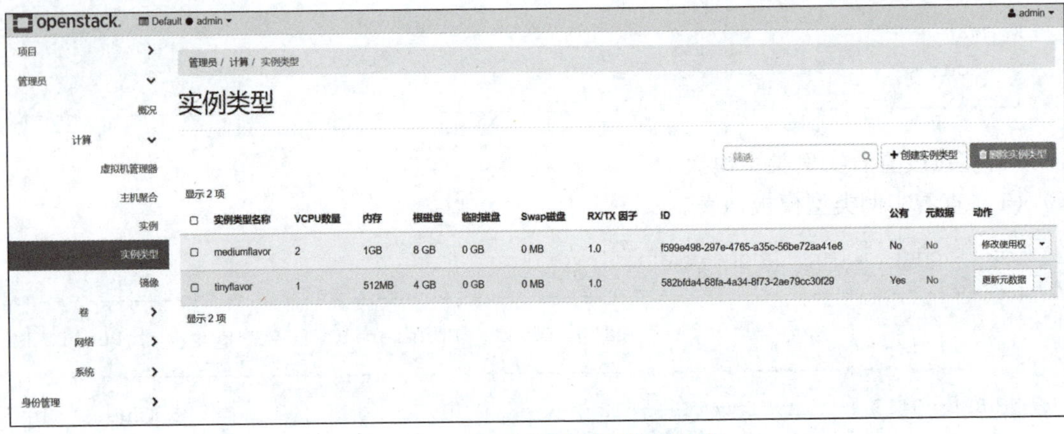

图 5-19　管理员实力类型模板列表

```
[root@ controller ~]# openstack flavor set --project myproject mediumflavor
[root@ controller ~]# openstack flavor show mediumflavor
+----------------------------+--------------------------------------+
| Field                      | Value                                |
+----------------------------+--------------------------------------+
| OS-FLV-DISABLED:disabled   | False                                |
| OS-FLV-EXT-DATA:ephemeral  | 0                                    |
| access_project_ids         | e598a75ab1434d4ea2cfa4e0da5ae9ee     |
| disk                       | 8                                    |
| id                         | f599e498-297e-4765-a35c-56be72aa41e8 |
| name                       | mediumflavor                         |
| os-flavor-access:is_public | False                                |
| properties                 |                                      |
| ram                        | 1024                                 |
| rxtx_factor                | 1.0                                  |
| swap                       |                                      |
| vcpus                      | 2                                    |
+----------------------------+--------------------------------------+
```

此时，access_project_ids 对应的值为 myproject 项目的 id；在 dashboard 界面，私有实例类型模板具有修改使用权选项，公共实例类型模板没有修改使用权选项，如图 5-20 所示。

单击 mediumflavor 实例类型模板的“修改使用权”，在“编辑实例类型”界面，单击加号就可以增加该实例类型在该项目的可见性，如图 5-21 所示。

（3）设置 mediumflavor 的属性，size = 'medium', visibility = 'private'。

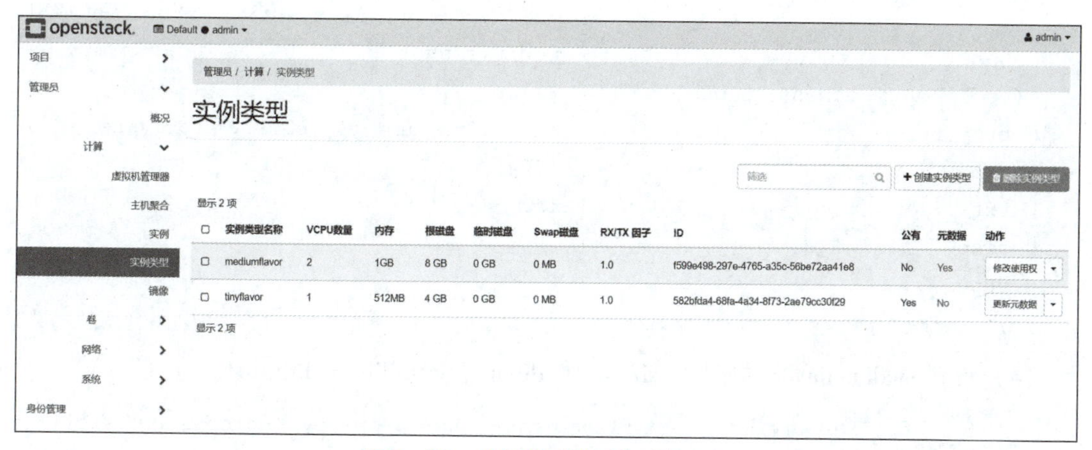

图 5－20　实例类型模板修改使用权

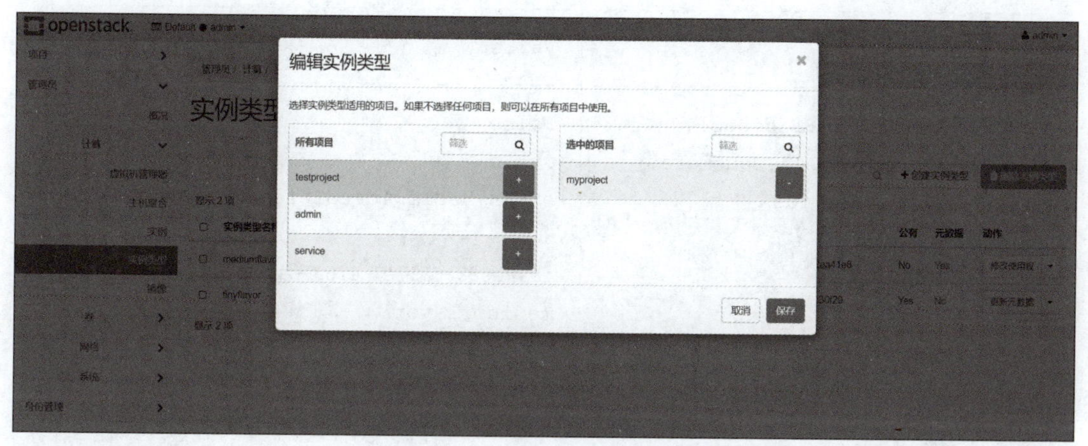

图 5－21　“编辑实例类型”界面

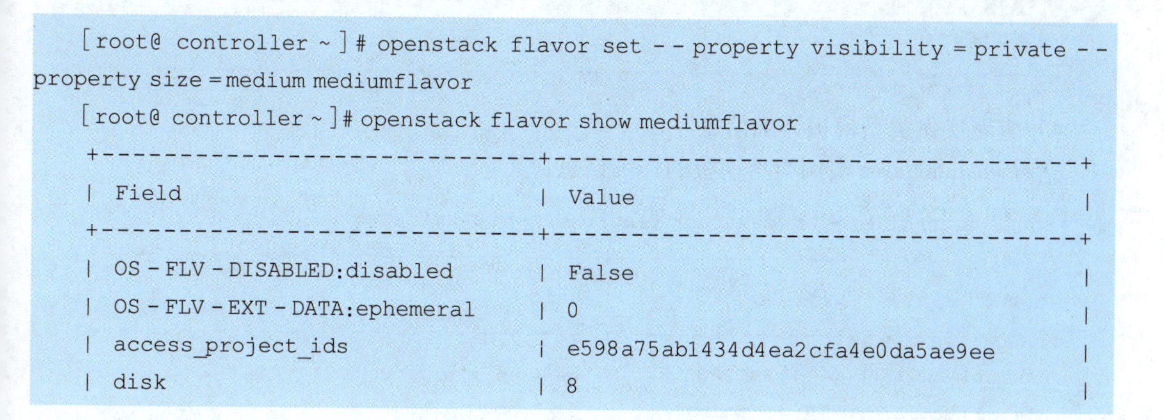

```
[root@ controller ~]# openstack flavor set --property visibility=private --property size=medium mediumflavor
[root@ controller ~]# openstack flavor show mediumflavor
+------------------------------+--------------------------------------+
| Field                        | Value                                |
+------------------------------+--------------------------------------+
| OS-FLV-DISABLED:disabled     | False                                |
| OS-FLV-EXT-DATA:ephemeral    | 0                                    |
| access_project_ids           | e598a75ab1434d4ea2cfa4e0da5ae9ee     |
| disk                         | 8                                    |
```

```
| id                             | f599e498-297e-4765-a35c-56be72aa41e8|
| name                           | mediumflavor                        |
| os-flavor-access:is_public     | False                               |
| properties                     | size='medium',visibility='private'  |
| ram                            | 1024                                |
| rxtx_factor                    | 1.0                                 |
| swap                           |                                     |
| vcpus                          | 2                                   |
+--------------------------------+-------------------------------------+
```

（4）设置 mediumflavor 的属性，size = 'Medium'，visibility = 'Private'。

```
[root@ controller ~]# openstack flavor set --no-property --property visibility
=Private --property size=Medium mediumflavor
[root@ controller ~]# openstack flavor show mediumflavor
+--------------------------------+-------------------------------------+
| Field                          | Value                               |
+--------------------------------+-------------------------------------+
| OS-FLV-DISABLED:disabled       | False                               |
| OS-FLV-EXT-DATA:ephemeral      | 0                                   |
| access_project_ids             | e598a75ab1434d4ea2cfa4e0da5ae9ee    |
| disk                           | 8                                   |
| id                             | f599e498-297e-4765-a35c-56be72aa41e8 |
| name                           | mediumflavor                        |
| os-flavor-access:is_public     | False                               |
| properties                     | size='Medium',visibility='Private'  |
| ram                            | 1024                                |
| rxtx_factor                    | 1.0                                 |
| swap                           |                                     |
| vcpus                          | 2                                   |
+--------------------------------+-------------------------------------+
```

4）查看实例类型模板详细信息

查看 mediumflavor 实例类型模板的详细信息：

```
[root@ controller ~]# openstack flavor show mediumflavor
+--------------------------------+-------------------------------------+
| Field                          | Value                               |
+--------------------------------+-------------------------------------+
| OS-FLV-DISABLED:disabled       | False                               |
| OS-FLV-EXT-DATA:ephemeral      | 0                                   |
| access_project_ids             | e598a75ab1434d4ea2cfa4e0da5ae9ee    |
| disk                           | 8                                   |
| id                             | f599e498-297e-4765-a35c-56be72aa41e8|
| name                           | mediumflavor                        |
```

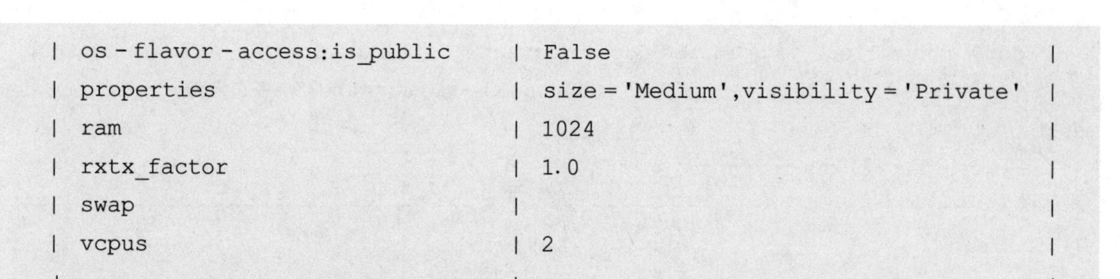

```
| os-flavor-access:is_public   | False                                   |
| properties                   | size='Medium',visibility='Private'      |
| ram                          | 1024                                    |
| rxtx_factor                  | 1.0                                     |
| swap                         |                                         |
| vcpus                        | 2                                       |
+------------------------------+-----------------------------------------+
```

5）取消实例类型模板信息设置

取消 mediumflavor 实例类型模板在 myproject 项目的可见性，并取消 property 下的 size 属性。

```
[root@ controller ~]# openstack flavor unset --project myproject --property size mediumflavor
[root@ controller ~]# openstack flavor show mediumflavor
+------------------------------+-----------------------------------------+
| Field                        | Value                                   |
+------------------------------+-----------------------------------------+
| OS-FLV-DISABLED:disabled     | False                                   |
| OS-FLV-EXT-DATA:ephemeral    | 0                                       |
| access_project_ids           |                                         |
| disk                         | 8                                       |
| id                           | f599e498-297e-4765-a35c-56be72aa41e8    |
| name                         | mediumflavor                            |
| os-flavor-access:is_public   | False                                   |
| properties                   | visibility='Private'                    |
| ram                          | 1024                                    |
| rxtx_factor                  | 1.0                                     |
| swap                         |                                         |
| vcpus                        | 2                                       |
+------------------------------+-----------------------------------------+
```

查看 mediumflavor 详细信息可知，此时的 access_project_ids 为空，也就是取消了在 myproject 项目的可见性，同时，properties 属性中，size 的属性没有了。

6）删除实例类型模板

删除 tinyflavor 和 mediumflavor 两个实例类型模板：

```
[root@ controller ~]# openstack flavor delete tinyflavor mediumflavor
```

2. 服务器管理

1）服务器创建

创建服务器实例 instance1，服务器以 cirros 镜像启动，规格为 s1.tiny（64 MB 内存，1 GB 硬盘，1 个 vCPU），创建网卡并连接到 int-net 网络（id 为 2682f5bc-e03e-4038-a883-c0b8ac46be85），使用 default 安全组，注册 mykey 密钥对。

```
[root@ controller ~]# openstack server create --flavor s1.tiny --image cirros \
>--nic net-id=2682f5bc-e03e-4038-a883-c0b8ac46be85 --security-group de-
fault \
>--key-name mykey instance
+-----------------------------------+--------------------------------------+
| Field                             | Value                                |
+-----------------------------------+--------------------------------------+
| OS-DCF:diskConfig                 | MANUAL                               |
| OS-EXT-AZ:availability_zone       |                                      |
| OS-EXT-STS:power_state            | NOSTATE                              |
| OS-EXT-STS:task_state             | scheduling                           |
| OS-EXT-STS:vm_state               | building                             |
| OS-SRV-USG:launched_at            | None                                 |
| OS-SRV-USG:terminated_at          | None                                 |
| accessIPv4                        |                                      |
| accessIPv6                        |                                      |
| addresses                         |                                      |
| adminPass                         | vCNnP3Hyz6K8                         |
| config_drive                      |                                      |
| created                           | 2022-12-27T05:00:19Z                 |
| flavor                            | s1.tiny(0)                           |
| hostId                            |                                      |
| id                                | d3b0d790-f39a-47f4-acd8-e5411c888967 |
| image                             | cirros(a58a…33d)                     |
| key_name                          | mykey                                |
| name                              | instance                             |
| progress                          | 0                                    |
| project_id                        | e598a75ab1434d4ea2cfa4e0da5ae9ee     |
| properties                        |                                      |
| security_groups                   | name='55bc7a2…4e0731'                |
| status                            | BUILD                                |
| updated                           | 2022-12-27T05:00:19Z                 |
| user_id                           | 1ecd571c56b34ff78dbae47cd7b20c89     |
| volumes_attached                  |                                      |
+-----------------------------------+--------------------------------------+
```

2）查看服务器列表

查看当前云平台中的服务器列表：

```
[root@ controller ~]# openstack server list
+--------+------------+-------+---------------------+------+-------+
| ID     | Name       | Status | Networks           | Image | Flavor |
+--------+------------+-------+---------------------+------+-------+
| d3b…967 | instance  | ACTIVE | int-net=172.16.1.11 | cirros| s1.tiny|
```

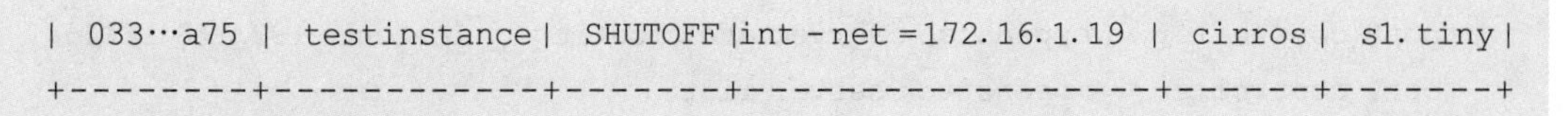

```
| 033…a75 | testinstance | SHUTOFF |int-net=172.16.1.19 | cirros | s1.tiny |
+--------+------------+-------+-----------------+------+-------+
```

3）查看服务器详细信息

查看 instance 服务器的详细信息：

```
+--------------------------------+--------------------------------------+
| Field                          | Value                                |
+--------------------------------+--------------------------------------+
| OS-DCF:diskConfig              | MANUAL                               |
| OS-EXT-AZ:availability_zone    |                                      |
| OS-EXT-STS:power_state         | NOSTATE                              |
| OS-EXT-STS:task_state          | scheduling                           |
| OS-EXT-STS:vm_state            | building                             |
| OS-SRV-USG:launched_at         | None                                 |
| OS-SRV-USG:terminated_at       | None                                 |
| accessIPv4                     |                                      |
| accessIPv6                     |                                      |
| addresses                      |                                      |
| adminPass                      | vCNnP3Hyz6K8                         |
| config_drive                   |                                      |
| created                        | 2022-12-27T05:00:19Z                 |
| flavor                         | s1.tiny(0)                           |
| hostId                         |                                      |
| id                             | d3b0d790-f39a-47f4-acd8-e5411c888967 |
| image                          | cirros(a58a…33d)                     |
| key_name                       | mykey                                |
| name                           | instance                             |
| progress                       | 0                                    |
| project_id                     | e598a75ab1434d4ea2cfa4e0da5ae9ee     |
| properties                     |                                      |
| security_groups                | name='55bc7a2…4e0731'                |
| status                         | BUILD                                |
| updated                        | 2022-12-27T05:00:19Z                 |
| user_id                        | 1ecd571c56b34ff78dbae47cd7b20c89     |
| volumes_attached               |                                      |
+--------------------------------+--------------------------------------+
```

4）修改服务器信息

将 instance 服务器实例名称改为 instance-cirros：

```
[root@controller ~]# openstack server set --name instance-cirros instance
```

5）为服务器添加浮动 IP

为 instance 服务器实例添加 192.168.16.25（已创建）浮动 IP：

```
[root@ controller ~]# openstack server add floating ip instance 192.168.16.25
[root@ controller ~]# openstack server list
+------+--------+-------+-----------------------------------+-----+----+
|ID    |Name    |Status |Networks                           |Image |Flavor |
+------+--------+-------+-----------------------------------+-----+----+
|d3…67 |instance |ACTIVE |int-net=172.16.1.11,192.168.16.25 |cirros|s1.tiny|
|03…75 |testins  |SHUTOFF |int-net=172.16.1.19,192.168.16.24 |cirros|s1.tiny|
+------+--------+-------+-----------------------------------+-----+----+
```

6）解除服务器的浮动 IP

解除 instance 服务器实例的 192.168.16.25 浮动 IP：

```
[root@ controller ~]# openstack server remove floating ip instance 192.168.16.25
[root@ controller ~]# openstack server list
+------+--------+------+------------------------------------+------+---+
|ID    |Name    |Status |Networks                           |Image |Flavor |
+------+--------+------+------------------------------------+------+---+
|d3…67 |instance |ACTIVE |int-net=172.16.1.11               |cirros|s1.tiny|
|03…75 |testins  |SHUTOFF|int-net=172.16.1.19,192.168.16.24 |cirros|s1.tiny|
+------+--------+------+------------------------------------+------+---+
```

7）删除服务器

删除 instance 服务器实例：

```
[root@ controller ~]# openstack server delete instance
```

【任务工单】

工单号：5 -4

<table>
<tr><td colspan="2">项目名称：OpenStack 云平台运维</td><td colspan="2">任务名称：计算服务 Nova 管理</td></tr>
<tr><td colspan="2">班级：</td><td>学号：</td><td>姓名：</td></tr>
<tr><td>任务安排</td><td colspan="3">□实例类型模板的管理，包括实例类型模板的创建、修改、查看和删除等
□服务器的管理，包括服务器的创建、修改、删除、绑定与解除浮动 IP 等</td></tr>
<tr><td>成果交付</td><td colspan="3">实验案例整理成操作指导交付文档</td></tr>
<tr><td>任务实施总结</td><td colspan="3">任务自评（0～10 分）：
任务收获：
改进点：</td></tr>
<tr><td>成果验收</td><td colspan="3">□完全满足任务要求
□基本满足任务要求
要求全部完成，能够对实例类型和服务器进行管理，但操作不够熟练：
□不能满足需求
要求不能独立完成，对服务器的管理比较混乱：</td></tr>
</table>

【知识巩固】

实例类型模板设置中，--project 只能对私有实例类型模板使用。(　　)

A. 对　　　　　　B. 错

【小李的反思】

工欲善其事，必先利其器。

出自《论语·卫灵公》，意思就是工匠想要使他的工作做好，一定要先让工具锋利。比喻要做好一件事，准备工作非常重要。在本任务中，实例类型模板（flavor）管理和服务器管理比较复杂，使用 Nova 组件对相关计算服务资源进行管理，使管理工作变得简单许多。

说到计算资源，神威·太湖之光超级计算机是中国自主研发，由国家并行计算机工程技术研究中心研制，安装在国家超级计算无锡中心的超级计算机。神威·太湖之光超级计算机安装了 40 960 个中国自主研发的“申威 26010”众核处理器，该众核处理器采用 64 位自主神威指令系统，峰值性能为 12.5 亿亿次/s，持续性能为 9.3 亿亿次/s，核心工作频率 1.5 GHz。2020 年 7 月，中国科大在“神威·太湖之光”上首次实现千万核心并行第一性原理计算模拟。2022 年，中国的神威·太湖之光全球超级计算机 500 强排名位列前十。

党的二十大报告指出，教育、科技、人才是全面建设社会主义现代化国家的基础性、战略性支撑。必须坚持科技是第一生产力，加快建设科技强国。作为新时代青年，要与时代同行，开拓思维，加强创新，从书本中去汲取力量，开阔自己的视野，为之后的发展做好铺垫，不断从百年党史中汲取应对风险、迎接挑战的智慧和力量，成为新时代中华民族伟大复兴的中国梦的坚定践行者！

任务5　存储服务 Cinder 管理

【任务描述】

小李已经学会了如何通过命令对云主机的身份认证、镜像、网络、计算等的管理，按照计算机的结构，还需要有硬盘的管理，相对于云平台，小李还需要学习云平台的存储管理，小李决定通过 Cinder 组件对云主机的块存储进行管理。

【知识要点】

块存储管理

1. 创建卷

```
openstack volume create
[--size <size>]
[--type <volume-type>]
[--image <image> | --snapshot <snapshot> | --source <volume>]
[--description <description>]
[--user <user>]
[--project <project>]
[--property <key=value>[...]]
[--bootable | --non-bootable]
[--read-only | --read-write]
<name>
```

--size <size>：指定卷的大小，单位是 GB，在未指定 --source 或 --snapshot 选项时，为必选项。

--type <volume-type>：指定卷类型。

--image <image>：指定卷的镜像源，一般制作服务器的启动卷时使用。

--snapshot <snapshot>：使用快照作为卷的源。

--source <volume>：根据已有卷克隆。

--description <description>：指定卷的描述信息。

--user <user>：指定卷所属的用户。

--project <project>：指定卷所属的项目。

--property <key=value>：设定卷的属性信息。

--bootable | --non-bootable：标记卷是可启动的 | 不可启动的。

--read-only：设定卷为只读模式。

--read-write：设定卷为读写模式。

2. 查看卷列表

```
openstack volume list
```

```
[ --project <project> ]
[ --user <user> ]
[ --name <name> ]
[ --status <status> ]
[ --all-projects]
[ --long]
```

--project <project>：根据所属项目过滤。

--user <user>：根据所属用户过滤。

--name <name>：根据名称过滤。

--status <status>：根据状态过滤。

--all-projects：查看所有项目的卷列表。

--long：查看更多属性。

3. 查看卷详细信息

```
openstack volume show
<volume>
```

4. 设置卷信息

```
openstack volume set
[ --name <name> ]
[ --size <size> ]
[ --description <description> ]
[ --no-property]
[ --property <key=value> [...]]
[ --state <state> ]
[ --type <volume-type> ]
[ --bootable | --non-bootable]
[ --read-only | --read-write]
<volume>
```

--name <name>：设置卷的名字。

--size <size>：设置卷的大小。

--description <description>：设置卷的描述信息。

--no-property：移除所有卷的 property 属性信息，一般将 --no-property 和 --property 一起使用，用来在设置新的属性信息前删除所有当前属性信息。

--property <key=value>：设定卷的属性信息。

--state <state>：设定卷的状态。

--type <volume-type>：设定卷的类型。

--bootable | --non-bootable：设定卷为可启动的 | 不可启动的。

--read-only | --read-write：设定卷是只读模式 | 读写模式。

5. 取消卷信息设置

```
openstack volume unset
[--property <key>]
<volume>
```

--property <key>：取消卷属性信息设置。

6. 卷迁移

```
openstack volume migrate
--host <host>
[--lock-volume|--unlock-volume]
<volume>
```

--host <host>：指定迁移的目标主机。

--lock-volume：锁定卷的状态，迁移过程中不可以中止。

--unlock-volume：不锁定卷的状态，迁移过程中可以中止。

7. 卷删除

```
openstack volume delete
[--force|--purge]
<volume>[<volume>...]
```

--force：强制删除卷，不考虑卷的状态。

--purge：删除卷的同时删除所有的快照。

【任务实施】

块存储管理

1. 创建卷

创建卷 vol，大小为 1 GB，类型为 nfs，所属项目为 myproject，描述信息为“test volume for myproject”。

```
[root@ controller ~]# openstack volume create --size 1 --type nfs --description "
test volume for myproject" vol1
+---------------------+--------------------------------------+
| Field               | Value                                |
+---------------------+--------------------------------------+
| attachments         | []                                   |
| availability_zone   | nova                                 |
| bootable            | false                                |
| consistencygroup_id | None                                 |
| created_at          | 2022-12-27T13:34:03.000000           |
| description         | test volume for myproject            |
| encrypted           | False                                |
```

```
| id                    | 9a7197f3-7071-4417-9a70-e4efdb2d25e9    |
| migration_status      | None                                    |
| multiattach           | False                                   |
| name                  | vol1                                    |
| properties            |                                         |
| replication_status    | None                                    |
| size                  | 1                                       |
| snapshot_id           | None                                    |
| source_volid          | None                                    |
| status                | creating                                |
| type                  | nfs                                     |
| updated_at            | None                                    |
| user_id               | 974ca43e23234f0b91847c162982961d        |
+-----------------------+-----------------------------------------+
```

2. 查看卷列表

查看当前云平台的所有卷列表：

```
[root@ controller ~]# openstack volume list -long
+-------+-----+--------+-----+-----+--------+----------+--------+
|ID      |Name  |Status     |Size  |Type  |Bootable  |Attached to  |Properties |
+-------+-----+--------+-----+-----+--------+----------+--------+
|9a7…5e9 |vol1  |available |1     |nfs   |false     |             |           |
+-------+-----+--------+-----+-----+--------+----------+--------
```

3. 查看卷的详细信息

查看 vol1 卷的详细信息：

```
[root@ controller ~]# openstack volume show vol1
+--------------------------------+--------------------------------------+
| Field                          | Value                                |
+--------------------------------+--------------------------------------+
| attachments                    | []                                   |
| availability_zone              | nova                                 |
| bootable                       | false                                |
| consistencygroup_id            | None                                 |
| created_at                     | 2022-12-27T13:34:03.000000           |
| description                    | test volume for myproject            |
| encrypted                      | False                                |
| id                             | 9a7197f3-7071-4417-9a70-e4efdb2d25e9 |
| migration_status               | None                                 |
| multiattach                    | False                                |
| name                           | vol1                                 |
| os-vol-host-attr:host          | compute@ lvm#LVM                     |
| os-vol-mig-status-attr:migstat | None                                 |
| os-vol-mig-status-attr:name_id | None                                 |
| os-vol-tenant-attr:tenant_id   | 62b92970ed2a42cb9ea0e8f013004062     |
| properties                     |                                      |
```

```
| replication_status                | None                              |
| size                              | 1                                 |
| snapshot_id                       | None                              |
| source_volid                      | None                              |
| status                            | available                         |
| type                              | nfs                               |
| updated_at                        | 2022-12-27T13:34:03.000000        |
| user_id                           | 974ca43e23234f0b91847c162982961d  |
+-----------------------------------+-----------------------------------+
```

4. 设置卷的信息

设置 vol1 卷的名字为 vol2，大小为 2 GB，可启动的，添加 creater = ly 的属性信息。

```
[root@ controller ~]# openstack volume set --name vol2 --size 2 --bootable --no-property --property creater=ly vol1
[root@ controller ~]# openstack volume show vol2
+-----------------------------------+-----------------------------------+
| Field                             | Value                             |
+-----------------------------------+-----------------------------------+
| attachments                       | []                                |
| availability_zone                 | nova                              |
| bootable                          | true                              |
| consistencygroup_id               | None                              |
| created_at                        | 2022-12-27T13:34:03.000000        |
| description                       | test volume for myproject         |
| encrypted                         | False                             |
| id                                | 9a7197f3-7071-4417-9a70-e4efdb2d25e9|
| migration_status                  | None                              |
| multiattach                       | False                             |
| name                              | vol2                              |
| os-vol-host-attr:host             | compute@ lvm#LVM                  |
| os-vol-mig-status-attr:migstat    | None                              |
| os-vol-mig-status-attr:name_id    | None                              |
| os-vol-tenant-attr:tenant_id      | 62b92970ed2a42cb9ea0e8f013004062  |
| properties                        | creater='ly'                      |
| replication_status                | None                              |
| size                              | 2                                 |
| snapshot_id                       | None                              |
| source_volid                      | None                              |
| status                            | available                         |
| type                              | nfs                               |
| updated_at                        | 2022-12-27T13:38:45.000000        |
| user_id                           | 974ca43e23234f0b91847c162982961d  |
+-----------------------------------+-----------------------------------+
```

5. 取消卷信息设置

取消 vol2 卷的 creater = ly 的属性信息：

```
[root@ controller ~]# openstack volume unset --property creater vol2
[root@ controller ~]# openstack volume show vol2
+--------------------------------+-------------------------------------+
| Field                          | Value                               |
+--------------------------------+-------------------------------------+
| attachments                    | []                                  |
| availability_zone              | nova                                |
| bootable                       | true                                |
| consistencygroup_id            | None                                |
| created_at                     | 2022-12-27T13:34:03.000000          |
| description                    | test volume for myproject           |
| encrypted                      | False                               |
| id                             | 9a7197f3-7071-4417-9a70-e4efdb2d25e9|
| migration_status               | None                                |
| multiattach                    | False                               |
| name                           | vol2                                |
| os-vol-host-attr:host          | compute@ lvm#LVM                    |
| os-vol-mig-status-attr:migstat | None                                |
| os-vol-mig-status-attr:name_id | None                                |
| os-vol-tenant-attr:tenant_id   | 62b92970ed2a42cb9ea0e8f013004062    |
| properties                     |                                     |
| replication_status             | None                                |
| size                           | 2                                   |
| snapshot_id                    | None                                |
| source_volid                   | None                                |
| status                         | available                           |
| type                           | nfs                                 |
| updated_at                     | 2022-12-27T13:38:45.000000          |
| user_id                        | 974ca43e23234f0b91847c162982961d    |
+--------------------------------+-------------------------------------+
```

6. 卷删除

删除 vol2 卷：

```
[root@ controller ~]# openstack volume delete vol2
```

【任务工单】

工单号：5－5

<table>
<tr><td colspan="2">项目名称：OpenStack 云平台运维</td><td colspan="2">任务名称：存储服务 Cinder 管理</td></tr>
<tr><td colspan="2">班级：</td><td>学号：</td><td>姓名：</td></tr>
<tr><td>任务安排</td><td colspan="3">□完成卷的管理，包括卷的创建、查看、修改、迁移及删除等</td></tr>
<tr><td>成果交付</td><td colspan="3">实验案例整理成操作指导交付文档</td></tr>
<tr><td>任务实施总结</td><td colspan="3">任务自评（0～10 分）：
任务收获：

改进点：</td></tr>
<tr><td>成果验收</td><td colspan="3">□完全满足任务要求
□基本满足任务要求
要求全部完成，能够对卷进行管理，但操作不够熟练：

□不能满足需求
要求不能独立完成，对卷的迁移等稍难的操作无法完成：</td></tr>
</table>

【知识巩固】

使用 openstack volume delete 命令在删除卷的时候，--purge 参数会在删除卷的同时删除所有的快照。(　　)

A. 对　　　　B. 错

【小李的反思】

居安思危，思则有备，有备无患。

出自春秋时期左丘明所作的《左传·襄公十一年》，意思就是处于安全环境时，要考虑到可能出现的危险，考虑到危险就会有所准备，事先有了准备就可以避免祸患。对数据存储服务也是同样的道理，要求有居安思危的意识，对数据安全做好技术预案，防止磁盘存储出现故障，出现数据丢失的现象。

2020 年 2 月 23 日晚，微盟集团旗下 SaaS 业务服务突发故障，系统崩溃，基于微盟 SaaS 业务的小程序宕机，受波及商户线上生意几乎“停摆”。得知消息的 300 万商户非常焦灼，“我们所有的数据都存在微盟商城，包括底价、售卖价格等，数据没了，无法销售”。经过一周之后，微盟发出公告称，于 2020 年 3 月 3 日上午 6 点完成 SaaS 业务数据恢复上线。然而不断有商家向媒体反映“商品信息恢复了，但客户信息严重丢失”“商店原本 5 000 多人的会员，现在只有 694 人恢复了信息”“说是恢复了，实际上很多细节并没有恢复，对接库房的 ERP 系统和微盟小程序的对接也没有恢复，还是用不了”……网络时代，一个以数据服务为主营业务的商家应该有备份的系统，然而其恢复时间竟然需要一周之久，并且修复后仍然 bug 频出，令人费解。

党的二十大报告指出，我们必须增强忧患意识，坚持底线思维，做到居安思危、未雨绸缪，准备经受风高浪急甚至惊涛骇浪的重大考验。新时代青年要居安思危，练就过硬本领，坚定信心、锐意进取，不断夺取全面建设社会主义现代化国家新胜利！

项目评价

项目名称：OpenStack 云平台运维					
班级：		学号：	姓名：		
评价指标		评价等级及分值	学生自评	组内互评	教师评分
素质目标达成情况（30%）	精益求精的工匠精神（10%）	A（10 分）：云平台运维过程中能不断修正问题，精益求精，追求完美 B（7 分）：能够完成云平台运维的任务，但对过程中存在的问题修改积极性不高 C（3 分）：对云平台运维过程中存在的问题放任不管			
	乐于探索的学习精神（10%）	A（10 分）：自我学习热情高涨，积极和同学探讨学习问题 B（7 分）：自我学习热情较好，能够自主完成学习 C（3 分）：自我学习热情一般，学习积极性不高			
	云计算工程师职业素养（10%）	A（10 分）：云平台运维过程中，细致认真，积极主动，具备团队协作精神和很强的责任心，交付文档书写规范 B（7 分）：云平台运维过程中，不够细致，有一定的团队协作精神和很强的责任心，交付文档能够按时提交，但缺少规范性 C（3 分）：云平台运维过程中，由于粗心大意出现了一些本不该出现的错误，缺少团队合作，交付文档不能按时提交			
知识目标达成情况（40%）	任务实施完成情况（20%）	A（20 分）：云平台运维完成，且满足使用要求 B（16 分）：云平台运维任务基本完成，但存在一些小的问题，且操作不够熟练 C（10 分）：云平台运维任务部分完成，完成率低于 60%，且存在一些影响任务实施结果的问题			
	测验作业完成情况（10%）	A（10 分）：测验作业全部完成，知识理解透彻 B（7 分）：测验作业大部分完成，能基本完成知识的理解 C（3 分）：测验作业部分完成，对知识的理解较为片面			

续表

评价指标		评价等级及分值	学生自评	组内互评	教师评分
知识目标达成情况（40%）	课上活动（10%）	A（10 分）：积极参与课上抢答、提问、主题讨论等 B（7 分）：能够参与课上抢答、提问、主题讨论等，但积极性不够高 C（3 分）：很少参与课上抢答、提问、主题讨论等			
能力目标达成情况（30%）	任务实施完成质量（20%）	A（20 分）：任务实施完成质量优秀 B（16 分）：任务实施完成质量良好 C（10 分）：任务实施完成质量一般			
	超凡脱俗（10%）	A（10 分）：能够帮助同组同学解决云平台运维过程中存在的问题，并能整理问题解决手册 B（7 分）：能够帮助同组同学解决云平台运维过程中存在的问题 C（3 分）：能够规定时间内完成学习任务			

项目总结

本项目主要讲述通过命令行端对 OpenStack 云平台的 Keystone、Glance、Neutron、Nova 和 Cinder 组件的管理。任务 1 主要讲述用户、项目、域、角色、服务的创建、修改、删除等基本管理操作；任务 2 主要讲述镜像服务的管理，包括镜像的创建、共享、关联、保存等操作，并通过制作 CentOS 7 镜像加强对镜像管理的掌握；任务 3 的网络管理模块主要讲述对网络、子网、路由器和浮动 IP 等的管理，通过网络的管理云主机实例到外网的访问；任务 4 主要讲述计算服务管理，包括实例模板的管理和服务器的管理、为服务器绑定与解除浮动 IP；任务 5 主要讲述块存储的管理，包括卷的创建、修改、迁移和删除等操作。通过本项目的学习，让学生掌握如何通过命令行操作对云平台的各组件进行管理，构建功能更加强大的云主机实例，并能够解决云平台运行中出现的问题。